John Simonich
ME Departm...
Stanford

MW00844383

MOMENTUM TRANSFER IN BOUNDARY LAYERS

SERIES IN THERMAL AND FLUIDS ENGINEERING

JAMES P. HARTNETT and THOMAS F. IRVINE, JR., Editors
JACK P. HOLMAN, Senior Consulting Editor

MOMENTUM TRANSFER IN BOUNDARY LAYERS

Tuncer Cebeci
Douglas Aircraft Company
and
California State University
at Long Beach

Peter Bradshaw
Imperial College of
Science & Technology

**HEMISPHERE
PUBLISHING CORPORATION**

Washington   London

McGRAW-HILL BOOK COMPANY

New York   Montreal
St. Louis   New Delhi
San Francisco   Panama
Auckland   Paris
Bogotá   São Paulo
Düsseldorf   Singapore
Johannesburg   Sydney
London   Tokyo
Madrid   Toronto
Mexico

MOMENTUM TRANSFER IN BOUNDARY LAYERS

Copyright © 1977 by Hemisphere Publishing Corporation. All rights reserved.
Printed in the United States of America. No part of this publication may be
reproduced, stored in a retrieval system, or transmitted, in any form or by
any means, electronic, mechanical, photocopying, recording, or otherwise,
without prior written permission of the publisher.

1 2 3 4 5 6 7 8 9 0   KPKP   7 8 3 2 1 0 9 8 7

This book was set in Press Roman by Hemisphere Publishing Corporation. The
editors were Barbara Davis and Mary A. Phillips; the production supervisor
was Rebekah McKinney; and the compositor was Bernie Doenhoefer.
The printer and binder was The Kingsport Press, Inc.

Library of Congress Cataloging in Publication Data

Cebeci, Tuncer.
   Momentum transfer in boundary layers.

   (Series in thermal and fluids engineering)
   Includes bibliographical references and index.
   1. Boundary layer.   2. Momentum transfer.
I. Bradshaw, Peter, joint author.   II. Title.
TA357.C4   532'.053   76-57750
ISBN   0-07-010300-3

Surface flow directions (limiting streamlines) on a wind-tunnel wall on which a constant-chord lifting airfoil is mounted. Flow is left to right. (Top) wall boundary layer initially laminar, with transition about halfway across the picture. (Bottom) turbulent wall boundary layer. For an explanation, see Sec. 11.4. (We are indebted to Dr. V. A. S. L. de Brederode for these photographs, which were taken at Imperial College.)

Contents

             AND TURBULENT BOUNDARY LAYERS   235

    8.1   Introduction   235
    8.2   External Laminar Boundary Layers   237

           8.2.1   Two-Dimensional Similar Flows   237
           8.2.2   Two-Dimensional Nonsimilar Flows   239
           8.2.3   Axisymmetric Flows   242

    8.3   External Turbulent Boundary Layers   245

           8.3.1   Two-Dimensional Flows   248
           8.3.2   Axisymmetric Flows   250

    8.4   Internal Turbulent Flows   253
    8.5   A Transport-Equation Model of Turbulence   255

Appendix 8A: Fortran Program for Solving TSL Equations
                for Two-Dimensional Laminar and Turbulent
                Boundary Layers   258

    8A.1   General Description of the Method   258

            8A.1.1   Main   259
            8A.1.2   Subroutine Input   260
            8A.1.3   Subroutine Grid   261
            8A.1.4   Subroutine IVPL   262
            8A.1.5   Subroutine Growth   262
            8A.1.6   Subroutine Eddy   263
            8A.1.7   Subroutine CMOM   265
            8A.1.8   Subroutine Output   266

    8A.2   Description of Input   268

Appendix 8B: Solution of Third-Order Ordinary or Parabolic
                Partial Differential Equations   271

Chapter 9   STABILITY AND TRANSITION   281

    9.1   Introduction   281
    9.2   Small-Disturbance Equations   282
    9.3   Orr-Sommerfeld Equation   285

           9.3.1   Boundary Conditions   286

    9.4   Properties of the Orr-Sommerfeld Equation   288

Preface

This book is intended as an introduction to fluid flows controlled by viscous or turbulent stress gradients and to modern methods of calculating these gradients. It is nominally self-contained, but the explanations of basic concepts are intended as review for senior students who have already taken an introductory course in fluid dynamics, rather than for beginning students. Nearly all stress-dominated flows are shear layers, the most common being the boundary layer on a solid surface. Jets, wakes, and duct flows are also shear layers and are discussed in this volume. Nearly all modern methods of calculating shear layers require the use of digital computers. Computer-based methods, involving calculations beyond the reach of electro-mechanical desk calculators, began to appear around 10 years ago; for instance, they are mentioned briefly as a desirable possibility in the book edited by Rosenhead (1963, Section 1.7). With the exception of one or two specialist books (Patankar and Spalding, 1970; Cebeci and Smith, 1974) this revolution has not been noticed in academic textbooks, although the new methods are widely used by engineers. The background knowledge of shear-layer physics required by a user of modern computer methods is, if anything, more extensive than what sufficed in the days of desk machine methods, because a wider range of problems can be tackled. Certainly a more homogeneous treatment of physics and calculation methods is possible, and we have tried to produce one.

Rather than treat heat and momentum transfer together, with consequent confusing detail about the effect of density variations, we have chosen to present momentum transfer here, mainly for constant-property flow, and to postpone heat and mass transfer to a later volume.

This volume comprises three main sections. Chapters 1–4 introduce basic ideas about boundary layers and other shear layers and the mathematical and computational equipment needed to analyze them. Chapters 5–8 deal with the behavior of two-dimensional or axisymmetric, laminar or turbulent, shear layers with prescribed boundary conditions, presenting a variety of prediction methods from the very simple to the most advanced. The demonstration calculations of Chapter 8 can be taken (at least for qualitative purposes!) as equivalent to experimental data on shear-layer behavior. The last section, Chapters 9–11, deals with more complicated problems.

Chapter 1 demonstrates the importance of shear layers and some general examples of their behavior and their interaction with an external stream. This chapter is self-contained, but elementary concepts are explained only briefly, on the assumption that most students will have attended courses on one-dimensional flow and laminar pipe flow. In Chapter 2 the full equations of motion are derived by combining the momentum theorem for a control volume (Newton's second law of motion) first with a general stress tensor, representative of viscous or turbulent flow, and then with a Newtonian viscous stress tensor for compressible or incompressible flow. The emphasis is on the conceptual simplicity of these equations despite their algebraic complexity. In Chapter 3 the thin-shear-layer (boundary layer) equations are derived, carefully considering the cases of laminar and turbulent, plane and curved flow, and discussing the limits of validity of the equations more thoroughly than in most textbooks. Chapter 4 begins with a brief review of dimensional analysis, a powerful tool in the analysis of shear layers especially in the presence of heat transfer. This is followed by a discussion of differential equations with emphasis on the boundary conditions. At the senior or graduate level for which this book is intended, courses on numerical analysis rarely contain a general account of boundary conditions, and even at the research level confusion frequently arises.

Chapter 5 is an introduction to the quantitative behavior of shear layers. For simplicity the discussion is carried through for laminar flow although many of the principles recur in the later treatment of turbulent flow. Some readers may be surprised at the emphasis on "similar" flows, which originally became popular because they are so much less tedious to calculate on desk machines than are more general flows. One reason for this emphasis is that similar solutions are still the easiest to understand and therefore the most useful for an introduction to quantitative behavior. Also, the application of similarity transformations greatly reduces variation with the streamwise coordinate even in nonsimilar flows, permitting great economies in computing time in either laminar or turbulent flows. Chapter 6 is a brief introduction to turbulent shear layers, again making use of similarity or self-preservation concepts, and presenting results parallel to those in Chapter 5. Inner layer similarity, which has no counterpart in laminar flow but plays a large part in most calculations for turbulent wall flows, is treated in detail. We have mentioned only briefly empirical transport equations for Reynolds stress, with references to recent reviews; this is not a specialized book on turbulence. In Chapter 7 numerical methods for finite-difference solutions of the thin-shear-layer equations are introduced. There are several well-proven methods; for convenience of exposition we present a highly flexible method developed by Keller and Cebeci (1971), which has been applied to a wide range of external- and internal-flow problems. Chapter 8 presents some numerical solutions by various methods for several laminar and turbulent flows.

The main discussion of instability and transition is in Chapter 9. The physics and mathematics of this subject are rather different from what is needed in the case of fully laminar or fully turbulent flows, and only a spurious unity would be achieved by discussing it after Chapter 5. Chapter 10 is a discussion of three-dimensional and (two-dimensional) time-dependent flows, with emphasis on nearly plane shear layers. Except for the axisymmetric or fully developed cases already discussed, slender shear layers present rather severe problems of turbulence modeling and numerical calculation, for which the solutions are still controversial. Chapter 11 is a discussion of complex shear layers distorted by the boundary conditions or interaction with another shear layer.

A minimal course of instruction on shear layers, for senior undergraduate students or for graduate students with only a fringe interest, would include Sections 1.1, 1.2, 2.1–2.4, 3.1–3.4, and 1.3, in that order, followed by Chapters 5 and 6. In Chapters 4–8, complete computer programs for a variety of methods are given. Some of the homework problems require the use of these methods, while other, shorter problems test comprehension of physical principles. Fortran card decks for the computer programs are available at nominal reproduction and mailing cost. They can be obtained from either author.

The authors would like to express their appreciation to several people who have given thought, time, and financial assistance to the development of this book. In particular, we want to thank Herb Keller of the California Institute of Technology, Keith Stewartson of University College, London, and Eph Sparrow of the University of Minnesota for their many valuable comments and suggestions. Special thanks are due to Mr. D. S. Siegel, Director, Vehicles and Propulsion Program of the Office of Naval Research and to Dr. G. K. Lea, Director of Fluid Mechanics Program of the National Science Foundation for their support of work leading to some of the material presented in this book. Judy Ramsey, Kalle Kaups, and Sue Schimke of the Aerodynamics Research Group of Douglas Aircraft Company and K. C. Chang of the California State University at Long Beach have contributed significantly to the computer programs and the formulation of problems. Students at California State University at Long Beach and Imperial College, London, made helpful comments on the material covered in Chapters 1–6. Finally, we are also indebted to Chris Benge and Nancy Thomas for their excellent typing of the manuscript, and to Mary Phillips and the staff of Hemisphere Publishing for expert editing. The contribution made by our families is something that only another author can fully understand.

Tuncer Cebeci
Peter Bradshaw

MOMENTUM
TRANSFER IN
BOUNDARY LAYERS

Chapter 1

Introduction

1.1 WHAT ARE SHEAR LAYERS?

1.1.1 Idealized Examples of Shear Layers

Momentum is transferred to, from, or within a fluid by internal stresses, namely, the hydrostatic pressure and the viscous stresses. The behavior of fluids affected only by the pressure and not by viscous stresses is relatively easy to predict by standard inviscid flow computer programs, except perhaps for flows partly above and partly below the speed of sound where numerical difficulties exist. This book is concerned with the main class of flows in which viscous (or turbulent) stresses are important. Strictly speaking, the viscous stresses, as seen by a macroscopic observer who cannot detect individual molecules, are the equivalent of momentum transfer by random molecular motion. However, in this book only the macroscopic view will be considered and the fluid will be regarded as continuous.

A variation of velocity in the direction normal to the direction of the velocity itself is called a *shear*.* We shall see that strong shears are usually confined to thin layers. The simplest example of a shear layer is the *mixing layer* separating two

---

*Strictly speaking, it is a *rate of shear strain* rather than a shear strain in the sense of solid mechanics (i.e., a rate of deformation rather than a total deformation). The abbreviation shear is standard.

nearly parallel streams at different velocities (Fig. 1.1a). Note that the velocity tends smoothly to the external-stream values at each side. Within the layer, transfer of momentum by molecular (and possibly turbulent) processes retards the fluid near the high-velocity side and accelerates it near the low-velocity side, so that the width of the layer increases in the streamwise direction. Another example of a layer with

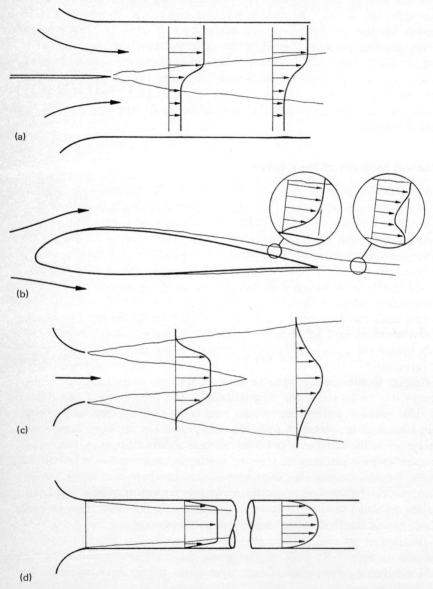

(a)

(b)

(c)

(d)

Fig. 1.1 Examples of shear layers, showing velocity profiles. (a) Mixing layer between parallel streams (of large extent normal to the plane of the paper); (b) boundary layer and wake of airfoil; (c) merging mixing layers in jet; (d) merging boundary layers in pipe flow.

a strong shear is the *boundary layer* between a stream and a solid surface (Fig. 1.1b), which is the commonest type of shear layer. On a solid surface the fluid velocity is reduced to zero (the "no-slip condition"), but there is no direct constraint on the velocity gradient at the surface. At the outer edge the velocity tends asymptotically to the free-stream value as in the mixing layer. A jet (Fig. 1.1c) also has a large shear; so does the flow in a pipe (Fig. 1.1d), where the velocity is again zero at the solid surface. Although regions with annular or circular cross section are not strictly layers, we still call jet or pipe flows *shear layers* because the physical processes are much the same and the equations of motion differ only in detail. Other slender, shear-dominated regions with noncircular cross sections are more complicated, but again they are commonly called shear layers. *Free* shear layers are those not adjacent to a solid surface, the jet being an example. Occasionally, the term boundary layer is used for the whole family of shear layers, but we avoid this usage here.

1.1.2 Practical Examples of Shear Layers

Shear layers can be quite complicated. For example, Fig. 1.2 shows a vertical-takeoff aircraft near the ground, supported by the thrust of its jet engines. There is a boundary layer on the inside circumference of each jet nozzle, and this boundary layer becomes an annular mixing layer as the exhaust leaves the nozzle. The mixing layer grows in thickness until it engulfs the whole cross section of the exhaust flow, which is therefore called a jet. When the jet hits the ground, it is sharply deflected (by pressure gradients; viscous or turbulent stresses are not as powerful as this) and spreads out radially in a flow called, logically enough, a wall jet. If there is a wind blowing (the earth has a boundary layer up to a kilometer thick), the part of the wall jet that blows upwind will eventually separate from the ground and be blown backwards toward the aircraft. Reingestion of hot exhaust gas is undesirable for jet engines, particularly if dust particles have been scoured off the ground, so the aircraft designer is interested in the transport of heat and foreign matter by the shear layer.

The flow inside a jet engine, or any other type of turbomachine, is very complicated because each blade has two boundary layers (one on each surface), which merge at the downstream end of the blade to form a wake (Fig. 1.1b). The wake of each blade then interacts with the flow over blades further downstream; boundary layers grow on the duct walls as well, further complicating the flow over the blades. Jet engine combustion chambers (and many other types of chemical reactors) use jet flows to mix the reactants; extremely complicated flow patterns result, and the amount of smoke and other pollutants emitted by the engine depends critically on the completeness of the mixing.

Once the aircraft of Fig. 1.2 is moving fast enough to be supported by its wings, the behavior of the boundary layers on wings and fuselage determines the drag and limits the lift. In Chap. 11 (especially Fig. 11.5) we shall see how greatly the boundary layer can affect the lift at given incidence, even at incidences well below the stall (by definition, the condition of maximum lift).

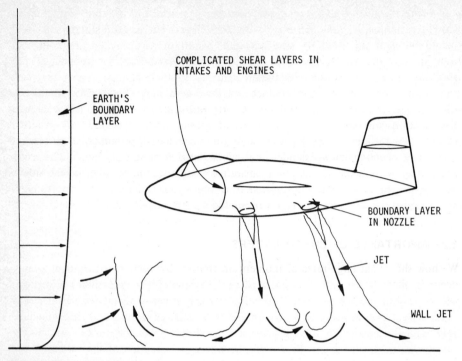

Fig. 1.2 Examples of shear layers.

Figure 1.2 is only one example of the importance of shear layers in engineering. Examples in the environmental sciences, additional to the planetary boundary layer mentioned above, include jet streams in atmosphere and ocean, river flows, and buoyant plumes. In many cases the practical interest is in the transport of heat or of a pollutant rather than of momentum, but because pollutant transport is determined by the momentum (velocity) field the latter must be known first. In this book we concentrate on momentum transfer; heat transfer will be covered elsewhere (Cebeci and Bradshaw, in preparation).

1.1.3 Laminar and Turbulent Flow

Shear layers in which the flow is steady or varies with time only in a simple way are called *laminar* shear layers. The name comes from the Latin *lamina*, meaning sheet (at one time it was thought that sheets of fluid slid over each other, and although the picture is erroneous, the name has survived). However, most shear layers of interest to engineers and environmental scientists are turbulent: that is, although the external stream may be steady or smoothly varying in time, the velocity within the shear layer fluctuates irregularly but continuously in magnitude and direction, both in time and in space. The earth's boundary layer is usually turbulent, and the motion of smoke from a chimney or garden fire on a windy day gives a good idea of the erratic eddying nature of turbulent flow.

Laminar and turbulent flows obey the same equations of motion at any instant. However, in Chap. 2 we shall see that the *mean* motion in turbulent flow is what would result if the fluid were to experience extra internal stresses acting in addition to the viscous stresses (and usually much larger than the latter). Strictly speaking, these stresses are the equivalent, seen by an observer who can sense only the mean flow, of the rates of momentum transfer by the turbulent velocity fluctuations.* If we ignore the details of the turbulent motion and deal only with the mean viscous stresses and the apparent mean turbulent stresses, we can treat laminar and turbulent shear layers in a common framework of formal equations and qualitative phenomena, which will be done in the first part of the book. The reader has merely to remember that the quantitative results depend on whether the flow is laminar or turbulent. Naturally, turbulence will be mentioned from time to time, but we postpone detailed consideration of the subject until Chap. 6.

1.2 IMPORTANCE OF SHEAR LAYERS

We now show that large rates of momentum transfer by viscous or turbulent stresses occur in shear layers and, for the most part, only in shear layers. Newton's second law of motion (the principle of conservation of momentum) applies to solids and fluids alike. This law can be stated for a moving *fluid element* (for instance, a small volume of fluid marked by smoke or dye as sketched in Fig. 1.3) as follows:

$$
\boxed{\begin{array}{l}\text{Rate of increase}\\\text{of momentum of}\\\text{fluid element}\end{array}} = \boxed{\begin{array}{l}\text{Net rate of}\\\text{transfer of}\\\text{momentum to}\\\text{fluid element}\end{array}} = \boxed{\begin{array}{l}\text{Net force}\\\text{applied to}\\\text{fluid element}\end{array}} \qquad (1.2.1)
$$

The net force is the difference between the forces on opposite sides of the element, so that it is proportional to the spatial gradients of the stresses (including the pressure), rather than to the stresses themselves. Figure 1.4 is a simple illustration of net force for the case of the pressure gradient in the x-direction, $\partial p/\partial x$. It will be shown in Chap. 2 that Eq. (1.2.1) can be rewritten as

$$
\boxed{\begin{array}{l}\text{Rate of increase}\\\text{of (momentum per}\\\text{unit volume)}\end{array}} = \boxed{\text{Acceleration}} = \boxed{\begin{array}{l}\text{Sum of relevant}\\\text{stress gradients}\end{array}} \qquad (1.2.2)
$$

Momentum is a vector, so that Eq. (1.2.2) really has three components, one for each spatial direction. Also, each component of momentum can be affected by viscous and turbulent stress gradients in all three coordinate directions, as well as by the pressure gradient in the direction of the component considered.

---

*This is mathematically analogous to the replacement of molecular momentum transfer by viscous stresses in the macroscopic view of laminar flow mentioned in Sec. 1.1.1. There are, however, great differences between the behavior of small, intermittently colliding molecules and that of the large-scale eddying motions of turbulent flow.

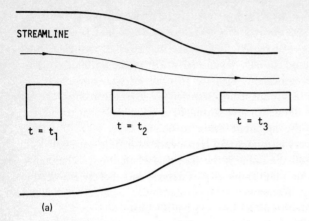

(a)

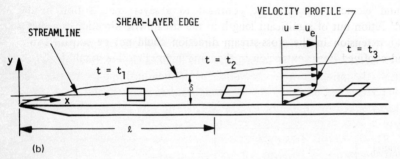

(b)

Fig. 1.3 Deformation of a fluid element by a rate of strain e. Here u and v are the velocity components in the x- and y-directions, respectively. (a) Fluid element in linear deformation ($e \simeq \partial u/\partial x$ or $-\partial v/\partial y$: flow in a nozzle); (b) fluid element in simple shear deformation ($e \simeq 1/2\, \partial u/\partial y$: flow in a boundary layer on a flat surface).

Beginners need to distinguish carefully between the rate of transfer of momentum *to* a fluid element as discussed here and the rate at which momentum is transported downstream *by* the fluid element. In a uniform flow without acceleration or viscous stresses, the former rate is zero but the latter is not. Fundamentally, there are three kinds of momentum transfer: transport by the mean motion of the fluid, transfer by random molecular motion (represented by the viscous stresses), and transfer by turbulent eddies (represented by the mean turbulent stresses). They are different aspects of the same process, and the main purpose of this book is to compare them and to find out where each is significant.

Viscous stresses are usually* small compared with the rate of momentum transfer by the mean motion of fluid elements. Therefore, viscous stress *gradients* will produce significant rates of change of momentum per unit volume only if the viscous stress changes by a large fraction over a small distance, δ, say. By "small" we mean small compared to the distance, l, say, over which large fractional changes

---

*The reader with some previous knowledge of fluid dynamics will recognize that "usually" means "at all but very low Reynolds numbers."

in momentum of a fluid element occur. We must now ask whether the viscous stress can change by a large fraction in a small streamwise distance (x in Fig. 1.3b) or only in a small cross-stream distance (y or z in Fig. 1.3b). Viscous stresses are proportional to the rate of deformation of a fluid element (see Fig. 1.3 for examples of rates of deformation, and for more details see Chap. 2) so that to change the viscous stresses rapidly in the streamwise direction requires the rate of deformation of a given fluid element to change rapidly as it moves downstream. Such a change in deformation rate can, in general, be accomplished only by large changes in pressure gradient (as in the nozzle of Fig. 1.3a), which would overwhelm the viscous stress gradients in Eq. (1.2.2). Therefore, the case of large streamwise gradients of viscous stresses is not important in practice (except in shock waves). Rapid changes of viscous stress in the cross-stream direction (Fig. 1.3b) will occur only if the region of significant viscous stress has a cross-stream thickness $\delta \ll l$.

By the process of elimination, it can be shown that the only kind of deformation of a fluid element that can be confined to a layer that is thin in the cross-stream direction but of significant length in the streamwise direction is a shear. (Briefly, linear extension in the cross-stream direction could not be confined to a thin layer, and confined linear extension in the streamwise direction would lead to a shear.) Therefore, the only kind of flow that can sustain significant viscous stress gradients is a shear layer (Fig. 1.3b) whose cross-stream thickness, δ, is small compared to a typical length in the streamwise direction, l. Figure 1.3b shows a shear layer with a large extent in the direction normal to the page. In contrast, Fig. 1.1, c and d, shows regions of roughly circular cross section: these are "slender" shear layers and the above arguments apply to them as well.

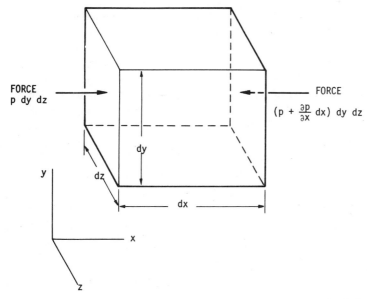

Fig. 1.4 Net force on fluid element due to pressure gradient is $- (\partial p / \partial x) \, dx \, dy \, dz \equiv - (\partial p / \partial x) \times$ (volume).

Outside the shear layer, in the "free stream," the rate of deformation is much smaller. Viscous stresses are therefore negligible, and the acceleration of a fluid element is equal to the pressure gradient.

Very similar arguments can be advanced for turbulent stresses, although unlike viscous stresses they are not directly proportional to local deformation rate. Similar arguments can be used to show that significant heat transfer in a fluid occurs mainly in thin layers separating two regions at different temperatures. Such thermal layers are usually coincident with a shear layer.

Recall the steps in the argument:

1. Viscous stresses are always small in high Reynolds-number flows.
2. Therefore, significant stress *gradients* are confined to thin regions.
3. In practice, regions thin in the streamwise direction are dominated by pressure gradients.
4. Therefore, regions with significant stress gradients are thin in the cross-stream direction.
5. The only kind of deformation that can be confined to such a layer is a shear.

1.3 BEHAVIOR OF SHEAR LAYERS—PRELIMINARY DISCUSSION

In this section we introduce shear-layer phenomena, using the qualitative concepts embodied in the equations of motion while avoiding the rather heavy algebra needed in quantitative discussion. Students who prefer rigorous mathematical analysis to qualitative discussion can proceed to Chap. 2, but they should return to this section after reading Chap. 3, at the latest, to acquire a grasp of the physics as well as the mathematics of shear layers. They should also review Probs. 1.1-1.3 at the end of this chapter.

The viscous stress produced by a rate of simple shear deformation (Fig. 1.3b) is, of course, a shear stress: in this example it acts in the streamwise direction. If the stream lines are not exactly parallel, there is in addition a rate of linear deformation (as in Fig. 1.3a) producing viscous normal stresses. In turbulent flow even a simple shear will produce normal stresses of the same order as the shear stress, but we shall see in Chap. 3 that normal-stress gradients do not usually play an important part in the momentum-conservation equations. The principle of conservation of momentum [Eq. (1.2.2)] for a laminar or turbulent shear layer is, therefore, approximately

$$
\boxed{\begin{array}{l}\text{Rate of increase}\\ \text{of } \textit{streamwise}\\ (x)\text{ component of}\\ \text{momentum per unit}\\ \text{volume}\end{array}} \approx - \boxed{\begin{array}{l}\text{Streamwise}\\ \text{gradient of}\\ \text{pressure } (p)\end{array}} + \boxed{\begin{array}{l}\text{Cross-stream}\\ \text{gradient of}\\ \text{shear stress}\end{array}} \qquad (1.3.1)
$$

and

$$
\boxed{\begin{array}{l} \text{Rate of increase} \\ \text{of } \textit{cross-stream} \\ (y) \text{ component of} \\ \text{momentum per unit} \\ \text{volume} \end{array}} \quad \approx \ - \quad \boxed{\begin{array}{l} \text{Cross-stream} \\ \text{gradient of} \\ \text{pressure } (p) \end{array}} \qquad\qquad (1.3.2)
$$

because the stress gradients in the cross-stream equation are small. If the properties of the shear layer change significantly in the direction, z, which is normal both to the cross-stream direction, y in Fig. 1.3b, and to the general direction of flow, x, a third equation, resembling Eq. (1.3.1), is needed, and the layer is called "three-dimensional" (z is hereafter called the "spanwise" direction, by analogy with the situation of an aircraft wing). If z-wise mean velocities are zero, the layer is "two-dimensional."

The cross-stream equation (1.3.2) is often trivial, because the curvature of the layer in the xy-plane is either negligible (Fig. 1.3b) or known because it is forced upon the shear layer by the external flow or boundary shape (Fig. 1.1b). This is a more powerful simplification than the smallness of some of the stress gradients.

If the ratio of the shear-layer thickness, δ, to the streamwise length of the flow, l, is small (or if the local growth rate $d\delta/dx$ is everywhere small), the effect of the shear layer on the external flow is small. For instance, the flow near an airfoil, but outside the boundary layer and wake, and even the pressure distribution on the airfoil, is close enough to the "inviscid" flow (with the same lift force, not necessarily the same incidence) for the latter to be a good enough approximation for design purposes. However, there are many cases in which δ/l, though small, is not small enough to convert the approximation signs in Eqs. (1.3.1) and (1.3.2) into acceptable equalities or to make the interaction between the shear layer and the external flow negligible. Some examples are given in Chap. 11. The simplification that results when δ/l or $d\delta/dx$ is very small—the thin-shear-layer (TSL) approximation—is so spectacular and so satisfying, aesthetically and mathematically, that the less tractable cases of moderately small δ/l receive little attention in textbooks. One of the novel contributions of this book is the attempt to put the TSL approximation in perspective for students. It appears to us that such an attempt is overdue because modern computers can do the heavy arithmetic required to calculate the interaction between a shear layer—or layers—and the external flow, whereas only a few years ago even the simplest laminar TSL problems were barely amenable to calculation unless crude numerical approximations were made.

Before going on to the equations of motion and the TSL simplifications of them, we shall outline the general behavior of shear layers. Special attention will be given to the boundary layer (Figs. 1.1b and 1.3b), because it is one of the most common types of shear layer and certainly the type whose accurate prediction has attracted most effort. The TSL approximation was first derived for boundary layers by Ludwig Prandtl in 1904 and is frequently called the "boundary-layer approximation." Indeed it is not uncommon to describe all shear layers as boundary layers although this is hardly a logical name for a wake or a jet. In this book a boundary layer is a fluid layer adjacent to a boundary, and the more general designation is shear layer. Prandtl not only developed the equations but also illustrated the behavior of the solutions with some flow-visualization photographs, which are still instructive today (Goldstein, 1965; Prandtl and Tietjens, 1954).

Consider the boundary layer on a smooth plate with an external stream of velocity u_e, independent of x and y (Fig. 1.3b). We shall assume that the flow is steady and laminar. If the nose of the plate is sharp, the only relevant length scale is the distance from the leading edge, x. The relevant fluid properties are the viscosity, μ, and the density, ρ (both assumed constant in this volume). Since no other quantities with mass dimensions enter the problem, we need consider only the ratio $\nu \equiv \mu/\rho$, the kinematic viscosity with the dimensions of (length)^2/time. In this context "kinematic" simply means "divided by ρ." The quantity $u_e x/\nu$, which is dimensionless, is a Reynolds number. Many other definitions of Reynolds number are possible; we denote this one by Re_x. In this particular flow there is no other *independent* definition of Reynolds number because there are no other independent velocity and length scales. Thus $u_e \delta/\nu \equiv Re_\delta$, where δ is the local thickness of the boundary layer defined explicitly below, is a unique function of $u_e x/\nu$. Obviously δ, rather than x, is the most useful length scale for discussing local conditions. Analysis and numerical solutions to be presented in Chap. 5 show that if we define δ as the distance from the surface at which the x-component of velocity, u, is $0.995u_e$, then

$$\frac{u_e \delta}{\nu} \equiv \frac{u_e \delta_{995}}{\nu} \simeq 5.3 \left(\frac{u_e x}{\nu}\right)^{1/2} \tag{1.3.3}$$

or

$$\delta \simeq 5.3 \left(\frac{\nu x}{u_e}\right)^{1/2} \tag{1.3.4}$$

or

$$\frac{d\delta}{dx} = \frac{1}{2}\frac{\delta}{x} \simeq 2.65 \left(\frac{u_e x}{\nu}\right)^{-1/2} \tag{1.3.5}$$

As an example of the thinness of a laminar shear layer, consider an air flow at atmospheric temperature and pressure ($\nu \simeq 1.5 \times 10^{-5}$ m^{2}s^{-1}) at a speed of 40 m s^{-1} (about 130 ft/s, a typical speed for university test rigs) over a plate of length $l = 1$ m, parallel to the stream. For these conditions, we have $u_e l/\nu \simeq 3 \times 10^6$, so that at the end of the plate, where according to Eq. (1.3.3) $u_e \delta/\nu \simeq 9 \times 10^3$, the layer thickness δ is about 3×10^{-3} m $= 3$ mm. It follows that δ/l is only about 1/300. This is about the smallest value of δ/l likely to be found in practice: δ/l decreases as $u_e l/\nu$ increases, and the chosen value of $u_e l/\nu$ is about the highest at which the boundary layer can be persuaded not to go turbulent, even in a high-quality test rig. If the boundary layer had been forced to go turbulent fairly close to the leading edge, its thickness at $x = 1$ m would be about 1.9 cm ($\delta/l \simeq$ 1/50). In turbulent jets in still air, $\delta/l \approx 1/5$, where δ is the jet thickness or for circular jets the diameter, defined as the distance between points at which $u = 0.005\, u_{max}$, say. Laminar shear layers are discussed in more detail in Chap. 5 and turbulent shear layers in Chap. 6.

Laminar shear layers become unstable at a Reynolds number that depends on the shape of the velocity profile. For the boundary layer on a flat surface adjacent to a stream of constant speed this "critical" Reynolds number is about $u_e \delta/\nu = 1600$. The most unstable type of disturbance is a sinusoidal fluctuation of the streamwise and cross-stream velocity components within the shear layer, moving downstream as a "traveling wave" (like an ocean wave) at a speed somewhere near the average speed of the shear-layer fluid (about $0.4u_e$ in the case of the flat-plate boundary layer) and having a wavelength of about 5–10δ. When the velocity fluctuations have reached an amplitude on the order of $0.01u_e$, secondary instabilities appear, generally involving spanwise variations. If disturbances of this order of amplitude are present in the free stream, the secondary instabilities can appear at once. In either case the fluctuations become more intense and more complicated, covering a wider and wider range of frequencies, until the amplitudes of fluctuation of all three velocity components in the central part of the layer are on the order of $0.1u_e$, when a limiting condition is reached* and the statistical properties of the fluctuations become nearly constant. The exact point at which the motion should be called "turbulence" is a matter of definition: the statistically steady state is called "fully developed turbulence." The kinetic energy supplied to the growing disturbances and the energy needed to maintain fully developed turbulence in the presence of the fluctuating viscous stresses that drain its kinetic energy come from the mean flow. The process of transition from laminar to turbulent flow can be quite prolonged, as witness the fact that the flow can still be effectively laminar at six times the critical value of $u_e \delta/\nu$, as in the example above. The "transition *point*" mentioned in elementary accounts is an oversimplification. However, transition takes place much more rapidly in flows with disturbances in the free stream, in free shear layers, and in boundary layers with u_e decreasing in the streamwise direction. Transition and its prediction are discussed briefly in Sec. 6.1 and in more detail in Chap. 9.

The flow external to a boundary layer or other shear layer is sometimes called inviscid. Of course, the viscosity in the external flow is the same as in a shear layer at the same temperature and pressure. What is really meant by the term is that viscous stresses are negligibly small, because the rate of deformation, rather than the viscosity, is negligible. Sometimes the inviscid flow external to a turbulent shear layer is called laminar. This term is even more confusing and should be avoided.

In the inviscid flow Bernoulli's equation applies, and so, just outside the edge of the shear layer (denoted by subscript e), we have

$$p_e + \frac{1}{2} \rho u_e^2 = \text{const} \tag{1.3.6}$$

*Not necessarily monotonically; the amplitudes often "overshoot."

where we use the constant-density form of Bernoulli's equation for simplicity. Therefore,

$$\frac{dp_e}{dx} = -\rho u_e \frac{du_e}{dx} \tag{1.3.7}$$

(this form is valid in variable-density flow also), and if the shear layer is thin enough the pressure will be closely equal to p_e throughout its thickness [see Eq. (1.3.2)]. Thus, we have related the streamwise pressure gradient, which appears in Eq. (1.3.1), to the streamwise variation of u_e. In the flow over a flat plate mentioned above, u_e is independent of x. This is "constant-pressure," "zero-pressure gradient," or simply "flat-plate" flow. Within the shear layer, viscous stress gradients affect the motion, and Bernoulli's equation is no longer correct. However, the motion may be usefully regarded as a response to Bernoulli's equation, modified by viscous stresses. In particular, the qualitative implication of Eq. (1.3.7) still holds: the smaller u is, the larger the streamwise change in u that corresponds to a given change in p_e (up to a point: viscous stresses constrain u to remain zero at the surface).

Figure 1.5 shows the velocity profile and the profile of the shear stress, τ (simple, $\tau = \mu \, \partial u / \partial y$), in a two-dimensional laminar boundary layer with a streamwise variation of u_e; this kind of distribution of u_e is called a "roof top," and airfoil sections are often designed to have such a distribution on the upper surface. At the surface Eq. (1.2.2) reduces to

$$\boxed{\begin{array}{l}\text{Sum of}\\ \text{stress}\\ \text{gradients}\end{array}} = 0 \tag{1.3.8}$$

or, in two-dimensional flow, to

$$\boxed{\begin{array}{l}\text{Pressure gradient}\\ \text{in } x\text{-direction}\\ \equiv dp_e/dx\end{array}} = \boxed{\begin{array}{l}\text{Shear stress gradient}\\ \text{in } y\text{-direction}\\ \equiv \partial\tau/\partial y\end{array}} \tag{1.3.9}$$

Therefore $\partial\tau/\partial y$ remains zero at the surface until u_e starts to decrease (and p_e starts to increase). Thereafter τ increases away from the surface before finally decreasing to zero at the outer edge. The shear stress in the central part of the layer usually increases above that found in a constant-pressure layer at the same Reynolds number, because the relation

$$\int \tau \, dy = \int \mu \frac{\partial u}{\partial y} \, dy = \mu u_e \tag{1.3.10}$$

must be satisfied. However, the increase is not enough to prevent the shear stress at the wall, τ_w, from falling below the constant-pressure value. The value of $d\tau_w/dx$

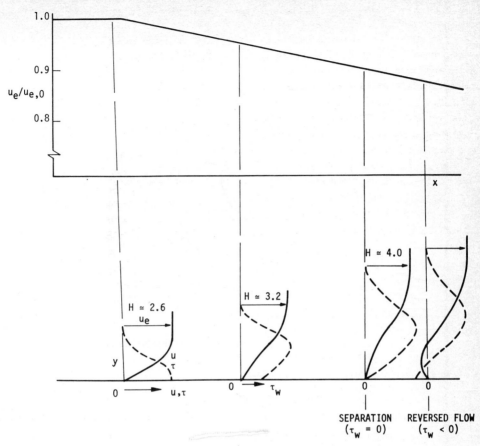

Fig. 1.5 Profiles of velocity u and shear stress $\tau \equiv \mu \, \partial u/\partial y$ in the laminar boundary layer on a "roof top" airfoil. The symbol H is explained on p. 18.

depends—in a complicated way—on du_e/dx, but for a linear and not-too-rapid decrease of u_e with x, τ_w passes through zero when u_e has fallen to roughly 0.9 of its original value. Negative τ_w means negative $\partial u/\partial y$ at and near the surface—that is, the flow near the surface is reversed, and the shear layer itself separates from the surface. The subsequent behavior depends on where the upstream-moving fluid comes from, but in most cases the reversed-flow region increases in thickness as x increases, as shown in Fig. 1.6. The exaggerated drawings seen in some textbooks, in which the shear layer separates from the surface so rapidly that the streamlines turn through a large angle, are not realistic. (Figure 1.6 is unrealistic in showing a steady flow: usually an oscillatory "vortex street" forms behind a body with a large separated region.) If the rate of decrease of u_e becomes less soon after separation and if the surface is not too highly convex, the shear layer may reattach to the surface, as shown in Fig. 1.7. This is because the rate at which the lower edge of the separated shear layers propagates toward the surface exceeds the rate at which the reversed flow region widens under the influence of the pressure gradient. A limited separated-flow region of this kind is called a "bubble."

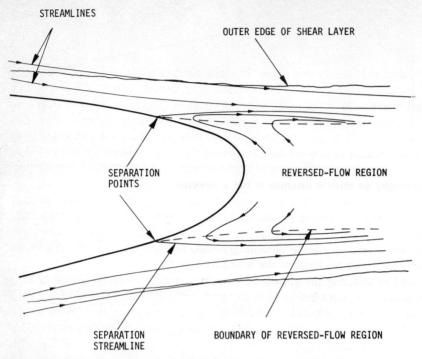

STREAMLINES

OUTER EDGE OF SHEAR LAYER

SEPARATION
POINTS

REVERSED-FLOW REGION

SEPARATION
STREAMLINE

BOUNDARY OF REVERSED-FLOW REGION

Fig. 1.6 Separation from rear of a bluff body.

The whole of the above qualitative description applies to turbulent boundary layers also, except that there is no unique relation between τ and $\partial u/\partial y$, and the profile shapes differ somewhat from those in laminar flow. Since turbulent stresses are much larger than viscous stresses, a turbulent boundary layer can withstand a greater decrease in u_e before separating. For a linear and not-too-rapid variation, u_e can fall to roughly 0.5–0.6 of its original value (note that according to Bernoulli's equation, the pressure change is proportional to the change in velocity squared so that a turbulent boundary layer can withstand an order of magnitude more pressure rise than a laminar boundary layer can).

Because a positive pressure gradient ($dp_e/dx > 0$, $du_e/dx < 0$) tends to cause separation, which is generally (but not always) undesirable, this is called an "adverse" pressure gradient. A negative ("favorable") pressure gradient increases τ_w above the value found in a constant-pressure boundary layer at the same Reynolds number and never causes separation. Positive pressure gradients do not always lead to separation, even if u_e decreases to a small fraction of its value. The question is whether pressure gradients are large compared to typical shear stress gradients.

If we consider laminar flow and write a typical value of the velocity gradient $\partial u/\partial y$ as u_e/δ, so that a typical viscous stress is $\mu u_e/\delta$, a typical shear stress gradient in the y direction can be written as $\mu u_e/\delta^2$. The ratio of the pressure gradient, $dp/dx \equiv -\rho u_e\, du_e/dx$, to this shear stress gradient is

$$-\Lambda \equiv -\frac{\delta^2}{\nu}\frac{du_e}{dx} \tag{1.3.11}$$

We shall frequently use typical values in this way. Any more careful definition of a typical viscous stress would just introduce a numerical factor in Eq. (1.3.11). Confusion might arise if the factor were an order of magnitude greater or less than unity, but one can usually guess typical values to better accuracy than this. Separation occurs when the dimensionless parameter $(-\Lambda)$ increases to about 10 or 12. If u_e is arranged to vary so that this parameter never becomes too large—that is, if du_e/dx decreases as δ^2 increases—u_e can be reduced to zero without separation, although clearly an infinite distance in the x direction is required.

In turbulent flow a typical shear stress is closer to a constant multiple of ρu_e^2 than to a constant multiple of $\mu u_e/\delta$, so that a more relevant pressure-gradient parameter in that case is $-(\rho u_e \, du_e/dx)/(\rho u_e^2/\delta) \equiv -(\delta/u_e) \, du_e/dx$. In both cases, however, the effect of a given pressure gradient—say, the roof top airfoil distribution—depends on the thickness of the boundary layer.

As well as affecting the growth of a laminar or a turbulent boundary layer, a pressure gradient affects the likelihood of transition from laminar to turbulent flow. The theory of hydrodynamic stability (Chap. 9) predicts that a shear layer with a point of inflexion in the velocity profile (i.e., a point where $\partial^2 u/\partial y^2 = 0$) is unstable to infinitesimal traveling-wave disturbances except at very low Reynolds numbers. This is why laminar free shear layers at moderate or high Reynolds numbers are always unstable. Since they have $\partial u/\partial y = 0$ at each edge (Fig. 1.1, a and c), there must be at least one point where $\partial^2 u/\partial y^2 = 0$. As mentioned above, a laminar boundary layer in an adverse (positive) pressure gradient has a point of inflexion; in zero pressure gradient the inflexion is at the surface.

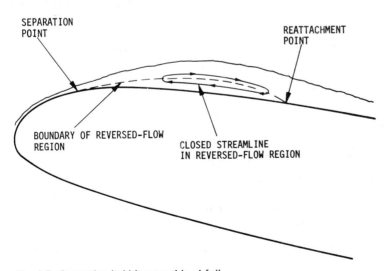

SEPARATION
POINT

REATTACHMENT
POINT

BOUNDARY OF REVERSED-FLOW
REGION

CLOSED STREAMLINE
IN REVERSED-FLOW REGION

Fig. 1.7 Separation bubble on a thin airfoil.

The critical Reynolds number, $u_e\delta/\nu$, for instability of a boundary layer to infinitesimal disturbances is, to a fair approximation, a function of the local value of $(-\Lambda) \equiv (\delta^2/\nu)\, du_e/dx$ only. A laminar boundary layer becomes unstable before separation if the value of $u_e\delta/\nu$ at separation exceeds about 170. As mentioned above, fully turbulent flow takes a finite streamwise distance to develop after the initial instability, but it is not possible to be precise about this because the process depends on the size of the disturbances introduced by slight unsteadiness in the external flow, as well as on the further development of the pressure gradient. Certainly laminar separation can precede the appearance of significant turbulent stresses, even at Reynolds numbers, $u_e\delta/\nu$, much higher than that at which instability appears. However, in such cases transition is usually complete soon after separation, and the growth rate of the shear layer therefore increases. The tendency for the shear layer to reattach is therefore increased. Indeed the behavior of separation bubbles (Fig. 1.7) depends critically on whether or when transition occurs, and this is one of the most unpredictable forms of "scale effect," the failure of wind-tunnel or towing-tank tests at low Reynolds number to reproduce the behavior of the flow over a fullsize aircraft, ship, or other body.

Uncertainty in the position of transition of the *attached* boundary layer can occasionally cause large differences between model and full-scale tests, and therefore test-rig operators normally force transition on a model at the approximate position where it is expected to occur at full scale. This can be done by producing a local separation at an artificial step in the surface, so that the profile becomes unstable even at the model Reynolds number. Three-dimensional disturbances such as scattered sand grains glued to the surface are particularly effective, because three dimensionality is a prerequisite of transition, but a spanwise wire attached to the surface has the advantage of precise geometry.

Two matters not fully discussed above are the effect of a boundary layer or other shear layers on the external inviscid flow and the definition of a physically meaningful thickness in place of the arbitrary $\delta \equiv \delta_{995}$. The main effect of a boundary layer on the external flow is to displace the streamlines away from the surface in the direction of the surface normal. This occurs because the fluid near the surface is slowed down by viscous effects. In a two-dimensional flow, the rate at which fluid mass passes the plane $x =$ constant between $y = 0$ and $y = h$, where h is slightly larger than δ, is

$$\int_0^h \rho u \, dy$$

per unit distance in the z- (spanwise) direction. In the absence of a boundary layer, $u = u_e$ and $\rho = \rho_e$ (we can usually neglect the very small value of $\partial u/\partial y$ in the inviscid flow). Therefore, the reduction in mass flow rate per unit span between $y = 0$ and $y = h$ caused by the presence of the boundary layer is

$$\int_0^h (\rho_e u_e - \rho u) \, dy$$

(note that the integrand is zero outside the boundary layer, so the upper limit of integration is not critical). The thickness in the y-direction of a layer of external stream fluid carrying this mass flow per unit span is

$$\delta^* \equiv \int_0^h \left(1 - \frac{\rho u}{\rho_e u_e}\right) dy \qquad (1.3.12)$$

This is the distance by which the external-flow streamlines are displaced in the y-direction by the presence of the boundary layer and is called the displacement thickness. For a laminar boundary layer in constant-property constant-pressure flow, $\delta^* = 1.73(vx/u_e)^{\frac{1}{2}}$ or about $0.33\delta_{995}$. As well as providing a measure of the effect of a boundary layer on the external flow, δ^* is clearly a physically meaningful thickness of the boundary layer. Thus both the matters discussed at the start of this paragraph have been covered.

However, δ^* is not the only physically meaningful thickness: we can also define a momentum-deficit thickness (or simply momentum thickness) as the thickness of a layer of external stream fluid carrying a momentum flow rate equal to the reduction in momentum flow rate caused by the presence of the boundary layer. Momentum flow rate is defined as mass flow rate times momentum per unit mass (i.e., velocity). It is convenient to define the momentum-deficit thickness in terms of actual mass flow rate times deficit in momentum per unit mass, rather than in terms of the deficit of mass flow rate times momentum per unit mass. This is done because the mass-flow deficit already appears in δ^*. Therefore, following the arguments used in defining δ^*, we get

$$\theta = \int_0^h \frac{\rho u}{\rho_e u_e}\left(1 - \frac{u}{u_e}\right) dy \qquad (1.3.13)$$

In constant-density flow

$$\delta^* = \int_0^h \left(1 - \frac{u}{u_e}\right) dy \qquad (1.3.14)$$

$$\theta = \int_0^h \frac{u}{u_e}\left(1 - \frac{u}{u_e}\right) dy \qquad (1.3.15)$$

(In Europe, symbols δ_1 and δ_2 are sometimes used instead of δ^* and θ.) Both δ^* and θ are useful definitions of thickness and can be applied to shear layers other than boundary layers. Of the two, θ is the more relevant for engineers concerned with momentum losses, and $u_e\theta/v$ is usually the most convenient Reynolds number for correlating shear-layer properties. However, we shall see later that the momentum-conservation equation written for the shear layer as a whole, expressible as an equation for $d\theta/dx$, still contains δ^* or the ratio δ^*/θ. This ratio δ^*/θ, written as H or H_{12}, has its own interpretation as a "shape parameter" of the velocity profile. If for illustrative purposes we assume that the velocity profile of a boundary layer is given, between $y = 0$ and $y = \delta$ only, by the "power law"

$$\frac{u}{u_e} = \left(\frac{y}{\delta}\right)^{1/n} \tag{1.3.16}$$

for some value of n, we find that in incompressible flow $H = (n + 2)/n$. If $n = 1$ [a linear profile, not too far from the real profile of a constant-pressure laminar boundary layer with a value of δ_{995} about 1.5 times the artificial δ used in Eq. (1.3.16)], then $H = 3$ (for the real constant-pressure laminar layer it is 2.6). Power-law profiles for $n \simeq 7$ are surprisingly good fits for constant-pressure turbulent boundary layers, except near the wall where Eq. (1.3.16) predicts an impossible infinite value of $\partial u/\partial y$ for $n > 1$, and near $y = \delta$ where $\partial u/\partial y$ does not asymptote to zero. For $n = 7$, $H = 1.286$. Figure 1.8 shows that the larger the value of H (the smaller the value of n), the less "full" is the profile. If there were no boundary layer, H would be 1.0.

Approximate values of H (for a laminar boundary layer) are shown in Fig. 1.5. A turbulent boundary layer has $H \simeq 1.2$-1.5 in constant-pressure flow, decreasing as the Reynolds number increases ($H = 1.286$ at $u_e\delta/\nu = 27,500$) but rising to $H \simeq 2$-3 at separation (depending on the details of the approach to separation). A full profile indicates that shear stresses, aided or resisted by pressure gradients, are causing a high rate of momentum transfer. The following rules of thumb are useful for constant-property boundary layers in mild pressure gradients:

$$\frac{\delta^*}{\delta} \simeq \begin{cases} \dfrac{1}{3} \text{ for laminar flow} \qquad h = 2.67 \\[2ex] \dfrac{1}{7} \text{ for turbulent flow} \qquad h = 1.43 \end{cases}$$

$$\frac{\theta}{\delta} \simeq \begin{cases} \dfrac{1}{8} \text{ for laminar flow} \\[2ex] \dfrac{1}{10} \text{ for turbulent flow} \end{cases}$$

$$\frac{d\delta}{dx} \simeq \begin{cases} 0.015 \text{ (outer edge inclined at } 1°) \text{ for turbulent flow} \\ 0.015 \text{ in laminar flow at } u_e x/\nu = 3 \times 10^4 \end{cases}$$

The numerical values for turbulent flow all decrease slowly with increasing Reynolds

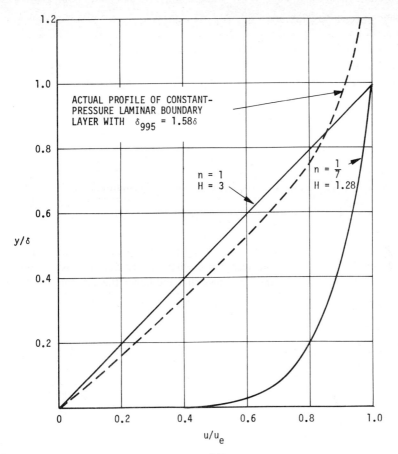

ACTUAL PROFILE OF CONSTANT-PRESSURE LAMINAR BOUNDARY LAYER WITH $\delta_{995} = 1.58\delta$

$n = 1$
$H = 3$

$n = \frac{1}{7}$
$H = 1.28$

y/δ

u/u_e

Fig. 1.8 Power-law profiles, $u/u_e = (y/\delta)^{1/n}$.

number, those given applying to typical laboratory conditions. A reader to whom the value $u_e\delta/\nu = 27{,}500$ quoted above for a turbulent boundary layer means nothing can now see that it corresponds roughly to $u_e x/\nu \simeq (u_e\theta/\nu) \times (\delta/\theta)/(d\delta/dx) = 27{,}500 \times 10/0.015 \simeq 2 \times 10^7$ or about six times the value in the example given on p. 10. Here x is the distance from the "effective origin" of the turbulent layer: because the thickness at transition is finite, the effective origin is some way upstream of transition (see Fig. 6.19 and the accompanying discussion).

Other thicknesses and other shape parameters can be defined. The thicknesses that appear in analysis of heat transfer (for instance, the enthalpy-deficit thickness analogous to θ) are discussed elsewhere (Cebeci and Bradshaw, in preparation).

Some care is needed in interpreting δ^*, θ, and other thicknesses in flows other than boundary layers because of difficulties with the boundary conditions. We shall deal with such difficulties as they arise. Here we remark only that significant values of $\partial p/\partial y$, which occur in strongly curved shear layers, invalidate not only the above definitions of δ^* and θ but also most of the equations in which one would like to use them. The above remarks about typical profile shapes apply mainly to boundary layers, although they are broadly correct for duct flows, which in longitudinal section look like pairs of (interacting) boundary layers. Free shear-layer profiles show less variety in profile shape, at least in the sample cases shown in Fig. 1.1, because they are constrained to $\partial u/\partial y = 0$ at each edge. Also, the marked difference in profile shape between laminar and turbulent shear layers is found only in wall flows. In wall flows the efficiency of turbulent transfer of momentum (which will not be defined more closely at present) falls rapidly near the wall because of the constraint on eddy size, while the efficiency of viscous transfer of momentum (the viscosity) is independent of y in isothermal flow. In free shear layers the efficiency of turbulent transfer is roughly independent of y, as is the viscosity.

EXAMPLE 1.1

Make a rough estimate of the boundary-layer thickness, δ, and the momentum-thickness Reynolds number, $u_e\theta/\nu$, near the back of a supertanker of length 300 m, moving at 10 m s^{-1}.

Solution The best approximation we can make is that the boundary layer is the same as on a smooth flat plate: this will be an underestimate because of the effects of pressure gradient, three dimensionality, and surface roughness. The boundary layer will thicken rapidly or even separate very near the stern. The boundary layer, of course, goes turbulent very near the bow. Therefore, take $d\delta/dx = 0.015$ (this is an overestimate; the figure of 0.015 applies to lower Reynolds numbers). Then $\delta = 4.5$ m. Take $\theta/\delta = 1/10$ (overestimate); then $\theta = 0.45$ m and $u_e\theta/\nu = 10 \times 0.45/1.3 \times 10^{-6} = 3.5 \times 10^6$, in water at 10°C. In fact, for $u_e x/\nu = 2.3 \times 10^9$, $u_e\theta/\nu \simeq 1.7 \times 10^6$ on a smooth flat plate: the rough estimate is probably nearer the truth for a real ship.

EXAMPLE 1.2

It is predicted that in inviscid flow a certain family of roof top airfoils (Fig. 1.5) would have constant external-stream velocity from the leading edge to $x = \alpha c$, where c is the chord, followed by a linear decrease to half the initial velocity at the trailing edge, $x = c$. Find the value of α for which a laminar boundary layer would just separate at the start of the region of adverse pressure gradient.

Solution $\delta = 5.3 \sqrt{\nu x/u_e}$ from Eq. (1.3.7). According to the rough criterion following Eq. (1.3.11) separation occurs when $(\delta^2/\nu) \, du_e/dx \simeq -10$. At $x = \alpha c$ with $du_e/dx = -0.5u_e/[(1-\alpha)c]$ we get $(\delta^2/\nu) \, du_e/dx = -0.5 \times 5.3^2\alpha/(1-\alpha)$, which equals -10 when $\alpha \simeq 0.42$. This is an unrealistic calculation because separation will considerably change the pressure distribution.

PROBLEMS

1.1. Does the TSL approximation $d\delta/dx \ll 1$ apply throughout each of the following flows, or only to part of the flow, or not at all?
   a. Boundary layer in zero pressure gradient on a flat plate.
   b. Boundary layer on an airfoil.
   c. Wake of an airfoil.
   d. Boundary layer and wake of a stalled airfoil.
   e. Free jet.
   f. Jet impinging on a solid surface (see Fig. 1.2).
1.2. Which of the following statements are true?
   a. Separation can occur only in an adverse pressure gradient (i.e., $dp/dx > 0$, or u_e decreasing in the streamwise direction).
   b. Transition can never occur in a favorable pressure gradient.
   c. The skin friction coefficient is always larger in turbulent flow than in laminar flow.
   d. Shear stress is always a maximum at the surface.
1.3. Assuming that the velocity profile in a turbulent boundary layer in mild pressure gradient can be approximated by a one-seventh power law, show that when the boundary layers in a two-dimensional duct flow are about to meet on the centerline (Fig. 1.1d can be regarded as the side view of the duct) the centerline velocity u_e is eight-sevenths times the value at the entry, u_0, say.

Chapter 2

The Continuity and Momentum Equations of Fluid Flow

2.1 CONSERVATION OR "TRANSPORT" EQUATIONS

Fluids, like solids, obey the principles of conservation of mass, momentum, and energy. In Chap. 1 we discussed the qualitative consequences of momentum conservation for a moving fluid element, analogous to a particle in solid-body mechanics. For quantitative purposes, however, it is more convenient to consider the flow of fluid through a small stationary control volume (CV). The conservation principle or transport principle (Bird et al., 1960) for any quantity q can be written for the CV as

$$
\boxed{\begin{array}{l}\text{Rate of increase}\\\text{of } q \text{ in CV}\end{array}} = \boxed{\begin{array}{l}\text{Rate at which } q\\\text{enters through}\\\text{the surface of CV}\end{array}} - \boxed{\begin{array}{l}\text{Rate at which } q\\\text{leaves through}\\\text{the surface of CV}\end{array}}
$$

$$
+ \boxed{\begin{array}{l}\text{Net sum of sources}\\\text{and sinks of } q\\\text{within CV}\end{array}} \qquad (2.1.1)
$$

The equations obtained by taking the limit as the CV shrinks to zero are the *Eulerian* equations for quantities measured at a fixed point. CV analysis is not the only way of deriving the equations, but it is the most appealing for teaching purposes because it demonstrates the physical processes. Therefore we shall derive all the required conservation equations by CV analysis at the risk of repeating results with which the reader is already familiar. In this book we shall not consider equations written for a moving fluid element (the Lagrangian equations): they are less tractable and are used only in mass-diffusion studies.

We shall use rectangular coordinates x, y, and z, along which the velocity vector **V** has components u, v, and w. Pressure will be denoted by symbol p and density by symbol ρ, and u, v, w, p, and ρ can be functions of time t as well as of spatial coordinates x, y, and z. In many cases the variables may be independent of time (steady flow) or independent of one coordinate, usually z (two-dimensional flow), or the viscosity and/or density may be constant.

In aeronautical engineering, flows with nearly constant fluid properties are called "incompressible" or "low-speed" flows. However, there are some low-speed flows with large density changes caused by foreign-fluid injection or by large temperature changes associated with heat transfer. In the case of liquids the viscosity will vary with temperature, and in the case of gases both density and viscosity will vary, so these flows share most of the complications of high-speed (compressible) gas flows. In this book we shall refer rigorously to "constant-density" flows (changes in viscosity being allowed) and to "constant-property" flows in which both density and viscosity must be constant. In both cases "constant" means "constant to the required engineering accuracy," and the effect on density of the small pressure changes needed to drive a low-speed flow is ignored. To treat a constant-density flow we can always use the simple constant-density form of the mass-conservation equation, but if the viscosity varies we need the full version of the momentum-conservation equation.

2.2 THE CONTINUITY EQUATION (PRINCIPLE OF CONSERVATION OF MASS)

Consider an elementary CV with mutually perpendicular sides of length dx, dy, and dz (Fig. 2.1): the shape, a rectangular parallelepiped, is chosen for convenience of analysis in Cartesian coordinates but is not critical for the final result. The continuity equation is obtained from Eq. (2.1.1) with $q \equiv$ mass $\equiv$ (density times volume): there are no sources or sinks of mass, so the equation states that the net rate at which mass is entering the CV equals the rate at which the mass of fluid in the CV is increasing.

To formulate this and other equations algebraically, we need to evaluate increments in variables between pairs of parallel volume-element faces. If we expand the velocity component in the x-direction, u, in a Taylor series in powers of the distance from the center of the left-hand face of the CV (the point P in Fig. 2.1), the value at the center of the right-hand face (the point Q) can be written as

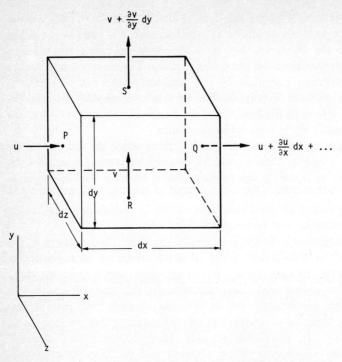

Fig. 2.1 CV for derivation of conservation equations (z-component of velocity, w, omitted for clarity).

$$u + \left(\frac{\partial u}{\partial x}\right) dx + \left(\frac{\partial^2 u}{\partial x^2}\right) \frac{(dx)^2}{2!} + \cdots$$

where all quantities are evaluated at the center of the left-hand face. According to the usual rules of differential calculus, we assume the CV to be so small that higher powers of dx, dy, or dz can be neglected and the x-component of the velocity at Q can be written simply as

$$u + \frac{\partial u}{\partial x} dx$$

Similarly, the mass flow rate per unit area at Q is

$$\rho u + \frac{\partial(\rho u)}{\partial x} dx$$

(The point on PQ at which $\partial u/\partial x$ is evaluated is now immaterial because terms in $\partial^2 u/\partial x^2$ are being neglected.) Similar results follow for directions other than x or quantities other than u. Neglect of squares and higher powers of dy and dz means that averages over the faces equal the respective values at the centers. Therefore the rate at which mass flows into the CV through the face whose center is P is

$\rho u \, dy \, dz$

Inspection of the dimensions of this quantity shows that they are (mass/time) as required. Similarly, the rate at which mass flows out of the CV through the face whose center is Q is

$$\left[\rho u + \frac{\partial(\rho u)}{\partial x} dx\right] dy \, dz$$

and therefore the net rate at which the x-component of velocity transfers mass out of the CV is

$$\frac{\partial(\rho u)}{\partial x} dx \, dy \, dz$$

Note that it does not matter if some fluid particles that enter through the face whose center is P leave through faces other than the one whose center is Q: they will be counted in the terms that follow.

Now—using an argument that will save much time and paper later on—we can derive the net rates of mass outflow through the other two pairs of faces by changing the symbols for velocity components *and* coordinates together (i.e., $x \rightarrow y \rightarrow z$ and $u \rightarrow v \rightarrow w$) so that the net rate at which the y-component of velocity transfers mass out of the CV is

$$\frac{\partial(\rho v)}{\partial y} dy \, dz \, dx$$

and, for the z-component,

$$\frac{\partial(\rho w)}{\partial z} dz \, dx \, dy$$

The rate at which the mass of fluid in the fixed CV, $\rho \, dx \, dy \, dz$, increases with time is

$$\frac{\partial}{\partial t}(\rho \, dx \, dy \, dz) = \frac{\partial \rho}{\partial t} dx \, dy \, dz$$

and the principle of conservation of mass states that the sum of the outflow and the accumulation must be zero, so after dividing by $dx \, dy \, dz$ we obtain the continuity equation for three-dimensional compressible flows:

$$\frac{\partial \rho}{\partial t} + \frac{\partial \rho u}{\partial x} + \frac{\partial \rho v}{\partial y} + \frac{\partial \rho w}{\partial z} = 0 \qquad (2.2.1)$$

In constant-density flow the continuity equation reduces to

$$\frac{\partial u}{\partial x} + \frac{\partial v}{\partial y} + \frac{\partial w}{\partial z} = 0 \qquad\qquad (2.2.2)$$

for both steady and unsteady flow. In two-dimensional flow (all quantities independent of z), Eq. (2.2.2) reduces to

$$\frac{\partial u}{\partial x} + \frac{\partial v}{\partial y} = 0 \qquad\qquad (2.2.3)$$

In vector notation, Eq. (2.2.2) becomes

$$\text{div } \mathbf{V} = 0 \qquad\qquad (2.2.4)$$

2.3 THE MOMENTUM EQUATION (NAVIER–STOKES EQUATION)

Newton's second law of motion (the principle of conservation of momentum) states that the rate of change (with respect to time) of the momentum of a body equals the sum of the forces applied to that body. Momentum, like force, is a vector, but we can resolve both into components and say, for instance,

$$\boxed{\begin{array}{l}\text{Rate of increase}\\ \text{of } x\text{-component}\\ \text{of momentum}\end{array}} = \boxed{\begin{array}{l}\text{Sum of } x\text{-}\\ \text{component of}\\ \text{applied forces}\end{array}} \qquad (2.3.1)$$

The y- and z-directions are treated in a similar manner. As in deriving the continuity equation, it is easiest to consider a small CV surrounding a point and then let the size of the CV tend to zero. For a CV, Newton's law can be rewritten in terms of momentum flux rate, which is the product of momentum per unit mass (i.e., velocity) and mass flow rate. The x-component (say) of Newton's law for a CV then amounts to a statement of Eq. (2.3.1) for the fluid element that occupies the CV at a given instant and becomes, in the formalism of Eq. (2.1.1),

$$\boxed{\begin{array}{l}\text{Rate of increase}\\ \text{of } x\text{-component}\\ \text{momentum of}\\ \text{fluid within CV}\end{array}} - \boxed{\begin{array}{l}\text{Rate of flux}\\ \text{of } x\text{-component}\\ \text{momentum}\\ \text{into CV}\end{array}} + \boxed{\begin{array}{l}\text{Rate of flux}\\ \text{of } x\text{-component}\\ \text{momentum out}\\ \text{of CV}\end{array}}$$

$$= \boxed{\begin{array}{l}\text{Sum of } x\text{-components of}\\ \text{forces applied to CV}\end{array}} \qquad (2.3.2)$$

The rate of change (with respect to time) of x-component momentum of the fluid in the CV of Fig. 2.1 is simply the rate of change of the product of the x-component velocity and the mass of the fluid in the CV:

$$\frac{\partial}{\partial t} (\rho u \, dx \, dy \, dz)$$

The rate of flow of x-component momentum into the CV through the face perpendicular to the y-direction containing the point R is the x-component momentum per unit mass (u again) times the rate of mass flow through the face, $\rho v \, dx \, dz$, that is,

$$\rho u v \, dx \, dz$$

The rate of flow of x-component momentum out of the opposite face, containing the point S, is

$$\left(\rho u v + \frac{\partial \rho u v}{\partial y} dy \right) dx \, dz$$

and so the net rate of flow of x-component momentum out of the CV via this pair of faces is

$$\frac{\partial}{\partial y} (\rho u v) \, dx \, dy \, dz$$

The net rate of outflow of x-component momentum via the pair of faces containing the points P or Q is

$$\frac{\partial}{\partial x} (\rho u^2) \, dx \, dy \, dz$$

The component of momentum being considered (the x-component here) should be distinguished from components of the velocity that transports that component of momentum. If you walk across the floor of a moving elevator, you have a horizontal component of momentum, which is being transported vertically by the motion of the elevator. Conversely, you are transporting horizontally the vertical component of momentum given you by the elevator: in the present context, $\rho u v$, the rate of flux in the y-direction of x-component momentum, is of course equal to $\rho v u$, the rate of flux in the x-direction of y-component momentum.

Finally, the net rate of outflow of x-component momentum via the pair of faces perpendicular to the z-direction is

$$\frac{\partial}{\partial z} (\rho u w) \, dx \, dy \, dz$$

The forces acting on the fluid in the CV are of two types, body forces and surface forces. The simplest example of a body force is gravity: the fluid in the CV experiences a downward force $\rho g\, dx\, dy\, dz$. Other body forces can be applied by magnetic fields acting on a current-carrying fluid, and the effects of rotating or accelerating the coordinate system can sometimes be represented by body forces. Clearly, a body force is a vector: it can have three components, which we will call f_x, f_y, and f_z per unit mass. If y is vertical, $f_y = -g$. Surface forces (i.e., forces on the imaginary surfaces of the CV) arise because of molecular stresses in the fluid. A surface force has two associated directions, the direction in which the force acts and the direction normal to the surface on which it acts (Fig. 2.2). If the two directions coincide, the force is a normal force; if they are perpendicular, it is a shear force (we always resolve surface forces so that only these two cases need be considered). We define the different components of normal stress (force per unit area) and shear stress as shown in Fig. 2.2: the first subscript represents the direction of the stress, and the second the direction of the surface normal. By convention, an *outward* normal stress acting on the fluid in the CV is positive, and the shear stresses are taken as positive on the faces furthest from the origin of coordinates. Thus σ_{xy} acts in the *positive* x-direction on the visible face perpendicular to the y-axis; a corresponding shear stress acts in the *negative* x-direction on the invisible face perpendicular to the y-axis.

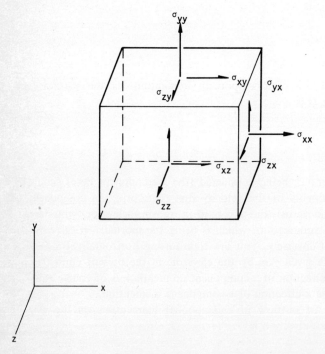

Fig. 2.2 Definitions of stress components applied to faces of the CV by surrounding fluid. Force components are stress components multiplied by areas of corresponding faces.

Pressure in a constant-density fluid is defined as minus one-third the sum of the three normal-stress components: invariably, the pressure is considered separately, so the quantities σ_{xx}, σ_{yy}, and σ_{zz} in Fig. 2.2 really represent the normal stresses with the pressure subtracted from each so that $\sigma_{xx} + \sigma_{yy} + \sigma_{yy} = 0$. The net force in the x-direction applied to the CV of Fig. 2.1 by the pressure is the difference in pressure times area between the face containing the point P and that containing Q:

$$-\left(\frac{\partial p}{\partial x} dx\right) dy\ dz$$

the minus sign arising because, by definition, a positive pressure acts inward. The other two faces do not contribute because the pressure on them is normal to the x-direction. Only a pressure gradient, not a constant pressure, produces a net force on an elementary CV.

By the same argument as that just used for the pressure, the net force in the x-direction applied by the normal stresses is proportional to a *stress gradient* and is

$$\left(\frac{\partial \sigma_{xx}}{\partial x}\right) dx\ dy\ dz$$

The normal stresses σ_{yy} and σ_{zz} have no components in the x-direction. The net force applied by the shear stress acting in the x-direction on the faces normal to the y-direction is

$$\left(\frac{\partial \sigma_{xy}}{\partial y} dy\right) dx\ dz$$

and the net force applied by σ_{xz} is

$$\left(\frac{\partial \sigma_{xz}}{\partial z} dz\right) dx\ dy$$

We can now write down the equation for conservation of x-component momentum in symbols by collecting all the above contributions to Eq. (2.3.1) and dividing by the volume $dx\ dy\ dz$. We obtain

$$\frac{\partial \rho u}{\partial t} + \frac{\partial \rho u^2}{\partial x} + \frac{\partial \rho uv}{\partial y} + \frac{\partial \rho uw}{\partial z} = -\frac{\partial p}{\partial x} + \frac{\partial \sigma_{xx}}{\partial x} + \frac{\partial \sigma_{xy}}{\partial y} + \frac{\partial \sigma_{xz}}{\partial z} + \rho f_x \qquad (2.3.3)$$

In this form the equation applies to *any* fluid, since we have not yet discussed any particular law for the surface stresses σ. It applies to *unsteady, three-dimensional compressible flows with body forces*. Two similar equations, beginning $\partial \rho v/\partial t \ldots$ and $\partial \rho w/\partial t \ldots$, can be derived for the y- and z-components of momentum in the same way as Eq. (2.3.3): see Eq. (2.3.11) and the problems at the end of the chapter. Equations (2.2.1), (2.3.3), and the two companions of the latter are the most general forms of the principles of conservation of mass and momentum for a fluid.

Now, we subtract u times the continuity equation (2.2.1) from Eq. (2.3.3). The right-hand side is unchanged because the right-hand side of Eq. (2.2.1) is explicitly zero; the left-hand side of Eq. (2.3.3) becomes

$$\rho\left(\frac{\partial u}{\partial t} + u\frac{\partial u}{\partial x} + v\frac{\partial u}{\partial y} + w\frac{\partial u}{\partial z}\right)$$

where we have used the rule for differentiation of a product to rewrite $\partial(\rho u v)/\partial y$ as $(\rho v\,\partial u/\partial y + u\,\partial\rho v/\partial y)$. The other terms on the left-hand side of Eq. (2.3.3) are treated in a similar manner. (Variations of this trick will be used below and described as "using the continuity equation.") Dividing through by ρ we get

$$\frac{\partial u}{\partial t} + u\frac{\partial u}{\partial x} + v\frac{\partial u}{\partial y} + w\frac{\partial u}{\partial z} = -\frac{1}{\rho}\frac{\partial p}{\partial x} + \frac{1}{\rho}\frac{\partial \sigma_{xx}}{\partial x} + \frac{1}{\rho}\frac{\partial \sigma_{xy}}{\partial y} + \frac{1}{\rho}\frac{\partial \sigma_{xz}}{\partial z} + f_x \qquad (2.3.4)$$

and the left-hand side is now the x-component of the acceleration following the motion of a fluid particle (e.g., in steady flow in the x-direction the left-hand side reduces to the familar expression $u\,du/dx$). The operator acting on u appears so frequently that it is given a special symbol

$$\frac{\partial}{\partial t} + u\frac{\partial}{\partial x} + v\frac{\partial}{\partial y} + w\frac{\partial}{\partial z} \equiv \frac{d}{dt} \qquad (2.3.5)$$

and is called the "substantial derivative" because it refers to the fluid substance rather than a fixed point. Equation (2.3.4) and its companion equations for v and w still represent Newton's second law, as applied to a fluid element, in the form

$$\text{Acceleration} = \frac{\text{force}}{\text{mass}} \qquad (2.3.6)$$

In this book we will consider only simple "Newtonian" fluids obeying Stokes' law for the surface stresses. In simple solids, stress is proportional to strain (Hooke's law); in simple fluids, stress is proportional to *rate of strain*. The rate of linear strain in the x-direction, e_{xx} (defined as the fractional increase, per unit time, of the x-dimension of a marked particle of fluid), is just $\partial u/\partial x$ (Fig. 1.3a). To verify this, consider an element of length l with its center at a point where u is zero, for convenience; then the left-hand end moves leftward at a speed $-(l/2)\,\partial u/\partial x$, the right-hand end moves rightward at a speed $(l/2)\,\partial u/\partial x$, and the result follows from the definition of e_{xx}. The rate of shear strain in the xy-plane, e_{xy} (the rate of deformation of an initially rectangular fluid element into a parallelogram), is $(1/2)[(\partial u/\partial y) + (\partial v/\partial x)]$. The other rates of linear (i.e., normal) and shear strain are defined by cyclic interchange of coordinates {e.g., $e_{yz} = (1/2)[(\partial v/\partial z) + (\partial w/\partial y)]$, x changing to y, u to v, and so on throughout}.

Stokes' law for the surface stresses is, like Hooke's law, empirical. Like Hooke's law it is a very accurate description for simple substances; it is valid up to the highest rate of strain ever achieved in practice, even in shock waves, as long as the fluid can be regarded as a continuum. In this book we pretend that it is exact. The changes in the book needed to include nonlinear relations between stress and strain rate would be small but confusing. Viscoelastic substances, in which stress depends on strain rate *and* initial strain, cause greater complications. Stokes' law is simply

$$\text{Stress} = 2\mu \times (\text{rate of strain}) \tag{2.3.7a}$$

for example,

$$\sigma_{xx} = 2\mu\left(\frac{\partial u}{\partial x}\right) \tag{2.3.7b}$$

$$\sigma_{xy} = \mu\left(\frac{\partial u}{\partial y} + \frac{\partial v}{\partial x}\right) \tag{2.3.7c}$$

where μ is the viscosity, a property of the fluid in the thermodynamic sense. Some other textbooks use double the present definition for rate of strain to avoid the factor 2 in Eq. (2.3.7a): our definition corresponds to the usual definition of strain in solid mechanics. Clearly, the full expressions for the viscous stress terms in Eq. (2.3.4) are quite complicated because μ depends on temperature (and weakly on pressure) and therefore varies with x, y, and z except in constant-property flow. We will not discuss details in this volume. Briefly, in a variable-density flow the sum of the three normal stresses as defined in Eq. (2.3.7), namely, $2\mu(\partial u/\partial x + \partial v/\partial y + \partial w/\partial z)$, is nonzero, as can be seen from the continuity equation, and one-third of that sum has to be subtracted from each normal stress to maintain the definition of pressure on p. 29. The bulk viscosity, β, which is often stated to be zero but is in fact of the same order as μ in diatomic gases (Chapman and Cowling, 1970), leads to a further apparent contribution to the pressure, $-\beta(\partial u/\partial x + \partial v/\partial y + \partial w/\partial z)$, which is left among the viscous terms because it can contribute to viscous dissipation. Writing $\partial u/\partial x + \partial v/\partial y + \partial w/\partial z$ in the vector form div **V** for short, the viscous-stress terms in Eq. (2.3.4) become

$$\frac{\partial}{\partial x}\sigma_{xx} + \frac{\partial}{\partial y}\sigma_{xy} + \frac{\partial}{\partial z}\sigma_{xz} \equiv \frac{\partial}{\partial x}\left[2\mu\frac{\partial u}{\partial x} + (\beta - \tfrac{2}{3}\mu)\text{ div }\mathbf{V}\right]$$
$$+ \frac{\partial}{\partial y}\left[\mu\left(\frac{\partial u}{\partial y} + \frac{\partial v}{\partial x}\right)\right] + \frac{\partial}{\partial z}\left[\mu\left(\frac{\partial u}{\partial z} + \frac{\partial w}{\partial x}\right)\right] \tag{2.3.8}$$

Equation (2.3.4) with the substitution of Eq. (2.3.8) is one of the three Navier-Stokes equations, the others being the analogous equations for v and w. With the continuity equation (2.2.1) they form four equations for the four variables u, v, w, and p. It is remarkable what complicated equations arise from the simple laws of

Newton [Eq. (2.3.1)] and Stokes [Eq. (2.3.7a)] —and even more remarkable how complicated the motion can be.

If μ is constant and div $\mathbf{V}$ is zero (constant-property flow), obvious simplifications ensue; a nonobvious simplification is that after rearranging the order of second derivatives a further term $\partial(\text{div }\mathbf{V})/\partial x$ can be identified and discarded, since it is zero. We finally get the x-component momentum equation for a Newtonian fluid in constant-property (but still unsteady and three-dimensional) flow as

$$\frac{du}{dt} \equiv \frac{\partial u}{\partial t} + u\frac{\partial u}{\partial x} + v\frac{\partial u}{\partial y} + w\frac{\partial u}{\partial z} = -\frac{1}{\rho}\frac{\partial p}{\partial x} + \nu\left(\frac{\partial^2 u}{\partial x^2} + \frac{\partial^2 u}{\partial y^2} + \frac{\partial^2 u}{\partial z^2}\right) + f_x \qquad (2.3.9)$$

where the operator $\partial^2/\partial x^2 + \partial^2/\partial y^2 + \partial^2/\partial z^2$ is often denoted by ∇^2. By interchanging coordinates in cyclic order the y-component equation for constant-property flow appears as

$$\frac{\partial v}{\partial t} + v\frac{\partial v}{\partial y} + w\frac{\partial v}{\partial z} + u\frac{\partial v}{\partial x} = -\frac{1}{\rho}\frac{\partial p}{\partial y} + \nu\left(\frac{\partial^2 v}{\partial y^2} + \frac{\partial^2 v}{\partial z^2} + \frac{\partial^2 v}{\partial x^2}\right) + f_y \qquad (2.3.10)$$

and it is usual to rearrange the terms so that derivatives appear in alphabetic order:

$$\frac{dv}{dt} \equiv \frac{\partial v}{\partial t} + u\frac{\partial v}{\partial x} + v\frac{\partial v}{\partial y} + w\frac{\partial v}{\partial z} = -\frac{1}{\rho}\frac{\partial p}{\partial y} + \nu\left(\frac{\partial^2 v}{\partial x^2} + \frac{\partial^2 v}{\partial y^2} + \frac{\partial^2 v}{\partial z^2}\right) + f_y \qquad (2.3.11)$$

Note that the final result is the same as if we had left the d/dt operator unchanged. It is left to the reader to write out the z-component equation for constant-property flow.

In steady flow, $\partial u/\partial t$ is zero, and in two-dimensional flow the velocity component, and all gradients, in one of the three coordinate directions are zero: in engineering this direction is taken as the z-direction and we get, for steady two-dimensional constant-property flow without body forces, the x-component momentum equation

$$u\frac{\partial u}{\partial x} + v\frac{\partial u}{\partial y} = -\frac{1}{\rho}\frac{\partial p}{\partial x} + \nu\left(\frac{\partial^2 u}{\partial x^2} + \frac{\partial^2 u}{\partial y^2}\right) \qquad (2.3.12)$$

and an analogous equation for the y-component.

Let us finally consider the case where the x-axis locally coincides with a streamline (i.e., $v = w = 0$): we can always choose the axis to achieve this at one point. We have, from Eq. (2.3.4) for steady constant-density flow without body forces,

$$\frac{\partial}{\partial x}(p + \tfrac{1}{2}\rho u^2 - \sigma_{xx}) = \frac{\partial \sigma_{xy}}{\partial y} + \frac{\partial \sigma_{xz}}{\partial z} \qquad (2.3.13)$$

where $p + (1/2)\rho u^2$ is the total pressure, P. This shows that the gradient of $P - \sigma_{xx}$

along a streamline (where σ_{xx} is the normal stress in the local flow direction) equals the sum of the shear-stress gradients along any two mutually perpendicular normals to the streamline: in many shear layers, σ_{xx} can be neglected in comparison to P. Clearly, if *all* stress gradients are negligible, we recover Bernoulli's equation, which states that total pressure is constant along a streamline in inviscid flow.

2.4 TURBULENT FLOW

The unsteady Navier-Stokes equations and the continuity equation are still instantaneously valid in turbulent flow because the smallest eddies are several orders of magnitude larger than the length scale of the molecular motion (the mean free path). Because the equations are nonlinear, the *time-mean* velocity components and pressure do *not* obey the Navier-Stokes equations. By the time mean of a quantity ϕ, usually denoted by $\bar{\phi}$ unless some special notation is being used, we imply

$$\bar{\phi} \equiv \lim_{T \to \infty} \frac{1}{T} \int_0^T \phi \, dt \qquad (2.4.1)$$

For simplicity, we omit overbars on the mean velocity components u, v, and w. Consider the x-component Navier-Stokes equation with its left-hand side in the "divergence" form [Eq. (2.3.3)]. For simplicity of exposition we use the constant-property expressions [Eq. (2.3.7)] for the σ stresses, as in Eq. (2.3.9). Add fluctuating (time-dependent) parts u', v', and w', with zero mean, to the mean (time-independent) u, v, and w components of velocity so that the instantaneous velocity components (the components of the vector $\mathbf{V} + \mathbf{V}'$) are $u + u'$, $v + v'$, and $w + w'$, and, similarly, let the instantaneous pressure and the x-component of instantaneous body force be $p + p'$ and $f_x + f_x'$. (See Fig. 2.3.) Noting that $\partial u / \partial t = 0$ by definition of the mean velocity u, we see that Eq. (2.3.9) becomes

$$\frac{\partial u'}{\partial t} + \frac{\partial}{\partial x}(u + u')^2 + \frac{\partial}{\partial y}(u + u')(v + v') + \frac{\partial}{\partial z}(u + u')(w + w')$$

$$= \frac{1}{\rho}\frac{\partial}{\partial x}(p + p') + \nu\left(\frac{\partial^2}{\partial x^2} + \frac{\partial^2}{\partial y^2} + \frac{\partial^2}{\partial z^2}\right)(u + u') + f_x + f_x' \quad (2.4.2)$$

The mean of $(u + u')(v + v')$, defined as in Eq. (2.4.1) and written as $\overline{(u + u')(v + v')}$, is just $uv + \overline{u'v'}$. The mean of uv' is zero because the mean of v' is zero and u is independent of time. The same principle applies for $u'v$. An overbar on uv would be superfluous because u and v are independent of time. By applying the same arguments to the other terms and noting that Eq. (2.4.1) implies that the mean of the time derivative of any finite quantity is zero, the mean of Eq. (2.4.2) becomes

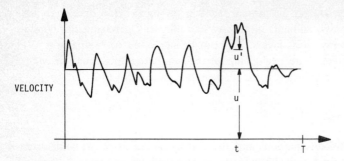

Fig. 2.3 The mean (u), fluctuating part (u'), and instantaneous part $(u + u')$ of the x-component velocity at a given point in turbulent flow. By definition $\overline{u'}$, obtained by putting $\phi = u'$ in Eq. (2.4.1), is zero.

$$\frac{\partial}{\partial x}(u^2 + \overline{u'^2}) + \frac{\partial}{\partial y}(uv + \overline{u'v'}) + \frac{\partial}{\partial z}(uw + \overline{u'w'}) = -\frac{1}{\rho}\frac{\partial p}{\partial x}$$

$$+ \nu\left(\frac{\partial^2}{\partial x^2} + \frac{\partial^2}{\partial y^2} + \frac{\partial^2}{\partial z^2}\right)u + f_x \quad (2.4.3)$$

which shows rather clearly how the velocity fluctuations produce mean rates of momentum transfer additional to those produced by the mean velocity components. A more conventional arrangement of the equation, reverting to the "acceleration" form for the mean velocity terms as in Eq. (2.3.4) and moving the terms containing the fluctuations to the right-hand side, is

$$u\frac{\partial u}{\partial x} + v\frac{\partial u}{\partial y} + w\frac{\partial u}{\partial z} = -\frac{1}{\rho}\frac{\partial p}{\partial x} + \nu\left(\frac{\partial^2}{\partial x^2} + \frac{\partial^2}{\partial y^2} + \frac{\partial^2}{\partial z^2}\right)u + f_x$$

$$- \frac{\partial \overline{u'u'}}{\partial x} - \frac{\partial \overline{u'v'}}{\partial y} - \frac{\partial \overline{u'w'}}{\partial z} \quad (2.4.4)$$

where the first line is the equation for steady viscous (laminar) flow [i.e., Eq. (2.3.9) with $\partial u/\partial t$ omitted] and the second line contains the gradients of the extra turbulent quantities divided by density. Referring back to the CV derivation of Eq. (2.3.3), we can see that $\rho\overline{u'u'}$ (usually written as $\rho\overline{u'^2}$), $\rho\overline{u'v'}$, and $\rho\overline{u'w'}$ are mean rates of transfer, *by the turbulence,* of u-component momentum through the faces of the CV perpendicular to x, y, and z, respectively. By Newton's second law these are equivalent to stresses $-\rho\overline{u'^2}$, $-\rho\overline{u'v'}$, and $-\rho\overline{u'w'}$ acting in the x-direction on the same three faces: $-\rho\overline{u'^2}$ is a normal stress and $-\rho\overline{u'v'}$ and $-\rho\overline{u'w'}$ are shear stresses; in our previous notation they represent the turbulent contributions to σ_{xx}, σ_{xy}, and σ_{xz}, respectively. The mean viscous contributions are still given by Eq. (2.3.7) based on the mean-velocity components.

Similar analyses can be done for the y- and z-component equations. The extra turbulent stresses (the Reynolds stresses) can be summarized by the following array, in which we have equated $\overline{v'u'}$ to $\overline{u'v'}$, and so on, for simplicity:

$$
\begin{vmatrix} \sigma_{xx} & \sigma_{xy} & \sigma_{xz} \\ \sigma_{yx} & \sigma_{yy} & \sigma_{yz} \\ \sigma_{zx} & \sigma_{zy} & \sigma_{zz} \end{vmatrix} \equiv -\rho \begin{vmatrix} \overline{u'^2} & \overline{u'v'} & \overline{u'w'} \\ \overline{u'v'} & \overline{v'^2} & \overline{v'w'} \\ \overline{u'w'} & \overline{v'w'} & \overline{w'^2} \end{vmatrix}
\tag{2.4.5a}
$$

or by the tensor* equation,

$$
\sigma_{ij} = -\rho\overline{u'_i u'_j}
\tag{2.4.5b}
$$

In tensor notation the complete formula for the "stress tensor," σ_{ij}, including the viscous contributions but neglecting the effects of div $\mathbf{V}$, is

$$
\sigma_{ij} = -\rho\overline{u'_i u'_j} + \mu\left(\frac{\partial u_i}{\partial x_j} + \frac{\partial u_j}{\partial x_i}\right)
\tag{2.4.5c}
$$

The first, second, and third rows of Eq. (2.4.5a) appear in the x-, y-, and z-component equations, respectively, and the first, second, and third columns appear within x-, y-, and z-derivatives, respectively. The stresses on the leading diagonal ($i = j$) are normal stresses, and the others are shear stresses.

Notice that the continuity equation, div $(\mathbf{V} + \mathbf{V}') = 0$, is linear, so its mean is div $\mathbf{V} = 0$ as usual. Subtracting the mean from the instantaneous equation, we get

$$
\text{div } \mathbf{V}' = \frac{\partial u'}{\partial x} + \frac{\partial v'}{\partial y} + \frac{\partial w'}{\partial z} = 0
\tag{2.4.6}
$$

In turbulent flows with mean density gradients, density fluctuations appear, and the time-mean equations become more complicated. However, the most important change is that the density no longer cancels out from the terms like $(1/\rho)\partial\sigma_{xx}/\partial x$. Experiments and theoretical arguments show that at nonhypersonic Mach numbers the density fluctuations are produced largely by temperature fluctuations rather than by pressure fluctuations and that temperature fluctuations are strongly anti-correlated with u-component velocity fluctuations. It is therefore possible to estimate the size of terms containing density fluctuations if the corresponding terms containing u-component velocity fluctuations are known. As a rough guide, if the density ratio across the width of the shear layer is less than about two, most of the terms involving fluctuating density are small, and if it is less than about seven, the terms can be satisfactorily approximated. The effects of density fluctuations can be removed from the equations, though not from the flow, by using mass-weighted averaging (see Cebeci and Smith, 1974).

---

*Tensor notation is used only rarely in this book, since we are dealing with flows with pre-ferred directions.

An alternative definition of a time mean to that given by the integral in Eq. (2.4.1) is the average of a large number of samples taken at independent times. If the flow has an orderly time dependence superimposed on the turbulent motion, as in the case of flow from a reservoir of finite size, we need to take an ensemble average over a large number of samples, each taken from a separate run of the experiment at a given time, t_1 say, after the start. This gives us an ensemble average *at the time* t_1. Strictly speaking, theoretical work should be based on ensemble averages. Clearly, if there is no orderly time dependence, we expect ensemble averages and the two kinds of time mean to coincide; the expectation is fulfilled in practice but is difficult to prove rigorously. Flows of this sort are called "statistically stationary" or, briefly but misleadingly, steady turbulent flows.

The Reynolds stresses appear as extra unknowns in the mean-motion equations. The central problem of turbulent flow is to find empirical expressions for them, perhaps by making use of more complicated weighted averages of the Navier-Stokes equations but necessarily making use of experimental data. We discuss the behavior of turbulent flow in Chap. 6.

2.5 VORTICITY

The above equations deal with the linear velocity components of a fluid element. In some cases it is convenient to treat the *angular* velocity of a fluid element explicitly, although it is simply related to the velocity components rather than being a separate variable. We define the *vorticity* vector in terms of the velocity vector **V** as $\Omega = \text{curl } \mathbf{V}$, so that its x-, y-, and z-components, ξ, η, and ζ, are

$$\xi \equiv \frac{\partial w}{\partial y} - \frac{\partial v}{\partial z} \qquad\qquad (2.5.1a)$$

$$\eta \equiv \frac{\partial u}{\partial z} - \frac{\partial w}{\partial x} \qquad\qquad (2.5.1b)$$

$$\zeta \equiv \frac{\partial v}{\partial x} - \frac{\partial u}{\partial y} \qquad\qquad (2.5.1c)$$

If a fluid element is in rigid or "solid body" rotation, its vorticity is a vector whose direction coincides with the axis of rotation and whose magnitude is twice the angular velocity. In a simple two-dimensional shear layer where the only velocity gradient is $\partial u/\partial y$, the vorticity vector is in the z-direction and its magnitude is $-\partial u/\partial y$. The equations for the rate of change of each component of vorticity of a fluid element as it moves along a mean streamline can be obtained by "taking the curl" of the Navier-Stokes equations, performing on the equations for each velocity component the operations that are performed on the velocity components to obtain Eq. (2.5.1). The equation for ζ in constant-property flow, obtained by subtracting the y-derivative of the x-component Navier-Stokes equation from the x-derivative of the y-component equation, is

$$\frac{\partial \zeta}{\partial t} + u \frac{\partial \zeta}{\partial x} + v \frac{\partial \zeta}{\partial y} + w \frac{\partial \zeta}{\partial z} \equiv \frac{d\zeta}{dt}$$

$$= \xi \frac{\partial w}{\partial x} + \eta \frac{\partial w}{\partial y} + \zeta \frac{\partial w}{\partial z} + \nu \left(\frac{\partial^2 \zeta}{\partial x^2} + \frac{\partial^2 \zeta}{\partial y^2} + \frac{\partial^2 \zeta}{\partial z^2} \right) \qquad (2.5.2)$$

The left-hand side and the viscous term have the same form as the corresponding terms in the Navier-Stokes equations, but the pressure term has disappeared and the first three terms on the right-hand side have no counterpart in the Navier-Stokes equation. These three terms are zero in two-dimensional flow, and the equation then reduces to

$$\frac{d\zeta}{dt} = \nu \nabla^2 \zeta \qquad (2.5.3)$$

so that the vorticity of a fluid element in two-dimensional flow is unaltered, as it moves downstream, except for viscous diffusion. In three-dimensional flow the extra terms on the right-hand side of Eq. (2.5.2) are nonzero: they represent skewing of the ξ and η components of vorticity into the z-direction by $\partial w / \partial x$ and $\partial w / \partial y$, respectively, and the intensification of ζ by the extensional rate of strain $\partial w / \partial z$. This intensification can be explained in physical terms by noting that if a fluid element in rigid rotation is stretched in the direction of the axis of rotation, conservation of mass (volume) requires the cross-sectional area to decrease while conservation of angular momentum requires the angular velocity (half the vorticity) to increase as the cross-sectional area decreases.

Equation (2.5.3) is a simple "diffusion" equation of a type to be discussed in Chap. 4 and applies to two-dimensional laminar shear layers with or without pressure gradient. In the external stream—that is, outside the shear layer—the vorticity is zero (by definition of an external stream), while at a solid surface $y = 0$ we have, with subscript w (for wall) to indicate conditions at $y = 0$,

$$\zeta_w = -\left(\frac{\partial u}{\partial y} \right)_w = -\frac{\tau_w}{\mu} \qquad (2.5.4)$$

and

$$\left(\frac{\partial \zeta}{\partial y} \right)_w = \left(\frac{\partial^2 v}{\partial x \, \partial y} - \frac{\partial^2 u}{\partial y^2} \right)_w = -\left(\frac{\partial^2 u}{\partial x^2} + \frac{\partial^2 u}{\partial y^2} \right)_w = -\frac{1}{\rho \mu} \frac{\partial p}{\partial x} \qquad (2.5.5)$$

using the continuity equation and the x-component equation of motion at the solid surface (where $du/dt = 0$). The outer elements of Eqs. (2.5.4) and (2.5.5) are the two boundary conditions for the second-order partial differential equation (2.5.3).

Unfortunately, fluid-flow problems cannot be formulated entirely in terms of vorticity rather than velocity: in essence this is because the boundary condition [Eq. (2.5.5)] includes the pressure, which is related to the velocity field. However, the concept of a shear boundary layer as a region in which vorticity is diffused in obedience to Eq. (2.5.3), after creation at a solid surface by Eq. (2.5.4), is helpful qualitatively. Qualitative or quantitative consideration of the skewing and stretching terms is almost essential in understanding three-dimensional shear layers. An elegant discussion of shear-layer vorticity is given by Lighthill in Chap. 2 of the book by Rosenhead (1963).

If we consider the very simplest case, the boundary layer on a thin flat plate with a uniform external velocity u_e, we see that in the limit of zero viscosity the flow would be unaltered by the presence of the plate. If the no-slip condition were imposed at the surface, an infinitesimal "vortex sheet" would separate the surface from the external flow. The velocity gradient $\partial u/\partial y$ (and therefore the spanwise component of vorticity, $\partial v/\partial x - \partial u/\partial y$) would be infinite, and the strength of the vortex sheet, defined as the velocity jump across it, would be constant everywhere on the plate and equal to u_e. The presence of viscosity causes the vorticity to diffuse away from the surface, so the vortex sheet is replaced by a growing boundary layer. The student may want to rephrase this argument for the mixing layer of Fig. 1.1a. In a way, vorticity is a more intrinsic property of a shear layer than velocity is: the equations governing the diffusion of vorticity do not explictly contain the pressure, so if a two-dimensional shear layer passes through a short region of strong pressure gradient, the vorticity on a given streamline remains nearly constant while the velocity changes rapidly. In such cases it may be preferable to rearrange the equations of motion so that the unknowns are the pressure and the vorticity vector rather than the pressure and the velocity vector. The pressure equation is derived in Prob. 2.5.

PROBLEMS

2.1. Write down the Navier-Stokes equation for the x-component of velocity in unsteady incompressible flow in the xy-plane. Write out in words the law of motion (due to Newton) that it represents. Check your answers from Sec. 2.3.

2.2. Derive the continuity equation for unsteady incompressible flow in the xy-plane, and then, by rewriting it in vector form, deduce the three-dimensional version [Eq. (2.2.2)].

2.3. Write out the viscous-stress terms in Eq. (2.3.8) for constant-property flow and show that they reduce to the form given in Eq. (2.3.9).

2.4. Write out the z-component Navier-Stokes equation for constant-property flow.

2.5. By "taking the divergence" of the constant-property Navier-Stokes equations [i.e., by adding the x-derivative of Eq. (2.3.9), the y-derivative of Eq. (2.3.11), and the z-derivative of the above equation] obtain an equation for the pressure.

2.6. Which of the following statements are true?

a. $-\rho \overline{u'v'}$ is the turbulent shear stress acting in the x-direction on an imaginary surface normal to the y-direction.

b. $-\rho \overline{u'v'}$ is the turbulent shear stress acting in the y-direction on an imaginary surface normal to the x-direction.

c. $-\rho \overline{w'^2}$ is the turbulent shear stress acting in the z-direction on an imaginary surface normal to the z-direction.

d. Viscous stresses are smaller in turbulent flow than in laminar flow.

Chapter 3

Thin-Shear-Layer Equations

3.1 THIN–SHEAR–LAYER APPROXIMATION IN TWO–DIMENSIONAL FLOW

It was mentioned in Chap. 1 that the equations of motion (Chap. 2) take a particularly simple form for a shear layer with sufficiently small δ/l because terms that are smaller than the main terms by a factor of δ/l can be neglected. It is very important to realize that neglecting a term in this way is an empirical procedure, no more and no less respectable than introducing an empirical data-correlation formula. Indeed, the equation "term = 0" can be regarded as a data-correlation formula, adequate for engineering purposes if $\delta/l \ll 1$. The laminar TSL equations have the same empirical status as, say, the full Navier-Stokes equation (2.4.5) with data-correlation formulas for the turbulent stresses. The difference is that the range of validity of the TSL equations is well known and simply stated ($\delta/l \ll 1$) while the range of validity of turbulence formulas is often doubtful. Therefore it is usually easy to check whether the TSL equations are valid for a particular problem. We now derive these simplified equations for the case of very small δ/l and then discuss the consequences of increasing δ/l. For simplicity, we begin with a two-dimensional steady constant-property flow without body forces and leave the stresses unspecified so that the results apply to laminar or turbulent flow. In this case, Eq. (2.3.4) and its y-component equivalent reduce to

$$u \frac{\partial u}{\partial x} + v \frac{\partial u}{\partial y} = -\frac{1}{\rho} \frac{\partial p}{\partial x} + \frac{1}{\rho} \frac{\partial \sigma_{xx}}{\partial x} + \frac{1}{\rho} \frac{\partial \sigma_{xy}}{\partial y} \tag{3.1.1}$$

$$u \frac{\partial v}{\partial x} + v \frac{\partial v}{\partial y} = -\frac{1}{\rho} \frac{\partial p}{\partial y} + \frac{1}{\rho} \frac{\partial \sigma_{yy}}{\partial y} + \frac{1}{\rho} \frac{\partial \sigma_{xy}}{\partial x} \tag{3.1.2}$$

and the continuity equation is

$$\frac{\partial u}{\partial x} + \frac{\partial v}{\partial y} = 0 \tag{2.2.3}$$

We now examine the relative sizes of the terms in these equations for a shear layer whose thickness is δ at a distance l from its origin (Fig. 1.3b) so that $d\delta/dx$ is of order δ/l and is small in comparison with unity. If the shear layer is perturbed so that, locally, $d\delta/dx$ is *not* of order δ/l, we should replace l by $\delta/(d\delta/dx)$, since it is the local rates of change that matter (the coordinates appear only in the derivative operators). However, the use of l is simpler algebraically and need cause no confusion. Again for simplicity, we will discuss the case of a boundary layer growing on the flat surface $y = 0$, but the results are applicable to any of the thin shear layers of Fig. 1.1.

For generality, we insert "typical" (order-of-magnitude) values. As well as replacing dependent variables by typical values, say u_e for the x-component velocity u, we replace derivatives by typical average values; that is, we replace the x or y derivative of a variable by the typical *change* in the variable over the distance l (or δ) divided by the distance l (or δ), so that a typical value of $\partial u/\partial y$ is u_e/δ, as was shown in the example on p. 14. We expect the typical changes in the velocity components to be of the same order[*] as typical values. Of course there may be special cases with smaller changes, but we want our equations to apply to as wide a variety of shear layers as possible. In the case of the pressure the typical change in the x-direction is of order ρu_e^2 (remembering that p and u_e are related by Bernoulli's equation), which is smaller than the absolute pressure p by a factor of the order of (Mach number)^2: it is not possible to make an a priori estimate of a typical change in pressure from $y = 0$ to $y = \delta$ (Bernoulli's equation does not help), so we will leave $\partial p/\partial y$ unaltered. Using the above rules and using the sign "$\sim$" to indicate a typical value of order-of-magnitude accuracy without regard to sign, we have

$$u \sim u_e \qquad \frac{\partial u}{\partial y} \sim \frac{u_e}{\delta} \qquad \frac{\partial u}{\partial x} \sim \frac{u_e}{l}$$

(from the last of these and from the continuity equation, $\partial v/\partial y \sim u_e/l$, leading to $v \sim u_e \delta/l$, since $v = 0$ at $y = 0$) and so on. Thus Eq. (3.1.1) gives

---

[*]Following the convention that an "order of magnitude" is a factor of about 10, the statement "A is of the same order as B," written "$A = O(B)$," means *roughly* that $A/\sqrt{10} < B < A\sqrt{10}$: the expression "$O(B)$" alone means "a quantity of the same order as that of B."

$$u \frac{\partial u}{\partial x} + v \frac{\partial u}{\partial y} = -\frac{1}{\rho} \frac{\partial p}{\partial x} + \frac{1}{\rho} \frac{\partial \sigma_{xx}}{\partial x} + \frac{1}{\rho} \frac{\partial \sigma_{xy}}{\partial y} \qquad (3.1.1)$$

$$\frac{u_e^2}{l} \qquad \frac{u_e^2}{l} \qquad \frac{u_e^2}{l} \qquad \frac{\sigma_{xx}/\rho}{l} \qquad \frac{\sigma_{xy}/\rho}{\delta} \qquad (3.1.3)$$

and Eq. (3.1.2) gives

$$u \frac{\partial v}{\partial x} + v \frac{\partial v}{\partial y} = -\frac{1}{\rho} \frac{\partial p}{\partial y} + \frac{1}{\rho} \frac{\partial \sigma_{yy}}{\partial y} + \frac{1}{\rho} \frac{\partial \sigma_{xy}}{\partial x} \qquad (3.1.2)$$

$$\frac{u_e^2 \delta}{l^2} \qquad \frac{u_e^2 \delta}{l^2} \qquad \frac{1}{\rho} \frac{\partial p}{\partial y} \qquad \frac{\sigma_{yy}/\rho}{\delta} \qquad \frac{\sigma_{xy}/\rho}{l} \qquad (3.1.4)$$

where the typical values have been written below the terms to which they correspond, arithmetic operators being omitted. We do not need to write out the continuity equation, since it has been used already. Note that the two left-hand-side terms in a given equation are of the same order. We consider three cases, of which the first is artificial but instructive.

The first case occurs when all σ stresses are locally negligible (inviscid flow). The first equation has all remaining terms of order u_e^2/l (justifying our choice for $\partial p/\partial x$), and no approximation is possible. The second equation implies that $\partial p/\partial y$ is of order $\rho u_e^2 \delta/l^2$ or δ/l times the order of $\partial p/\partial x$. It is very important to realize that the reason for this is *not* that δ/l is small but that the surface is *flat*. To see this, note that $\partial v/\partial x$ is zero at a flat surface, rising to order $u_e \delta/l^2$ if y is of the order of δ. On a flat surface $u \partial v/\partial x$ and $v \partial v/\partial y$ are of the same order, as is shown in Eq. (3.1.4). Now suppose that the surface is convex upward, like the top surface of the airfoil in Fig. 1.1b, with radius of curvature R, the x-axis remaining straight but tangential to the surface at the point considered. In this case $\partial v/\partial x$ is closely equal to $-u/R$. *Neglecting* contributions to $\partial p/\partial y$ of order $\rho u_e^2 \delta/l^2$, that is, the flat-surface terms, we get the "centrifugal force" equation

$$\frac{\partial p}{\partial y} \approx -\rho u \frac{\partial v}{\partial x} \approx \frac{\rho u^2}{R} \qquad (3.1.5)$$

The neglect of the flat-surface terms compared to the "centrifugal" terms is justified if $l/R \gg \delta/l$. If we take $\delta/l = 0.02$, a fairly generous value for a turbulent boundary layer, the centrifugal term dominates if $R < 50l$ (to an order of magnitude). $R = 50l$ corresponds to a thin circular-arc airfoil of chord l with a camber (distance from chord line to arc at midchord) of only 1/4% of the chord length. Because $p(y = \delta) - p(y = 0)$ is of order $\rho u_e^2/R$ from Eq. (3.1.5), the effect of $\partial p/\partial y$ on the difference between $\partial p/\partial x$ at $y = 0$ and $y = \delta$ depends on $d(1/R)/dx$; infinite values of the latter appear when a flat surface joins a curved one and the order-of-magnitude results become complicated.

The second case occurs when all σ stresses are of the same order. This case corresponds to turbulent flow. To a better—but still rough—approximation in a boundary layer, $0.4\sigma_{xx} = \sigma_{xy} = \sigma_{yy}$. In Eq. (3.1.3), $\partial\sigma_{xx}/\partial x$ is clearly an order δ/l smaller than $\partial\sigma_{xy}/\partial y$, and the latter must be of the same order as the remaining terms, $\rho u_e^2/l$, if stress gradients are to matter at all. Therefore a general stress σ must be of order $\rho u_e^2 \delta/l$. Equation (3.1.1) reduces to

$$u\frac{\partial u}{\partial x} + v\frac{\partial u}{\partial y} = -\frac{1}{\rho}\frac{\partial p}{\partial x} + \frac{1}{\rho}\frac{\partial \sigma_{xy}}{\partial y}\left[1 + O\left(\frac{\delta}{l}\right)\right] \qquad (3.1.6)$$

where $O(\delta/l)$ signifies a quantity of the order of magnitude of δ/l. For a flat surface the orders of magnitude of the terms are

$$u\frac{\partial v}{\partial x} + v\frac{\partial v}{\partial y} = -\frac{1}{\rho}\frac{\partial p}{\partial y} + \frac{1}{\rho}\frac{\partial \sigma_{yy}}{\partial y} + \frac{1}{\rho}\frac{\partial \sigma_{xy}}{\partial x}$$

$$\frac{u_e^2\delta}{l^2} \qquad \frac{u_e^2\delta}{l^2} \qquad \frac{1}{\rho}\frac{\partial p}{\partial y} \qquad \frac{u_e^2}{l} \qquad \frac{u_e^2\delta}{l^2} \qquad\qquad (3.1.7)$$

Therefore Eq. (3.1.2) reduces to

$$\frac{1}{\rho}\frac{\partial p}{\partial y} = \frac{1}{\rho}\frac{\partial \sigma_{yy}}{\partial y}\left[1 + O\left(\frac{\delta}{l}\right)\right] \qquad (3.1.8)$$

where the $O(\delta/l)$ term includes the terms from the left-hand side of Eq. (3.1.7) as well as the term in σ_{xy}. For a curved surface, Eq. (3.1.4) gives, using the results of the first case above and assuming that $l/R \gg \delta/l$,

$$u\frac{\partial v}{\partial x} + v\frac{\partial v}{\partial y} = -\frac{1}{\rho}\frac{\partial p}{\partial y} + \frac{1}{\rho}\frac{\partial \sigma_{yy}}{\partial y} + \frac{1}{\rho}\frac{\partial \sigma_{xy}}{\partial x}$$

$$\frac{u_e^2}{R} \qquad \frac{u_e^2\delta}{l^2} \qquad \frac{1}{\rho}\frac{\partial p}{\partial y} \qquad \frac{u_e^2}{l} \qquad \frac{u_e^2\delta}{l^2} \qquad\qquad (3.1.9)$$

so that if l/R is of order unity,

$$\frac{1}{\rho}\frac{\partial p}{\partial y} = \frac{u^2}{R} + \frac{1}{\rho}\frac{\partial \sigma_{yy}}{\partial y}\left[1 + O\left(\frac{\delta}{l}\right)\right] \qquad (3.1.10)$$

We shall see in Chap. 6 that the σ_{yy} term in Eq. (3.1.10) is negligible in weakly turbulent flows, so that Eq. (3.1.10) reduces to Eq. (3.1.5) and Eq. (3.1.8) reduces to $\partial p/\partial y = 0$.

The third case consists of stresses proportional to strain rate (laminar flow of Newtonian viscous fluid). We have

$$\sigma_{xx} = 2\mu \frac{\partial u}{\partial x}$$

$$\sigma_{xy} = \mu\left(\frac{\partial u}{\partial y} + \frac{\partial v}{\partial x}\right) \tag{2.3.7}$$

$$\sigma_{yy} = 2\mu \frac{\partial v}{\partial y}$$

so, with $\nu \equiv \mu/\rho$, Eq. (3.1.3) becomes

$$u\frac{\partial u}{\partial x} + v\frac{\partial u}{\partial y} = -\frac{1}{\rho}\frac{\partial p}{\partial x} + \frac{1}{\rho}\frac{\partial \sigma_{xx}}{\partial x} + \frac{1}{\rho}\frac{\partial \sigma_{xy}}{\partial y} \tag{3.1.11}$$

$$\frac{u_e^2}{l} \qquad \frac{u_e^2}{l} \qquad \frac{u_e^2}{l} \qquad \frac{\nu u_e}{l^2} \qquad \nu\left(\frac{u_e}{\delta^2}, \frac{u_e}{l^2}\right)$$

Note that the second element in the σ_{xy} term derived from Eq. (2.3.7) is smaller than the first. Therefore Eq. (3.1.11) becomes

$$u\frac{\partial u}{\partial x} + v\frac{\partial u}{\partial y} = -\frac{1}{\rho}\frac{\partial p}{\partial x} + \nu\frac{\partial^2 u}{\partial y^2}\left[1 + O\left(\frac{\delta}{l}\right)^2\right] \tag{3.1.12}$$

Also, Eq. (3.1.4) becomes, for a flat surface,

$$u\frac{\partial v}{\partial x} + v\frac{\partial v}{\partial y} = -\frac{1}{\rho}\frac{\partial p}{\partial y} + \frac{1}{\rho}\frac{\partial \sigma_{yy}}{\partial y} + \frac{1}{\rho}\frac{\partial \sigma_{xy}}{\partial x} \tag{3.1.13}$$

$$\frac{u_e^2\delta}{l^2} \qquad \frac{u_e^2\delta}{l^2} \qquad \frac{1}{\rho}\frac{\partial p}{\partial y} \qquad \frac{\nu u_e}{l\delta} \qquad \nu\left(\frac{u_e}{l\delta}, \frac{u_e\delta}{l^3}\right)$$

or, by writing all the viscous terms together,

$$\frac{u_e^2\delta}{l^2} \qquad \frac{u_e^2\delta}{l^2} \qquad \frac{1}{\rho}\frac{\partial p}{\partial y} \qquad \frac{u_e^2\delta}{l^2}\cdot\frac{\nu}{u_el}\left(\frac{l}{\delta}\right)^2\left[1, \left(\frac{\delta}{l}\right)^2\right] \tag{3.1.14}$$

Now, we know from Chap. 1 that $(\delta/l)^2 \sim \nu/(u_el)$ in laminar flow, so the viscous terms are also of order $u_e^2\delta/l^2$. This means that $\partial p/\partial y$ is of order $u_e^2\delta/l^2$. But the pressure difference between $y = 0$ and $y = \delta$ is of order δ times $\partial p/\partial y$, that is, $\rho u_e^2\delta^2/l^2$, and the difference in $\partial p/\partial x$ between $y = 0$ and $y = \delta$ will therefore be negligible in comparison to the external-stream dynamic pressure $\frac{1}{2}\rho u_e^2$. For practical purposes we can therefore put

$$\frac{\partial p}{\partial y} = 0 \tag{3.1.15}$$

for a laminar (or turbulent) layer on a flat surface. On a curved surface, the "centrifugal" approximation [Eq. (3.1.5)] should be adequate for practical purposes and may, of course, be close enough to Eq. (3.1.15) for the latter to be acceptable.

The above derivations are rather more detailed than those normally found in textbooks, which treat the difficulties of curved shear layers and turbulent flow cursorily if at all. Note that it is *not* valid to apply the TSL approximation to the instantaneous equations for turbulent flow, such as Eq. (2.4.2), although the resulting mean equations happen to be correct. In a turbulent TSL, *mean* velocity gradients in the x-direction are of order δ/l times those in the y-direction, but all fluctuating velocity components and their gradients are of the same order of magnitude. Also, the z-wise gradients of fluctuating quantities are not zero even if the flow is two-dimensional in the mean. Note further that the neglected terms in the x-component equation for turbulent flow [Eq. (3.1.6)] are of order δ/l (strictly speaking, $d\delta/dx$) times the retained ones, whereas in laminar flow [Eq. (3.1.12)] the ratio is $(\delta/l)^2$: the weakness of the TSL approximation in turbulent flow is accentuated because $d\delta/dx$ is usually larger than it is in laminar flow. The equations commonly used for the case $d\delta/dx \ll 1$ are

$$u\frac{\partial u}{\partial x} + v\frac{\partial u}{\partial y} = -\frac{1}{\rho}\frac{\partial p}{\partial x} + \frac{1}{\rho}\frac{\partial}{\partial y}\left(\mu\frac{\partial u}{\partial y} - \rho\overline{u'v'}\right) \tag{3.1.16}$$

$$\frac{\partial u}{\partial x} + \frac{\partial v}{\partial y} = 0 \tag{2.2.3}$$

$$\frac{\partial p}{\partial y} = 0 \tag{3.1.15}$$

The quantity in parentheses in Eq. (3.1.16) is the shear stress σ_{xy}, usually given the symbol τ. The stress term can be further simplified to $\partial/\partial y\,(\nu\,\partial u/\partial y - \overline{u'v'})$, since ρ is constant. These equations are the TSL equations or boundary-layer equations for constant-property flow. Most of the remainder of this book is based on them. Since $\partial p/\partial y$ is zero, $\partial p/\partial x$ (usually replaced by dp/dx to emphasize its independence of y) can be equated to the value at the surface or the value in the free stream, where Bernoulli's equation applies, $-\rho u_e\,du_e/dx$. Note that p is therefore no longer a variable but has been absorbed into the boundary conditions. The two-dimensional laminar TSL equation (3.1.16), with the simplification [Eq. (3.1.15)] and without the Reynolds stress term, and the continuity equation (2.2.3) are two equations for the two variables u and v. They can be shown to be "parabolic" (see Chap. 4) with disturbances propagating only downstream and not upstream. The two-dimensional Navier-Stokes equations (for u, v, and p) are elliptic, with disturbances propagating upstream as well as downstream. The change in type is caused by the elimination of p as a variable and by the neglect of $\partial\sigma_{xx}/\partial x$. Most attempts to approximate, rather than neglect, $\partial p/\partial y$ and $\partial\sigma_{xx}/\partial x$ change the type back to elliptic, and methods of solution intended for parabolic equations may not work. There is a strong temptation, for the sake of mathematical and computational simplicity, to use the TSL equations even when they are not accurate enough.

Equations (3.1.15) and (3.1.16) have been derived for boundary layers on a surface coincident with the x-axis for simplicity. The same equations apply to other types of shear layer with a sufficiently small value of δ/l or $d\delta/dx$, providing that the x-axis lies within, or close to, the shear layer. Equations (3.1.15) and (3.1.16) do not apply to shear layers flowing in, say, the y-direction because in deriving them we chose the x-axis as the general direction of flow. (We could of course have chosen the y axis instead.) In mathematical language the equations are not "rotationally invariant." In practice, difficulties arise only if the shear-layer direction is not known.

If the fluid properties vary, the viscous stress terms become more complicated, but the final result is to replace the stress gradients in Eq. (3.1.16) by

$$\frac{1}{\rho}\frac{\partial \sigma_{xy}}{\partial y} \equiv \frac{1}{\rho}\frac{\partial \tau}{\partial y} = \frac{1}{\rho}\frac{\partial}{\partial y}\left(\mu \frac{\partial u}{\partial y}\right) - \frac{1}{\rho}\frac{\partial}{\partial y}(\overline{\rho u'v'} + \overline{\rho'u'v'}) \tag{3.1.17}$$

where the last turbulent term is usually negligible at nonhypersonic Mach numbers, and the effect of the viscosity fluctuation μ' is almost always negligible. Orders of magnitude may be upset if the density itself changes by an order of magnitude across the layer, as happens in hypersonic flow.

It is generally assumed that unsteadiness does not affect the TSL equations, except for the appearance of $\partial u/\partial t$ on the right-hand side of Eq. (3.1.16), as long as $d\delta/dx$ remains small. Strong pressure gradients, in steady or unsteady flow, may overwhelm the viscous or turbulent stress gradients, except close to the surface where the stress gradient must balance the pressure gradient since the acceleration is necessarily small. However, this does not invalidate Eq. (3.1.16).

3.2 AXISYMMETRIC AND THREE-DIMENSIONAL FLOWS

When three-dimensional shear layers are considered, two classes must be distinguished. The first, sometimes called the class of "boundary sheets" and exemplified by the boundary layer on a swept wing not too near the root or tip, has a scale of order l in the z- (spanwise) direction, as well as in the x- (chordwise) direction, so the z-component equation can be simplified in the same way as the x-component equation. In some special cases the spanwise scale may be even larger, but as in Sec. 3.1 we want to derive widely applicable equations. An analysis similar to that of Sec. 3.1 gives, instead of the boxed group on p. 44, the equations

$$u\frac{\partial u}{\partial x} + v\frac{\partial u}{\partial y} + w\frac{\partial u}{\partial z} = -\frac{1}{\rho}\frac{\partial p}{\partial x} + \frac{1}{\rho}\frac{\partial}{\partial y}\left(\mu \frac{\partial u}{\partial y} - \overline{\rho u'v'}\right) \tag{3.2.1}$$

$$\frac{\partial p}{\partial y} = 0 \quad \text{or} \quad -\rho\left(u\frac{\partial v}{\partial x} + w\frac{\partial v}{\partial z}\right) \tag{3.2.2}$$

$$u\frac{\partial w}{\partial x} + v\frac{\partial w}{\partial y} + w\frac{\partial w}{\partial z} = -\frac{1}{\rho}\frac{\partial p}{\partial z} + \frac{1}{\rho}\frac{\partial}{\partial y}\left(\mu \frac{\partial w}{\partial y} - \overline{\rho v'w'}\right) \tag{3.2.3}$$

with the continuity equation

$$\frac{\partial u}{\partial x} + \frac{\partial v}{\partial y} + \frac{\partial w}{\partial z} = 0 \tag{2.2.2}$$

Note that the pressure is now a function of x and z, since the free-stream velocity has two components, u_e and w_e. The alternative expression for $\partial p/\partial y$ in Eq. (3.2.2) can be shown to be equivalent to $\mathbf{u}^2/\mathbf{R}$, where $\mathbf{u}$ is the vector whose components are (u, w) and $1/\mathbf{R}$ is the curvature in the plane of $\mathbf{u}$.

The second class of three-dimensional shear layers, boundary regions or "slender shear layers," is exemplified by the flow in a duct or a wing-body junction and has a scale of order δ in the z-direction as well as the y-direction. We discuss these flows further in Chap. 10.

If the flow in a slender shear layer is axisymmetric, so gradients around the circumference are zero, the exact equations of motion are similar to those for two-dimensional flow, and the shear-layer equations are also similar and of similar accuracy. These sets of equations can be derived in cylindrical coordinates either by transformation from rectangular Cartesian coordinates or by direct CV analysis (see Prob. 3.5). With a little extra effort they can be written in the special curvilinear (x, y) coordinates shown in Fig. 3.1, appropriate to a slowly growing annular shear layer such as the boundary layer on a body of revolution. Using the TSL approximation to simplify the stress-gradient terms and the y-component momentum equation, we have

$$\frac{\partial u}{\partial t} + u\frac{\partial u}{\partial x} + v\frac{\partial u}{\partial y} = -\frac{1}{\rho}\frac{dp}{dx} + \frac{1}{\rho r^k}\frac{\partial}{\partial y}\left[r^k\left(\mu\frac{\partial u}{\partial y} - \rho\overline{u'v'}\right)\right] \tag{3.2.4}$$

$$\frac{\partial}{\partial x}(r^k u) + \frac{\partial}{\partial y}(r^k v) = 0 \tag{3.2.5}$$

where k is a "flow index," equal to unity in axisymmetric flow and zero in two-dimensional flow, so r^k is r or unity, respectively. With $k = 0$ we recover Eqs. (3.1.16) and (2.2.3).

The y-component equation is usually taken as $\partial p/\partial y = 0$: if the streamwise curvature is significant, Eq. (3.1.5) can be used instead. Note that the slope of the surface $y = 0$ (the angle ϕ in Fig. 3.1) need *not* be small, nor need δ/r_0. If the flow field extends to the axis of symmetry as in a jet or pipe flow or the wake of a body of revolution, $r \equiv y$, the x-axis becomes the axis of symmetry, and the equations simplify to

$$\frac{\partial u}{\partial t} + u\frac{\partial u}{\partial x} + v\frac{\partial u}{\partial r} = -\frac{1}{\rho}\frac{dp}{dx} + \frac{1}{r}\frac{\partial}{\partial r}\left[r\left(\mu\frac{\partial u}{\partial r} - \rho\overline{u'v'}\right)\right] \tag{3.2.6}$$

and

$$\frac{\partial u}{\partial x} + \frac{1}{r}\frac{\partial rv}{\partial r} = 0 \tag{3.2.7}$$

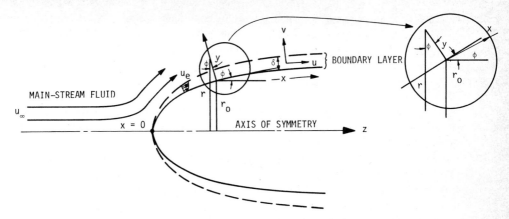

Fig. 3.1 The notation and the coordinate system for axisymmetric flows.

In general, r is related to r_0, the radius of the surface $y = 0$, by

$$r(x,y) = r_0(x) + y \cos \phi(x) \tag{3.2.8}$$

where

$$\phi = \tan^{-1} \frac{dr_0}{dz} \tag{3.2.9}$$

Defining

$$t \equiv \frac{y \cos \phi}{r_0} \tag{3.2.10}$$

we can write Eq. (3.2.8) as

$$\frac{r}{r_0} = 1 + t \tag{3.2.11}$$

Here t represents the deviation of r from r_0 and is called the transverse-curvature term. In many axisymmetric problems the body radius is quite large in relation to the boundary-layer thickness, so the transverse-curvature effect is negligible. In that case, Eqs. (3.2.4) and (3.2.5) simplify considerably and become

$$\frac{\partial u}{\partial t} + u \frac{\partial u}{\partial x} + v \frac{\partial u}{\partial y} = -\frac{1}{\rho} \frac{dp}{dx} + \frac{1}{\rho} \frac{\partial}{\partial y} \left(\mu \frac{\partial u}{\partial y} - \rho \overline{u'v'} \right) \tag{3.2.12}$$

$$\frac{\partial}{\partial x} (r_0{}^k u) + \frac{\partial}{\partial y} (r_0{}^k v) = 0 \tag{3.2.13}$$

We note that Eq. (3.2.12) now has exactly the same form as Eq. (3.1.16) for two-dimensional flow. In Sec. 3.3 we outline the Mangler transformation, which transforms the axisymmetric equations for *small t* into those of an equivalent two-dimensional flow.

In some problems, however, the body radius, while still finite, is of the same order of magnitude as the boundary-layer thickness, or even smaller. Typical examples are slender cylinders, the tail of a streamlined body of revolution, etc. In such cases the transverse-curvature effect can be quite important and must be accounted for in the equations. For this reason, for axisymmetric flows we shall consider the more general equations given by Eqs. (3.2.4) and (3.2.5).

In nonaxisymmetric slender shear layers the equations of motion can be simplified only by neglect of stress gradients in the x-direction. The y- and z-component equations again indicate that pressure differences in the yz-plane are small, but the effect of pressure gradients on the v and w components of velocity may be important. Moreover, the behavior of the pressure may be significantly affected by interaction between the shear layer and the external flow, the usefulness of the TSL concept thus being reduced. The v and w components are small but have a large effect on the u-component distribution. The motion in the yz-plane is called the "secondary flow," and in general the x-component of vorticity is nonzero. The laminar shear layer in a corner is discussed by Zamir and Young (1970); in turbulent flow the behavior of the Reynolds stresses is extremely complicated (Mojola and Young, 1971), and the behavior of the pressure has not been explored in detail.

3.3 THE MANGLER TRANSFORMATION

The boundary-layer equations for two-dimensional and axisymmetric flows differ from each other only by the appearance of the radial distance $r(x, y)$. For flows without transverse-curvature effect [i.e., negligible t in Eq. (3.2.11)] the momentum equation is the same in both cases, and only the continuity equations differ from each other. For such flows the axisymmetric flow equations can be placed in a two-dimensional form by using a transformation known as the Mangler transformation. When there is transverse-curvature effect, this transformation puts the equations into nearly two-dimensional form.

The Mangler transformation is defined by the equations

$$d\bar{x} = \left(\frac{r_0}{L}\right)^{2k} dx \tag{3.3.1}$$

and

$$d\bar{y} = \left(\frac{r}{L}\right)^{k} dy \tag{3.3.2}$$

$$\bar{\psi}(\bar{x}, \bar{y}) = \left(\frac{1}{L}\right)^{k} \psi(x, y) \tag{3.3.3}$$

so an axisymmetric flow with coordinates (x, y), as in Fig. 3.1, is transformed approximately into a two-dimensional flow with coordinates $(\bar{x}, \bar{y})$. From Eqs. (3.3.1) and (3.3.2) we can write

$$\frac{\partial}{\partial x} = \left(\frac{r_0}{L}\right)^{2k} \frac{\partial}{\partial \bar{x}} + \frac{\partial \bar{y}}{\partial x} \frac{\partial}{\partial \bar{y}} \tag{3.3.4a}$$

$$\frac{\partial}{\partial y} = \left(\frac{r}{L}\right)^{k} \frac{\partial}{\partial \bar{y}} \tag{3.3.4b}$$

We define a stream function $\psi(x, y)$ that satisfies the continuity equation (3.2.5) automatically:

$$r^k u = \frac{\partial \psi}{\partial y} \qquad r^k v = -\frac{\partial \psi}{\partial x} \tag{3.3.5}$$

Now, if in the barred plane [Eq. (3.2.5)] is written as

$$\frac{\partial \bar{u}}{\partial \bar{x}} + \frac{\partial \bar{v}}{\partial \bar{y}} = 0 \tag{3.3.6}$$

then the stream function $\bar{\psi}(\bar{x}, \bar{y})$ that satisfies Eq. (3.3.6) is given by

$$\frac{\partial \bar{\psi}}{\partial \bar{y}} = \bar{u} \qquad \frac{\partial \bar{\psi}}{\partial \bar{x}} = -\bar{v} \tag{3.3.7}$$

Therefore if the relations defined by Eqs. (3.3.3), (3.3.4), and (3.3.7) are used, Eqs. (3.3.5) become

$$u = \left(\frac{L}{r}\right)^{k} \frac{\partial \bar{\psi}}{\partial y} = \left(\frac{L}{r}\right)^{k} \left(\frac{r}{L}\right)^{k} \frac{\partial \bar{\psi}}{\partial \bar{y}} = \bar{u} \tag{3.3.8a}$$

$$v = -\left(\frac{L}{r}\right)^{k} \frac{\partial \bar{\psi}}{\partial x} = \left(\frac{L}{r}\right)^{k} \left(\frac{r_0}{L}\right)^{2k} \bar{v} - \left(\frac{L}{r}\right)^{k} \frac{\partial \bar{y}}{\partial x} \bar{u} \tag{3.3.8b}$$

By substituting from Eq. (3.3.8) into Eq. (3.2.4) and using the relations of Eqs. (3.3.4) it can be shown that we can write the Mangler transformed momentum equation as

$$\bar{u} \frac{\partial \bar{u}}{\partial \bar{x}} + \bar{v} \frac{\partial \bar{u}}{\partial \bar{y}} = -\frac{1}{\rho} \frac{dp}{d\bar{x}} + \frac{1}{\rho} \frac{\partial}{\partial \bar{y}} \left\{ (1 + t)^{2k} \left[\mu \frac{\partial \bar{u}}{\partial \bar{y}} - \left(\frac{L}{r}\right)^{k} \rho \overline{u'v'} \right] \right\} \tag{3.3.9}$$

Here

$$t = -1 + \sqrt{1 + \frac{2 \cos \phi}{L} \left(\frac{L}{r_0}\right)^{2} \bar{y}} \tag{3.3.10}$$

where r_0 and ϕ are the radius and slope of the surface as shown in Fig. 3.1. For small y/r_0, $t \approx (y/r_0) \cos \phi$. The boundary conditions are

$$\bar{y} = 0 \quad \bar{u} = \bar{v} = 0 \tag{3.3.11a}$$

$$\bar{y} = \bar{\delta}(\bar{x}) \quad \bar{u} = u_e(\bar{x}) \tag{3.3.11b}$$

Comparison of the Mangler transformed equation (3.3.9) with the momentum equation (3.1.16) for two-dimensional flows shows that Eq. (3.3.9) is identical with Eq. (3.1.16) except for the presence of the transverse-curvature term, t. If this term is negligible (i.e., if $\delta/r_0 \ll 1$), then we see that the Mangler transformation allows the momentum equation for axisymmetrical flows to be written exactly in the same form as the momentum equation for two-dimensional flows.

3.4 INTEGRAL EQUATIONS FOR THIN OR AXISYMMETRIC SHEAR LAYERS

The equations that result from integrating the momentum and continuity equations across the width of the shear layer (i.e., from $y = 0$ to $y = \delta$) are useful in qualitative discussion and in some simplified types of calculation method. We derive the equations for two-dimensional constant-property flow and then quote more general results without proof.

3.4.1 The Momentum Integral Equation

We start with Eqs. (3.1.16) and (2.2.3), replacing the total shear stress $\mu \, \partial u/\partial y - \rho \overline{u'v'}$ by the usual symbol τ, and $-(1/\rho) \, dp/dx$ by $u_e \, du_e/dx$, invoking Eq. (3.1.15). For the sake of simplicity we derive the result for a boundary layer on the surface $y = 0$. We have

$$u \frac{\partial u}{\partial x} + v \frac{\partial u}{\partial y} = u_e \frac{du_e}{dx} + \frac{1}{\rho} \frac{\partial \tau}{\partial y} \tag{3.4.1}$$

$$\frac{\partial u}{\partial x} + \frac{\partial v}{\partial y} = 0 \tag{2.2.3}$$

Now add u times Eq. (2.2.3) to Eq. (3.4.1)

$$\frac{\partial u^2}{\partial x} + \frac{\partial uv}{\partial y} = u_e \frac{du_e}{dx} + \frac{1}{\rho} \frac{\partial \tau}{\partial y} \tag{3.4.2}$$

and integrate with respect to y from $y = 0$ to $y = h > \delta$. Then

$$\int_0^h \frac{\partial u^2}{\partial x} \, dy + u_e v_h = \int_0^h u_e \frac{du_e}{dx} \, dy - \frac{\tau_w}{\rho} \tag{3.4.3}$$

where τ_w is the surface shear stress and

$$v_h = -\int_0^h \frac{\partial u}{\partial x}\, dy$$

from Eq. (2.2.3). Thus

$$\int_0^h \left(\frac{\partial u^2}{\partial x} - u_e \frac{\partial u}{\partial x} - u_e \frac{du_e}{dx} \right) dy = -\frac{\tau_w}{\rho} \qquad (3.4.4)$$

or, after rearranging,

$$\int_0^h \left\{ -\frac{\partial}{\partial x}\left[u(u_e - u)\right] - \frac{du_e}{dx}(u_e - u) \right\} dy = -\frac{\tau_w}{\rho} \qquad (3.4.5)$$

Now since $u_e - u = 0$ for $y \geqslant \delta$, both parts of the integrand contribute only for $y < \delta$ and so are independent of h. Therefore the first part of the integral can be written as

$$-\frac{d}{dx}\int_0^h u(u_e - u)\, dy$$

the x-derivative now being ordinary rather than partial because the definite integral is independent of y. With a little rearrangement and a change of sign we get

$$\frac{d}{dx}\left[u_e^2 \int_0^h \frac{u(u_e - u)}{u_e^2}\, dy \right] + u_e \frac{du_e}{dx} \int_0^h \left(\frac{u_e - u}{u_e} \right) dy = \frac{\tau_w}{\rho} \qquad (3.4.6)$$

We now introduce the displacement thickness δ^* and the momentum thickness θ, as defined by Eqs. (1.3.14) and (1.3.15), and obtain

$$\frac{d}{dx}(u_e^2 \theta) + \delta^* u_e \frac{du_e}{dx} = \frac{\tau_w}{\rho} \qquad (3.4.7)$$

or

$$\frac{d\theta}{dx} = \frac{\tau_w}{\rho u_e^2} - (H + 2)\frac{\theta}{u_e}\frac{du_e}{dx} \qquad (3.4.8)$$

where

$$H \equiv \frac{\delta^*}{\theta}$$

This is the momentum integral equation for two-dimensional constant-density laminar or turbulent flows.

In the form

$$\frac{d}{dx}(\rho_e u_e^2 \theta) = \tau_w + \delta^* \frac{dp}{dx} \tag{3.4.9}$$

the momentum integral equation applies to compressible or incompressible flows. It also applies to free shear layers if we take $y = 0$ at or below the lower edge of the layer and put $\tau_w = 0$. In axisymmetric flow the definitions of the *areas* δ^* and θ are

$$\delta^* = \int_0^h r^k \left(1 - \frac{u}{u_e}\right) dy \tag{3.4.10}$$

$$\theta = \int_0^h r^k \frac{u}{u_e}\left(1 - \frac{u}{u_e}\right) dy \tag{3.4.11}$$

using the "flow index" convention again ($k = 0$ for plane flow and $k = 1$ for axisymmetric flow). The momentum-integral equation becomes

$$\frac{d}{dx}(\rho_e u_e^2 \theta) = r_0^k \tau_w + \delta^* \frac{dp}{dx} \tag{3.4.12a}$$

(since τ_w acts on the surface $y = 0$ whose radius is r_0 in the notation of Fig. 3.1), or

$$\frac{d\theta}{dx} = r_0^k \frac{\tau_w}{\rho u_e^2} - (H + 2)\frac{\theta}{u_e}\frac{du_e}{dx} \tag{3.4.12b}$$

EXAMPLE 3.1

Derive Eq. (3.4.12).

Solution   Multiply Eq. (3.2.4) by r and Eq. (3.2.5) by u, setting $k = 1$ and $dp/dx = - \rho u_e \, du_e/dx$, and add. We get

$$\frac{\partial}{\partial x}(ru^2) + \frac{\partial}{\partial y}(ruv) = ru_e \frac{du_e}{dx} + \frac{\partial}{\partial y}\left[r\left(\mu \frac{\partial u}{\partial y} - \rho\overline{u'v'}\right)\right]$$

analogous to Eq. (3.4.2). Formally integrating, we get

$$\int_0^h \frac{\partial (ru)^2}{\partial x} \, dy + u_e(rv)_h = \int_0^h ru_e \frac{du_e}{dx} \, dy - \frac{\tau_w r_0}{\rho}$$

The rest of the analysis follows the derivation of Eq. (3.4.7) except that an r appears in each integral and must therefore be included in the definition of δ^* and θ. Equation (3.4.12) is an exact consequence of Eqs. (3.2.4) and (3.2.5). The momentum-integral equation quoted in elementary accounts of axisymmetric flow is

$$\frac{d\theta}{dx} = \frac{\tau_w}{\rho u_e^2} - (H + 2) \frac{\theta}{u_e} \frac{du_e}{dx} - \frac{\theta}{r_0} \frac{dr_0}{dx}$$

where δ^* and θ are defined, as in two-dimensional flow, without the r factor. This equation results from putting $r = r_0$ in the above analysis and is therefore a good approximation only if $\delta/r_0 \ll 1$.

3.4.2 The Entrainment Equation

From the continuity equation we get

$$v_e = -\int_0^\delta \frac{\partial u}{\partial x} \, dy = -\frac{d}{dx} \int_0^\delta u \, dy + u_e \frac{d\delta}{dx} \tag{3.4.13}$$

where v_e is v at $y = \delta$ and we have used Leibnitz's rule for interchange of differentiation and integration (we can see that increasing δ by an amount $d\delta$ increases the integral on the right by $u_e \, d\delta$). Also, using the definition of δ^*, Eq. (1.3.14), and ignoring the difference between h and δ, we have

$$\frac{d}{dx}(u_e \delta^*) = \frac{d}{dx} \int_0^\delta (u_e - u) \, dy \tag{3.4.14}$$

Splitting up the integral in Eq. (3.4.14) and using the definition of δ^* again, we get

$$v_e = \frac{d}{dx}(u_e \delta^*) - \delta \frac{du_e}{dx} \tag{3.4.15}$$

(a result that is obviously correct if there is no boundary layer and $u = u_e$ everywhere so that $\delta^* = 0$).

Now, the definition of δ^* can be rearranged to give

$$\int_0^\delta u \, dy = u_e(\delta - \delta^*) \tag{3.4.16}$$

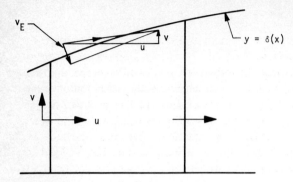

Fig. 3.2 Definition of entrainment velocity.

The quantity on the left is the volume flow rate per unit span between $y = 0$ and $y = \delta$. We define the "entrainment velocity," v_E, as the component of velocity normal to the edge of the boundary layer (Fig. 3.2). Then v_E is the rate at which the volume flow rate per unit span changes with x, so that by using Eqs. (3.4.15) and (3.4.16) or by applying the continuity equation by inspection to the CV of Fig. 3.2,

$$v_E \equiv \frac{d}{dx} \int_0^\delta u \, dy \tag{3.4.17}$$

$$v_E = \frac{d}{dx}[u_e(\delta - \delta^*)] = u_e \frac{d\delta}{dx} - v_e \tag{3.4.18}$$

This is the entrainment equation or volume-flow integral equation. In compressible flow,

$$\int_0^\delta \rho u \, dy = \rho_e u_e (\delta - \delta^*) \tag{3.4.19}$$

the definition of δ^*, also including a density term [see Eq. (1.2.17)], and so

$$v_E = \frac{1}{\rho_e} \frac{d}{dx}[\rho_e u_e (\delta - \delta^*)] \tag{3.4.20}$$

In axisymmetric flow, v_E is given by

$$v_E = \frac{1}{r_\delta} \frac{d}{dx} \int_0^\delta ru \, dy = \frac{1}{r_\delta} \frac{d}{dx} \left[u_e \left(\delta \frac{r_0 + r_\delta}{2} - \delta^* \right) \right] \tag{3.4.21}$$

where $r_\delta = r_0 + \delta \cos \phi$.

3.5 "FAIRLY THIN" SHEAR LAYERS

As we have seen, the common assumption $\partial p/\partial y = 0$ may not be very accurate on a curved surface even for small $d\delta/dx$. Also, the relatively large values of $d\delta/dx$ found in some turbulent flows, particularly turbulent jets, degrade the TSL approximation. A common but ill-codified practice is to *approximate*, rather than neglect, some of the terms whose smallness is in question. This provides an interpolation between the TSL approximation and the full Navier-Stokes equations and may be called the "fairly-thin-shear-layer (FTSL) approximation." The most obvious example is the replacement of $\partial p/\partial y = 0$ by the centrifugal formula [Eq. (3.1.5)] or some approximation to it. In turbulent flow, where the normal-stress gradients are of order δ/l smaller than the shear-stress gradient, normal stresses are sometimes included, for instance, in the momentum integral equation. The FTSL approximation is used most in turbulent flows because of the greater need and because any perturbation of a laminar flow leading to large $d\delta/dx$ is likely to lead to transition. Its use therefore merges with the approximations necessary in modeling the major turbulent stress terms. Unfortunately, the main body of mathematical work on higher-order boundary-layer theory has been done in laminar flows because the relation between the velocity field and the stresses is known. The use of the FTSL approximation in turbulent flow is reviewed by Bradshaw (1975).

3.6 LOW-REYNOLDS-NUMBER FLOWS

In a laminar boundary layer below a constant-speed stream $d\delta/dx \sim (u_e x/\nu)^{-1/2}$, so if $u_e x/\nu$ is small, the assumption $d\delta/dx \ll 1$ fails (and so does this formula for $d\delta/dx$). The FTSL approximations can be used to interpolate between the TSL equations and the full Navier-Stokes equations as long as the Reynolds number is large in comparison with unity, although, as was mentioned in Sec. 3.5, these approximations are necessarily case oriented and unsystematic. For Reynolds numbers of the order of unity the full Navier-Stokes equations are needed, but for Reynolds numbers much *smaller* than unity the acceleration terms in the equations become small in comparison with the viscous terms (the ratio of the two being of the same order as the Reynolds number). Thus the equations reduce to

$$\boxed{\text{Pressure gradient} \quad = \quad \begin{array}{l}\text{Sum of viscous}\\ \text{stress gradients}\end{array}}$$

The continuity equation is unaltered.

In constant-property flow we therefore have

$$\frac{\partial p}{\partial x} = \mu\left(\frac{\partial^2 u}{\partial x^2} + \frac{\partial^2 u}{\partial y^2} + \frac{\partial^2 u}{\partial z^2}\right) \equiv \mu\nabla^2 u \qquad (3.6.1)$$

and similar equations for the y- and z-directions, where the identity defines the

"Laplacian" or "harmonic" operator, ∇^2. The vorticity equation, obtainable by taking the curl of Eq. (2.3.9) and its companions or by taking the low-Reynolds-number limit of Eq. (2.5.2), can be written as

$$\nabla^2 \Omega = 0 \tag{3.6.2}$$

and the pressure equation, obtainable by taking the divergence of Eq. (3.6.1) and its companions, is

$$\nabla^2 p = \mu \nabla^2 \text{ div } \mathbf{u} = 0 \tag{3.6.3}$$

using the continuity equation.

In two-dimensional or axisymmetric flow we can define a stream function ψ, satisfying the continuity equation, by

$$r^k u = \frac{\partial \psi}{\partial y} \qquad r^k v = -\frac{\partial \psi}{\partial x} \tag{3.6.4}$$

Then the vorticity can be shown to satisfy

$$\nabla^2 \psi = -r^k \zeta \tag{3.6.5}$$

for the spanwise component of vorticity in two-dimensional flow or the circumferential component in axisymmetric flow. Combining with Eq. (3.6.2) gives

$$\nabla^4 \psi = 0 \tag{3.6.6}$$

which is the *biharmonic equation*. It is rather difficult to solve numerically, partly because the finite-difference scheme is necessarily complicated and partly because the boundary conditions are also complicated. We note that if the vorticity is zero, this equation [or, strictly speaking, Eq. (3.6.5)] reduces to $\nabla^2 \psi = 0$, as in inviscid irrotational flow. This is the basis of the Hele-Shaw analogy, in which inviscid flow is simulated by the low-Reynolds-number flow between closely spaced plates. Viewed normal to the plates, the streamlines in flow around an obstacle spanning the gap are the same as in inviscid flow past a two-dimensional body of the same cross section.

The practical importance of the so-called "Stokes flow" described by Eq. (3.6.2) applies to flow in porous media ("Darcy flow") or lubrication ducts or around small particles. Dust or silt particles and small raindrops are natural examples, while particle transport is important in many branches of engineering. Solutions for arbitrary particle shapes, or groups of particles whose flow fields overlap, must be obtained numerically, but a famous analytic solution exists for the important case of flow at a speed u_∞ past an isolated spherical particle of diameter d and gives Stokes' law for the drag D:

$$D = 3\pi \, d\mu u_\infty \tag{3.6.7}$$

For further discussions of Stokes flow and of the Oseen approximation in which the acceleration terms are retained but *linearized*, see Schlichting (1968, Chap. 6) and Rosenhead (1963, Chap. 4).

There is, of course, no turbulent analog of flow at very low Reynolds numbers: if the flow is turbulent at all, the Reynolds number based on the apparent "eddy" viscosity (defined as the ratio of Reynolds stress to rate of strain), instead of on the molecular viscosity, must be large in comparison with unity.

PROBLEMS

3.1. Which of the stress-gradient terms in Eq. (2.4.4) are negligible according to the boundary-layer approximation? If $d\delta/dx$ (or δ/x) is 10^{-2}, will the ratio of a typical neglected term to a typical remaining term be of the order of 10^{-2}, 10^{-4}, or 10^{-6}? Is this answer valid for laminar flow and turbulent flow or for laminar flow only?

3.2. Show that the shear stress, $\tau = \mu(\partial u/\partial y) - \overline{\rho u'v'}$, near the axis of a circular jet is proportional to the distance from the axis.

3.3. Show that at a solid surface in the xz-plane the tangential pressure gradient is equal to the derivative with respect to y of the component of shear stress in the direction of the pressure gradient. What is the corresponding result for two-dimensional laminar flow over a uniformly porous surface with transpiration velocity v_w? State any assumptions that you make.

3.4. Find the mean part u, the fluctuating part u', and the mean-square fluctuation $\overline{u'^2}$ for the following variations of instantaneous velocity with time.
    a. $u + u' = a + b \sin \omega t$.
    b. $u + u' = a + b \sin^2 \omega t$.
    c. $u + u' = at + b \sin \omega t$ (careful).

3.5. Derive Eq. (3.2.4) by CV analysis, using the CV shown below.

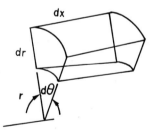

3.6. Derive Eq. (3.3.9).

3.7. Is $d\theta/dx$ always positive?

3.8. Calculate an average value of $(\delta^*/\tau_w) dp/dx$ for Prob. 1.3. Assume that the pressure gradient term in the momentum integral equation is fairly small.

3.9. Derive Eq. (3.4.12).

3.10. Derive Eq. (3.4.21).

Chapter 4

Tools

The three most powerful tools of fluid dynamics are experiment, the numerical solution of partial differential equations (PDEs), and dimensional analysis. In some ways, dimensional analysis is the most powerful of all because its skillful use greatly reduces the volume of experiment or computation needed. Dimensionless variables will be used extensively below. We shall not discuss shear-layer experimental techniques in detail because this is done in other books (Bradshaw, 1970; Merzkirch, 1974). Shear-layer experiments rely heavily on the use of pitot tubes to measure mean-velocity profiles and on hot-wire or laser-Doppler anemometers (Bradshaw, 1971a; Durst et al., 1976) for fluctuations. Flow visualization is quicker than quantitative measurement for determining general flow behavior, especially the location of transition or separation, while a stethoscope (a pitot tube connected to an earpiece) is a useful tool for mapping the boundaries of a low-speed turbulent flow.

This chapter is a brief review of dimensional analysis, familiar to most readers, and an introduction to numerical solution of ordinary differential equations (ODEs) or PDEs. The introduction is intended merely to help the reader through the book, and details of numerical methods are not given. Not all engineering courses have been reoriented to the new world in which we can compute solutions to most of the equations that interest us: shear layers may provide readers with their first contact with PDEs, other than the classical Laplace, heat-conduction, or wave equations, and possibly with their very first contact with the complexities of boundary conditions. Readers who have more extensive experience with numerical solution of differential equations may skip to Sec. 4.3.3. Because the shear layer is only part of the flow, shear-layer solutions must be matched to solutions for the outer flow, which is not always the trivial procedure described in elementary books. We therefore discuss boundary conditions for fluid-flow equations in general as well as for the TSL equations in particular in preparation for the discussion of viscous/inviscid interactions in Chap. 11.

4.1 DIMENSIONAL ANALYSIS

Any equation connecting physical quantities should be the same whatever units are employed (SI, metric, foot-pound-second, etc.). It follows that each term in an equation must have the same dimensions of mass, length, time, etc., and it is often convenient to rearrange the equation so that each term is dimensionless. The argument of a transcendental function (e.g., $\log x$, $\sin x$) must be dimensionless, so the above principle is obeyed by the equation expressing the function as a power series. To identify dimensionless variables, we do not need to know the equations but only the original (dimensional) variables: sometimes, though by no means always, the requirement of dimensional consistency specifies the functional form of the equation, though unknown constants still appear.

In fluid dynamics the mass dimension necessarily disappears when we consider the equations for acceleration: all the force terms are expressed as force/unit mass, which has the dimensions of acceleration; in particular, pressure or viscous stress appears only as (stress/ρ). Heat transfer rates per unit area always appear divided by ρc_p and have the dimensions of (velocity $\times$ temperature). This greatly simplifies dimensional analysis in most cases. Moreover, most of the dimensionless groups in any given case are identifiable as special cases of well-known parameters, like Reynolds number or Mach number, or as dimensionless coefficients whose form is easy to derive by inspection. It is therefore most inappropriate to use the traditional formal derivation of dimensionless groups by partial solution of indicial equations, which starts from an assumption of perfect ignorance. We strongly recommend the "matrix elimination" method (e.g., Taylor, 1973), which allows the use of prior knowledge but incorporates the necessary formality of checking that all the required groups have been obtained. Mathematically, it is equivalent to partial solution of the indicial equations by successive elimination.

For most purposes in fluid dynamics only a few well-known dimensionless groups need be borne in mind: their algebraic form may differ from case to case, but the physical meaning is the same. They are the following:

The Reynolds number, proportional to the ratio of a typical dynamic pressure to a typical viscous stress, for example, $\rho u_e^2/(\mu u_e/\delta) \equiv u_e\delta/\nu = \mathrm{Re}_\delta$.

The Mach number, proportional to the square root of the ratio of a typical dynamic pressure to the absolute pressure, for example, $\sqrt{(\rho u^2/p)} \equiv \sqrt{\gamma}\, u/a = \sqrt{\gamma}\, M$, where a is speed of sound and γ is ratio of specific heats: the interpretation of Mach number as a velocity ratio is less useful in shear layers.

The Prandtl number, proportional to the ratio of molecular diffusivity of momentum (i.e., viscosity) to molecular diffusivity of heat (i.e., conductivity), $\mu/k \equiv$ $\Pr/c_p$, where c_p is specific heat at constant pressure: note that the Prandtl number is a property of the fluid, being very nearly constant at 0.72 in air and decreasing in water from about 7 at room temperature to about 1.7 near boiling point.

The pressure coefficient, equal to the ratio of a typical pressure difference (including a typical shear stress) to a typical dynamic pressure: the skin-friction coefficient $c_f \equiv \tau_w/(1/2\rho u_e^2)$ is an example.

The Stanton number, a dimensionless heat-transfer coefficient analogous to the skin-friction coefficient, defined as $\mathrm{St} \equiv (Q_w/\rho c_p)/[u_e(T_w - T_e)]$.

4.2 ORDINARY DIFFERENTIAL EQUATIONS AND THEIR BOUNDARY CONDITIONS

4.2.1 Basic Theory

Equations like the momentum integral equation (3.4.8) are ODEs with derivatives in one direction only. The momentum integral equation, with u_e specified as a function of x and with c_f and H related to the other variables by further ODEs or algebraic equations, is the basis for many simple calculation methods for laminar or turbulent shear layers (see Chaps. 5 and 6). Other kinds of ODE derived from the NS or TSL equations will be introduced below.

The "order" of a single ODE (or PDE) is the order of the highest derivative occurring in it. The order of a system of several differential equations (DEs) is the least order of a single DE into which the system can be combined (if it can be so reduced). However, every single DE or system can be broken down, by defining extra dependent variables, into a system of first-order DEs. The number of equations in this system is the order (so that the order equals the number of dependent variables in this case). For example, the ODE

$$au''' + bu'' + cu' + du + e = 0 \tag{4.2.1}$$

(where primes denote differentiation with respect to the independent variable, x say, and a to e are, in general, functions of x and possibly of u) is equivalent to

$$v = u'$$
$$av'' + bv' + cv + du + e = 0 \tag{4.2.2}$$

or to

$$v = u'$$
$$w = v' \tag{4.2.3}$$
$$aw' + bw + cv + du + e = 0$$

A system of first-order equations, like Eq. (4.2.3), can generally be written as

$$\mathbf{u}' = \mathbf{f}(x, \mathbf{u}) \tag{4.2.4}$$

where $\mathbf{u}$ (the usual symbol for the unknowns, not necessarily velocities) and $\mathbf{f}$ are column vectors whose number of components equals the order of the system, n say: in the above example, $(n = 3)$, $\mathbf{u} = (u, v, w)$, and generally $\mathbf{u} = (u_1, u_2, u_3, \ldots, u_n)$.

Many standard computer programs for solving initial value problems for ODEs are arranged to operate on a system of first-order equations like Eq. (4.2.4), with $\mathbf{f}$ defined by a subroutine that must be rewritten by the user for each new problem. Numerical methods for solving equations like Eq. (4.2.4) can be devised by approximating the following obvious integral relation:

$$\mathbf{u}(x_0 + \Delta x) = \mathbf{u}(x_0) + \int_{x_0}^{x_0 + \Delta x} \mathbf{f}(x, \mathbf{u}) \, dx \tag{4.2.5}$$

For an interval Δx that is small enough, the integral is approximated, and then the procedure is continued starting now from $x_0 + \Delta x$. For example, an obvious crude approximation ("Euler's method") is to approximate the integral by $\mathbf{f}[x_0, \mathbf{u}(x_0)] \, \Delta x$.

In Sec. 4.2.3 we use more accurate stable methods for ODEs. Before a numerical solution can be started, from $x = x_0$ say, we must know the n values of the components of $\mathbf{u}$ at $x = x_0$. These starting values are called the "initial conditions." Boundary-layer calculation methods using the momentum integral equation and other ODEs start at $x = x_0$ with given values of θ, H, and other variables.

The so-called "initial value problem," that is, Eq. (4.2.4) subject to the requirement that $\mathbf{u}(x_0) = \mathbf{u}^0$ say, generally has one and only one solution. Furthermore, small changes in $\mathbf{u}^0$ or $\mathbf{f}(x, \mathbf{u})$ usually cause only small changes in the solution. This behavior is typical of the behavior of many boundary layer (or indeed other) flows. Problems having these features are called "well-posed problems."

There are many situations, however, in which it does not make sense physically to be able to specify a full set of *initial* conditions. For example, as we shall see in Sec. 5.4, for certain types of flows the velocity profile at given x in a laminar boundary layer is governed by a third-order ODE (the Falkner-Skan equation), but only two conditions (zero velocity) are imposed at the surface and one condition is imposed in the free stream (i.e., at the edge of the boundary layer). In general, a "boundary-value problem" or "two-point problem" for an nth order ODE or system of ODEs is one in which, say, p conditions are imposed at x_0 and $q = n - p$ conditions are imposed at $x_{\max}$ (here $pq \neq 0$). It may easily happen that two-point problems are not well posed in the above sense. In particular, a solution may not even exist, or if one does, it may not be unique. However, in most physically meaningful problems there will be some hint or reason to suspect a lack of "well posedness" if it indeed occurs. Again, we may consider the Falkner-Skan equation of Sec. 5.4. When the dimensionless pressure gradient parameter m is negative, corresponding to a decelerating flow, the flow might separate, and we should be alerted for possible trouble. It turns out that the two-point boundary-value problem, Eqs. (4.2.6)-(4.2.7) below, has *no solution* for $m < -0.0904 \ldots$, it has nonunique solutions for $-0.0904 \ldots < m < 0$, and it has one solution for each $m > 0$ (i.e., accelerating flows are well posed).

To solve nonlinear two-point boundary-value problems, some form of iteration or successive approximation is invariably used. For example, the well-named "shooting" method proceeds by adjusting q parameters in n initial conditions at $x = x_0$ so that the q boundary conditions at $x = x_{max}$ are satisfied (see, for example, Sec. 4.2.2). For linear problems an iterative procedure is not required, since superposition can be used. That is, n linearly independent solutions of the homogeneous DEs are combined with one particular solution of the inhomogeneous system. Then the n coefficients in this combination are chosen so that the boundary conditions are satisfied.

General reviews of numerical methods for solving ODEs are given by Keller (1968), and most computer center libraries will have general-purpose methods available, with program descriptions citing textbooks or research papers for the basic theory. The readers of this book do not *have* to understand the details of these methods (though it is desirable that they should). Here we give a fairly practical method for solving a real ODE that describes an important class of laminar boundary layers. The details can be skipped at a first reading.

4.2.2 A Shooting Method Using the Runge[*]-Kutta Integration Procedure

In Chap. 5 we shall discuss the Falkner-Skan equation for a laminar boundary layer,

$$f''' + \frac{m+1}{2} ff'' + m[1 - (f')^2] = 0 \tag{4.2.6}$$

Here $f(\eta)$ is a dimensionless stream function and m is a (constant) dimensionless pressure gradient, $-0.0904 < m < \infty$. Equation (4.2.6) is a nonlinear third-order ODE. It is to be solved subject to the following two-point boundary conditions:

$$\eta = 0 \quad f = 0 \quad f' = 0 \tag{4.2.7a}$$

$$\eta = \eta_\infty \quad\quad\quad f' = 1 \tag{4.2.7b}$$

Here $\eta = \eta_\infty$ corresponds to the edge of the boundary layer: for computational purposes, η_∞ is chosen arbitrarily to be larger than δ. If appropriate boundary conditions are applied at $\eta = 0$ (initial value problem), we can integrate outward once only to obtain a solution, using (say) the Runge-Kutta program given in Sec. 4.2.3. To satisfy Eq. (4.2.7) for the nonlinear equation (4.2.6), we need to iterate using a shooting method or otherwise.

A shooting method that can be used to solve the Falkner-Skan equation or other ordinary nonlinear DEs was developed by Keller (1968). One of the features of this method is the systematic way by which new values of $f''(0)$ are determined. The

[*]Pronounced "roonga."

traditional trial-and-error searching technique [e.g., Hartree (1937)] is replaced by Newton's method (see Isaacson and Keller, 1966). This generally provides quadratic convergence of the iterations and decreases the computation time.

According to Keller's shooting method, we first replace Eq. (4.2.6) by a system of three first-order ordinary DEs. If the unknowns f, f', and f'' are denoted by f, u, and v, respectively, the system of three first-order equations can be written as

$$f' = u \tag{4.2.8}$$

$$u' = v \tag{4.2.9}$$

$$v' = -\frac{m+1}{2}fv - m(1 - u^2) \tag{4.2.10}$$

Here u is related to the x-component velocity usually denoted by u, but v is related to the velocity gradient $\partial u/\partial y$ and *not* to the y-component velocity. The boundary conditions given by Eq. (4.2.7) are replaced by

$$f(0) = 0 \quad u(0) = 0 \tag{4.2.11a}$$

$$u(\eta_\infty) = 1 \tag{4.2.11b}$$

We denote $v(0)$, related to the wall shear stress $\mu(\partial u/\partial y)_0$, by

$$v(0) = s \tag{4.2.11c}$$

The problem is to find s such that the solution of the initial value problem, Eqs. (4.2.8)-(4.2.10), (4.2.11a), and (4.2.11c), satisfies the outer boundary condition (4.2.11b). That is, if we denote the solution of this initial value problem by $[f(\eta,s), u(\eta, s), v(\eta, s)]$, then we seek s such that

$$u(\eta_\infty,s) - 1 \equiv \phi(s) = 0 \tag{4.2.12a}$$

To solve Eq. (4.2.12a), we employ Newton's method. This widely used method for finding the root of an equation by successive approximation is most simply explained by reference to Fig. 4.1. If s^0 is a *guess* for a root of the equation $\phi(s) = 0$, a better guess, s^1, is (usually) obtained by extrapolation to the axis of the tangent to $y = \phi(s)$ at $s = s^0$ and so on. This yields the iterates s^ν defined by

$$s^{\nu+1} = s^\nu - \frac{\phi(s^\nu)}{(d\phi/ds)(s^\nu)} \equiv s^\nu - \frac{u(\eta_\infty,s^\nu) - 1}{(\partial u/\partial s)(\eta_\infty,s^\nu)} \quad \nu = 0, 1, 2, \ldots \tag{4.2.12b}$$

Obviously, $s^{\nu+1} = s^\nu$ only if $\phi(s^\nu) = 0$, and then Eq. (4.2.11b) is satisfied exactly. In general, this will not occur for any finite ν; instead we iterate until $|s^{\nu+1} - s^\nu| \leqslant \epsilon$ for some sufficiently small ϵ. Then the condition (4.2.11b) is also approximately satisfied.

In order to obtain the derivative of u with respect to s, we take the derivatives

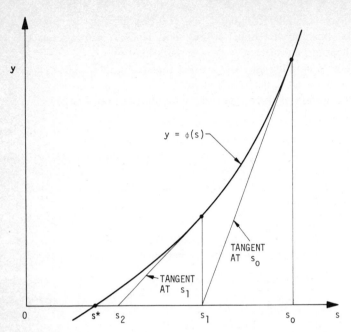

Fig. 4.1 Newton's method for improving an initial guess, s_0, for the root, s^*, of an equation, $\phi(s) = 0$.

of Eqs. (4.2.8)–(4.2.10), (4.2.11a), and (4.2.11c). This leads to the following linear DEs, known as the variational equations, for Eqs. (4.2.8)–(4.2.10):

$$F' = U \tag{4.2.13}$$

$$U' = V \tag{4.2.14}$$

$$V' = -\frac{m+1}{2}(fV + vF) + 2muU \tag{4.2.15}$$

and to the initial conditions, $\eta = 0$:

$$F(0) = 0 \quad U(0) = 0 \quad V(0) = 1 \tag{4.2.16}$$

Here

$$F(\eta,s) \equiv \frac{\partial f}{\partial s} \quad U(\eta,s) \equiv \frac{\partial u}{\partial s} \quad V(\eta,s) \equiv \frac{\partial v}{\partial s} \tag{4.2.17}$$

Note that the boundary condition at η_∞ has disappeared, being replaced by the last of Eq. (4.2.16).

Once the initial-value problem given by Eqs. (4.2.8)–(4.2.10), (4.2.11a), and (4.2.13)–(4.2.16) is solved, $u(\eta_\infty, s^\nu)$ and $U(\eta_\infty, s^\nu)$ are known, and consequently the next approximation to $v(0)$, namely, $s^{\nu+1}$, can be computed from Eq. (4.2.12b). A number of integration methods can be used to solve the initial value problem. Here, because of its simplicity, we use a fourth-order Runge-Kutta method.

The theory behind the Runge-Kutta method (Isaacson and Keller, 1966) is rather complicated and is not necessary for the present discussion. Its virtue is that it is "self starting": that is, a forward step of integration, evaluating the integral in Eq. (4.2.5) by numerical approximation, requires only the initial conditions already given for the ODE. Other, non-self-starting methods require details of the solution for several previous steps before a new step can be executed by integrating polynomial fits to the previous values of the derivatives, but the Runge-Kutta method can "pull itself up by its bootstraps." As might be expected, non-self-starting methods are faster to run, and where many calculations are to be done, it is common to start with a Runge-Kutta method and then switch over to a non-self-starting method such as the predictor-corrector method described by Isaacson and Keller (1966, Chap. 8).

4.2.3 Fortran Program for Solving Falkner–Skan Equation

In this section we describe a Fortran program for solving the Falkner-Skan equation by the shooting method discussed in Sec. 4.2.2. It is programed in Fortran IV for the IBM 370/165; it can be used on other computers with minor changes. Typical solutions are given in Chap. 5.

The program consists of a main program containing the logic of the computations and has three subroutines: STNDRD, VARS, and RKM. The following is a description of the computer program and its subroutines.

MAIN Contains the input data, the boundary-layer grid, the logic of the numerical scheme, the number of iterations performed, and the output, which consists of the profiles f, u, v as a function of η and the parameters δ_1^* and θ_1. See Eqs. (5.3.15), (5.3.16), and (5.5.2).

STNDRD Defines the boundary and initial conditions given by Eqs. (4.2.11a) and (4.2.11c) at the wall and the system of equations given by Eqs. (4.2.8)–(4.2.10). Also, $\phi(s)$ in Eq. (4.2.12a) is computed.

VARS Defines the variational equations and their initial conditions given by Eqs. (4.2.13)–(4.2.16). $U(\eta_\infty, s^\nu)$ in Eq. (4.2.12b) is computed here.

RKM Contains the Runge-Kutta method, which is of fourth order in $\Delta\eta$. It integrates any system of first-order quasi-linear simultaneous ordinary DEs and is not specialized to the Falkner-Skan system. As can be seen from the program, RKM must be called four times for each step in η.

The arguments X to IS have the following explanation:

A or X The independent variable η. The initial value for X must be input.

B A floating point array that is dimensioned N. The B(I) are the current values of the dependent variables. For example,

$$\begin{pmatrix} f \\ u \\ v \end{pmatrix} = \begin{pmatrix} B(1) \\ B(2) \\ B(3) \end{pmatrix}$$

Initial values $f(0)$, $u(0)$, and $v(0)$ must be input.

C   A floating point array that is dimensioned N. The (CI) are the current values of derivatives. These values are supplied to RKM through the argument list. For example,

$$\begin{pmatrix} f' \\ u' \\ v' \end{pmatrix} = \begin{pmatrix} C(1) \\ C(2) \\ C(3) \end{pmatrix}$$

DX   The increment in η. DX must be input.

N   The number of simultaneous equations to be integrated.

F   A floating point array used by the subroutine to store values of the array B. F is dimensioned N.

G   A floating array that contains intermediate values computed by the subroutine. Four entries of G are used to compute one entry of B. G is dimensioned 4*N.

IS   A code variable that must be set to zero to initialize the subroutine. IS is automatically stepped through the values 1, 2, 3, and 4 and is reset to zero by the subroutine after user variables for X and the array B are computed. The four entries to RKM are, in effect, four improved estimates of the required integral over the interval DX.

Definitions of symbols used in the computer program are as follows:

Symbol	Fortran name
f, u, v	F, U, V
m	P
$-\dfrac{m + 1}{2}$	PNX1
ϕ	PHI
$-(s^{v+1} - s^v) = \dfrac{\phi}{d\phi/ds}$	TERM
$\dfrac{\partial \phi}{\partial s}(s^v) \equiv U(\eta_\infty, s^v)$	DERWRS
η	X or ETA
$\delta_1{}^*$	DELST1
θ_1	THETA1

```
      COMMON/SHARE/ F(101),U(101),V(101),PNX1,NP,DX,NX,ETA(101),P(15),
     1              S(15),PHI,ETAINF,DERWRS
C - - - - - - - - - - - - - - - - - - - - - - - - - - - - - - - - -
C INPUT
      READ(5,8000) NXT,ITMAX,DX,S(1),ETAINF,EPS
      READ(5,8100) (P(I),I=1,NXT)
C TOTAL NUMBER OF GRID POINTS, NP
      NP    = ETAINF/DX + 1.1
      NX    = 1
C INITIAL VALUE OF INDEPENDENT VARIABLE
   60 ETA(1)= 0.0
      WRITE(6,9300) NX,P(NX),S(NX)
      WRITE(6,9050)
      IT    = 0
      PNX1  = -0.5*(P(NX)+1.0)
C
   90 IT    = IT+1
      IF(IT .LE. ITMAX) GO TO 95
      WRITE(6,9000)
      GO TO 400
C
   95 CALL STNDRD
      CALL VARS
      TERM  = PHI/DERWRS
      WRITE(6,9060) S(NX),TERM
      S(NX) = S(NX)-TERM
C
C  CHECK FOR CONVERGENCE
      IF(ABS(TERM) .GT. EPS) GO TO 90
      CALL STNDRD
      S(NX+1)=S(NX)
C         OUTPUT
  400 WRITE(6,9070)
      DO 2 J=1,NP,4
    2 WRITE(6,9100) J,ETA(J),F(J),U(J),V(J)
      DELST1= ETA(NP)-F(NP)
      THETA1=(S(NX)-P(NX)*DELST1)/(1.5*P(NX)+0.5)
      WRITE(6,9200) DELST1,THETA1
C
      NX    = NX+1
      IF(NX.LE.NXT.AND.IT.LE.ITMAX)GO TO 60
      STOP
C - - - - - - - - - - - - - - - - - - - - - - - - - - - - - - - - -
 8000 FORMAT(2I3,4F10.0)
 8100 FORMAT(F10.0)
 9000 FORMAT(1H0,25HITERATIONS EXCEED MAXIMUM)
 9050 FORMAT(1H0,11X,1HS,13X,3H-DS)
 9060 FORMAT(1H ,5X,2E14.6)
 9070 FORMAT(1H0,2X,1HJ,4X,3HETA,9X,1HF,13X,1HU,13X,1HV/)
 9100 FORMAT(1H ,I3,F10.4,4E14.6)
 9200 FORMAT(1H0,7HDELST1=,E14.6,3X,7HTHETA1=,E14.6//)
 9300 FORMAT(1H1,3HNX=,I3,5X,2HP=,F10.6,5X,2HS=,F10.6)
      END
```

```
      SUBROUTINE STNDRD
      COMMON/SHARE/ F(101),U(101),V(101),PNX1,NP,DX,NX,ETA(101),P(15),
     1              S(15),PHI,ETAINF,DERWRS
      DIMENSION C(3),B(3),Z(3),G(12)
C - - - - - - - - - - - - - - - - - - - - - - - - - - - - - - - - - - - -
C  DEFINITION OF FIRST-ORDER SYSTEM
C  INITIAL VALUES OF DEPENDENT VARIABLES
      N       = 3
      X       = 0.0
      IS      = 0
      B(1)    = 0.0
      B(2)    = 0.0
      B(3)    = S(NX)
      F(1)    = B(1)
      U(1)    = B(2)
      V(1)    = B(3)
C
      DO 100 J=2,NP
      DO 110 LL=1,4
      C(1)    = B(2)
      C(2)    = B(3)
      C(3)    = PNX1*B(1)*B(3)-P(NX)*(1.0-B(2)**2)
  110 CALL RKM(X,B,C,DX,N,Z,G,IS)
C  SAVE CURRENT VALUES
      F(J)    = B(1)
      U(J)    = B(2)
      V(J)    = B(3)
  100 ETA(J)= X
      PHI     = U(NP)-1.0
      RETURN
      END
```

```
      SUBROUTINE VARS
      COMMON/SHARE/ F(101),U(101),V(101),PNX1,NP,DX,NX,ETA(101),P(15),
     1              S(15),PHI,ETAINF,DERWRS
      DIMENSION C(3),B(3),Z(3),G(12)
C - - - - - - - - - - - - - - - - - - - - - - - - - - - - - - - - - - - -
C  INITIAL VALUES OF DEPENDENT VARIABLES
      N       = 3
      IS      = 0
      X       = 0.0
      B(1)    = 0.0
      B(2)    = 0.0
      B(3)    = 1.0
      FF=F(1)
      UU=U(1)
      VV=V(1)
C  SOLUTION FOR GIVEN F,FP,FPP PROFILES
      DO 100 J=2,NP
      FJ=F(J)
      UJ=U(J)
      VJ=V(J)
      DO 10 LL=1,4
      GO TO(4,2,4,3),LL
    2 FF=(FF+FJ)/2.0
      UU=(UU+UJ)/2.0
      VV=(VV+VJ)/2.0
      GO TO 4
    3 FF=FJ
      UU=UJ
      VV=VJ
    4 C(1)    = B(2)
      C(2)    = B(3)
      C(3)    = PNX1*(VV*B(1)+FF*B(3))+2.0*P(NX)*B(2)*UU
   10 CALL RKM(X,B,C,DX,N,Z,G,IS)
  100 CONTINUE
C
      DERWRS= B(2)
      RETURN
      END
```

```
      SUBROUTINE RKM(A,B,C,DX,N,F,G,IS)
      DIMENSION B(1),C(1),F(1),G(1)
C - - - - - - - - - - - - - - - - - - - - - - - - - - - - - - - - - -
      IS    = IS+1
      GO TO (10,30,60,80), IS
C  FIRST ENTRY
   10 E     = A
      DO 20 I=1,N
      F(I)  = B(I)
      G(4*I-3) = C(I)*DX
   20 B(I)  = F(I)+G(4*I-3)/2.0
      GO TO 50
C  SECOND ENTRY
   30 DO 40 I=1,N
      G(4*I-2) = C(I)*DX
   40 B(I)  = F(I)+G(4*I-2)/2.0
   50 A     = E+DX/2.0
      GO TO 10
C  THIRD ENTRY
   60 DO 70 I=1,N
      G(4*I-1) = C(I)*DX
   70 B(I)  = F(I)+G(4*I-1)
      A     = E+DX
      GO TO 100
C  FOURTH ENTRY
   30 DO 90 I=1,N
      G(4*I)= C(I)*DX
      B(I)  = G(4*I-3)+2.0*(G(4*I-2)+G(4*I-1))
   90 B(I)  = (B(I)+G(4*I))/6.0+F(I)
      IS    = 0
  100 RETURN
      END
```

4.3 PARTIAL DIFFERENTIAL EQUATIONS AND THEIR BOUNDARY CONDITIONS

4.3.1 Types of Partial Differential Equations

All shear layers, except those in a constant-diameter pipe or between parallel plates of constant spacing, have properties that vary in at least two coordinate directions and possibly in time. Thus they must be described by the solution of PDEs. We have already derived the three-dimensional unsteady NS equations and various simplified versions such as the TSL equations. In later chapters the solution of these equations for particular problems will be discussed; here we discuss generalities from the viewpoint of an engineer or physicist rather than that of a mathematician.

Note that in the general theory of DEs the distinction between space and time variables is often blurred: the unsteady boundary layer on an infinite plate below an accelerating stream and the steady boundary layer on a semi-infinite plate are mathematically almost the same problem, but in one case the independent variables are t and y and in the other case, x and y. The (t, y) problem is often called one-dimensional, meaning one *space* dimension: the downstream or x-direction in a unidirectional flow described by the TSL equations is "timelike" because the equations do not permit conditions at given x to affect what happens at smaller x (i.e., upstream). The full NS equations *do* permit this kind of upstream influence. The presence or absence of upstream influence, and the boundary and initial conditions required, depend in part on the "type" of the PDEs but are best derived on the basis of physical considerations. Unfortunately, most of the systems of PDEs used to describe complicated fluid motion do not have any standard type or other useful classification.

To understand some of the basic ideas underlying the classification into types, we examine the general linear second-order PDE in two independent variables:

$$Lu \equiv a\frac{\partial^2 u}{\partial x^2} + b\frac{\partial^2 u}{\partial x \bar{y}} + c\frac{\partial^2 u}{\partial y^2} + d\frac{\partial u}{\partial x} + e\frac{\partial u}{\partial y} + fu \equiv h(x,y) \tag{4.3.1}$$

By a simple rotation of the xy-plane into the $\overline{xy}$-plane, given by the coordinate transformation,

$$\bar{x} = \cos\theta x + \sin\theta y$$
$$\tag{4.3.2}$$
$$\bar{y} = -\sin\theta x + \cos\theta y$$

we can *always* transform Eq. (4.3.1) into the form

$$\bar{L}u \equiv \bar{a}\frac{\partial^2 u}{\partial \bar{x}^2} + \bar{c}\frac{\partial^2 u}{\partial \bar{y}^2} + \bar{d}\frac{\partial u}{\partial \bar{x}} + \bar{e}\frac{\partial u}{\partial \bar{y}} + fu = h \tag{4.3.3}$$

This removal of the mixed derivative term, by picking θ so that $\bar{b} = 0$, is similar to the transformation applied to the general quadratic equation so that the new coordinate axes will be aligned with those of the corresponding conic section defined by the quadratic. Indeed, we adopt the same terminology for the conics to classify the DEs.

Thus if $\bar{a}\bar{c} > 0$ (i.e., $\bar{a}$ and $\bar{c}$ have the same sign), we call Eq. (4.3.3) an *elliptic* equation; if $\bar{a}\bar{c} < 0$, we call Eq. (4.3.3) a *hyperbolic* equation; and finally if $\bar{a}\bar{c} = 0$ (but both do not vanish), we call Eq. (4.3.3) a *parabolic* equation. As with the conics we note that $b^2 - 4ac = \bar{b}^2 - \bar{a}\bar{c}$ is an invariant of the rotation. Thus we say that Eq. (4.3.1) is elliptic, hyperbolic, or parabolic if $b^2 - 4ac < 0, > 0$, or $= 0$, respectively.

In the elliptic case we can simply stretch (or compress) the coordinates by introducing

$$\bar{\bar{x}} = \frac{\bar{x}}{\sqrt{|\bar{a}|}} \qquad \bar{\bar{y}} = \frac{\bar{y}}{\sqrt{|\bar{c}|}} \tag{4.3.4}$$

to get finally

$$\bar{\bar{L}}u = \frac{\partial^2 u}{\partial \bar{\bar{x}}^2} + \frac{\partial^2 u}{\partial \bar{\bar{y}}^2} + \bar{\bar{d}}\frac{\partial u}{\partial \bar{\bar{x}}} + \bar{\bar{e}}\frac{\partial u}{\partial \bar{\bar{y}}} + fu = h \tag{4.3.5}$$

The leading term here is the well-known Laplacian in the $\overline{\overline{xy}}$-plane. So we see that by a simple rotation and stretching of the coordinate system any elliptic equation in the plane can be reduced to the Laplacian with possibly lower-order terms.

Suppose, in the hyperbolic case, that $\bar{a} > 0$ and $\bar{c} < 0$. Then, using Eq. (4.3.4), we obtain from Eq. (4.3.3)

$$\bar{\bar{L}}u = \frac{\partial^2 u}{\partial \bar{\bar{x}}^2} - \frac{\partial^2 u}{\partial \bar{\bar{y}}^2} + \bar{\bar{d}}\frac{\partial u}{\partial \bar{\bar{x}}} + \bar{\bar{e}}\frac{\partial u}{\partial \bar{\bar{y}}} + fu = h \tag{4.3.6}$$

Here the leading terms form the wave operator (i.e., $\partial^2 u/\partial \bar{\bar{x}}^2 - \partial^2 u/\partial \bar{\bar{y}}^2 = 0$ is the wave equation). Thus any hyperbolic equation in the plane is equivalent to the wave equation with possibly lower-order terms.

Assume, in the parabolic case, that $\bar{a} > 0$, $\bar{c} = 0$, and also $\bar{e} < 0$. Then, in place of Eq. (4.3.4) we use

$$\bar{x} = \frac{\bar{x}}{\sqrt{\bar{a}}} \qquad \bar{\bar{y}} = \frac{\bar{y}}{\bar{e}} \tag{4.3.7}$$

and Eq. (4.3.3) now becomes

$$\bar{\bar{L}}u = \frac{\partial^2 u}{\partial \bar{x}^2} - \frac{\partial u}{\partial \bar{y}} + \bar{\bar{d}}\frac{\partial u}{\partial \bar{x}} + fu = h \tag{4.3.8}$$

The leading terms here form the heat or diffusion operator (i.e., $\partial^2 u/\partial \bar{x}^2 - \partial u/\partial \bar{\bar{y}} = 0$ is the heat equation).

One of the basic differences between the various types of PDEs is in their "domains of dependence" and their "regions (or domains) of influence." Suppose each Eq. (4.3.1) is to be solved in some region R of xy-space with boundary B. Then, roughly speaking, the domain of dependence of a *point P* in R consists of all those points on B at which given values of the solution (and perhaps some of its derivatives) are required to uniquely determine the solution at the point in question (see Fig. 4.2). Conversely, the "region of influence" of a point P consists of all those points in R at which the solution is altered when a change in the solution at the point P occurs (see Fig. 4.3). In general, elliptic equations have the property that the domain of dependence of any point is a curve (or surface) completely enclosing the point as in Fig. 4.2a. In parabolic and hyperbolic equations this is not the case. The extent of the domain of dependence of a point is determined by the intersection of the so-called characteristic curves through that point with the "physical" boundary B. The characteristics also form part of the boundary of the domains of influence (see Fig. 4.3). The total number of characteristics is equal to the number of dependent variables. For a parabolic system they all coincide with a line normal to the timelike direction (Fig. 4.3a), whereas for a hyperbolic system they are distinct. An elliptic equation has no real characteristics. For further details see Garabedian (1964).

The two-dimensional steady NS equations can be reduced to a fourth-order nonlinear PDE that turns out to be elliptic. It does not resemble any of the classical types listed above, but the domain of dependence of any point is as in Fig. 4.2a. The equations for steady inviscid irrotational flow are second-order PDEs. They are

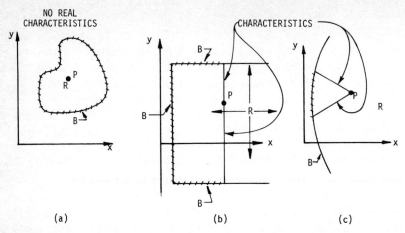

Fig. 4.2 Domains of dependence of the point *P* for the three classical equations to be solved in *R*. (a) Elliptic. (b) Parabolic. (c) Hyperbolic. The crosshatching shows the part of *B*, the boundary of *R*, that is the domain of dependence for *P*.

elliptic in subsonic flow and hyperbolic in supersonic flow, the Mach lines being the characteristics. In a partly subsonic, partly supersonic flow these equations are said to be of "mixed type" or of elliptic-hyperbolic type. The plane steady TSL equations are said to be parabolic, since the domain of influence is usually like that in Fig. 4.3b. However, this may not be the case when reverse flow is present. Thus these TSL equations may be of some mixed or improper type. The three-dimensional TSL equations have more complicated domains of influence, and their type cannot be easily classified (see pp. 77-78).

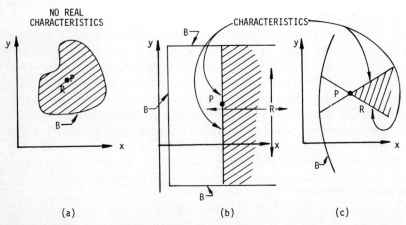

Fig. 4.3 Regions of influence of three classical equations. (a) Elliptic. (b) Parabolic. (c) Hyperbolic. Shading shows regions of influence. Strictly speaking, the boundary *B* in case (a) is at infinity.

4.3.2 Boundary Conditions—Generalities

The boundaries (curves or surfaces) on which conditions are to be imposed are normally determined by the physical formulation of the problem under study. These conditions and the boundaries on which they are imposed are related to the type of the equation. Together the equations and boundary conditions should form (for most physical problems) what is known as a *well-posed problem*. This is one that has a unique solution, which should depend continuously on the inhomogeneous data (i.e., boundary data as well as forcing terms in the equations). For example, in standard TSL problems a small change in the mass transfer at the boundary, or in the pressure distribution, or even a small perturbation in the wall contour should cause only a correspondingly small change in the solution.

A general theory ensuring well posedness for the classical equations (4.3.1) under appropriate boundary conditions is well known (see Courant-Hilbert, 1961). However, such a theory is not known for the full NS equations in most practical configurations or for the TSL equations in general. Thus in imposing boundary conditions we must first rely on physical arguments, then on known mathematical results, and finally on heuristic considerations.

The mathematical arguments that can be used to test or suggest boundary conditions are based on studying related ODEs in the neighborhood of the boundary. The procedure is roughly to introduce coordinates normal and tangential to the boundary, say (η, τ_1, τ_2), in a three-dimensional steady problem with η the normal coordinate. Then remove, by Fourier transform, the tangential variables (τ_1, τ_2) both in the DEs and the boundary conditions. There remains an ODE or system of ODEs in η as an independent variable, subject to some boundary conditions. This reduced ODE problem should be well posed for all values of the transform variables.

While the indicated procedure is far from complete (i.e., nonlinear equations and even linear equations with nonconstant coefficients are not easily included), it does give much useful information. For example, the proper number of required conditions is clearly suggested.

In counting boundary conditions, care should be taken to exclude any condition that can be deduced from the PDE and previously included boundary constraints. For instance, if we specify $u = $ constant along $y = 0$ in the solution of either Laplace's equation or a heat equation, we must not count $\partial^2 u/\partial y^2 = 0$ as another boundary condition because it follows from substituting $\partial^2 u/\partial x^2 = 0$ or $\partial u/\partial x = 0$ into those equations. This could also be seen by our transform method, since the resulting ODE problems would not be well posed (i.e., have bounded solutions) for all values of the transform variables.

4.3.3 Boundary Conditions for Fluid Flow Problems

The NS equations for steady laminar flow in (x, y, z) space, together with the continuity equation, are elliptic and of fourth order, with variables u, v, w, and p. The velocity component normal to a solid boundary is zero and if the viscous terms are included, the velocity component tangential to any solid boundary is zero also (the no-slip condition). Occasionally, porous boundaries occur: in such cases the normal component of velocity is assumed known, and the tangential component is usually taken as zero, although this cannot be exactly correct. It is usually convenient, and frequently even necessary, to idealize the problem by replacing an infinite domain by a finite one. In this process, fictitious boundaries are introduced upstream and downstream, say. Then we are faced with the problem of specifying conditions on these boundaries. It is usually reasonable to specify the velocity components on the upstream boundary. On the downstream side the correct conditions are less obvious. The alternative to imposing such idealized conditions would be to specify conditions on all the solid boundaries in the flow rig (including, for instance, the blades of the pump or blower). If the normal and tangential components of velocity are properly specified on all the real and fictitious boundaries in an isothermal flow, the pressure is thereby determined to within an arbitrary constant. (For example, consider flow in a pipe: if the mass flow rate is given, the pressure drop can be found.) Conversely, the pressure may be specified on part of the boundary and the velocity on other parts.

One can easily think of impossible boundary conditions. In incompressible flow the mass flow rate at entry to a pipe cannot be different from the mass flow rate at exit: in unsteady compressible flow, however, the entry mass flow can—in principle—exceed the exit mass flow because the density of the fluid in the pipe can increase (but not decrease) indefinitely with time. Often, as in this case, given boundary conditions can be satisfied by an unsteady flow but not a steady one, even if the boundary conditions are steady. The most difficult aspect of NS problems, particularly for numerical solutions in domains of limited size, is choosing the right downstream boundary conditions. The most common boundary geometry is like the wind-tunnel working section shown in Fig. 4.4: the upper and lower boundaries may be fictitious slip boundaries (no constraint on tangential velocity component) or solid walls; there may or may not be a central body. If the flow is to be steady, the exit mass flow rate must equal the entry mass flow rate. However, it is obvious that the exit momentum flux rate will differ from that at entry by an amount equal to the force on the solid boundaries and that the exit velocity will be nonuniform unless the exit boundary is a porous surface with special properties.

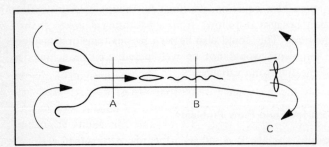

Fig. 4.4 Artificial boundary conditions for an NS problem: conditions on line *B* depend on solutions of NS equations in the domain inside *C*.

Therefore the exit boundary condition appears to depend on the solution, although the equations seem to be elliptic. There are two reasons, most easily discussed for the NS equations than for an abstract elliptic system. One reason is that equations can impose integral constraints on their own boundary conditions; these are often derivable by inspection from the physical principles represented by the equations (e.g., conservation of mass or momentum). The other reason is the exit boundary condition one would like to impose on the NS equations is that which provides a solution that would change as slowly as possible downstream of the exit boundary, simulating a larger domain of integration. The point is most easily grasped by reference to the wind tunnel in Fig. 4.4. The only "exit" boundary conditions that one can specify accurately are those on the laboratory walls, but instead one tries to guess the *properties of the solution* for these conditions and apply them as boundary conditions at the ends of the working section. A more detailed discussion of NS problems is outside the scope of a book on shear layers, but it is necessary to remember that a shear layer is usually only part of a flow and must be imbedded in a domain governed by NS or inviscid-flow equations.

The boundary conditions on the (parabolic) laminar TSL equations are more straightforward (Fig. 4.5, a repetition of Fig. 1.1). Let us first consider the two-dimensional flow ($w = 0$) of a boundary layer over a solid surface (Fig. 4.5b). We must obviously specify the normal (v) and tangential (u) components of velocity at the solid surface, the u component at a distance from the surface chosen to be outside the shear layer (giving the pressure within the shear layer as a function of x) and the u component chosen to be on the initial line $x = 0$. Are these boundary conditions sufficient? We do *not* need to specify v on $x = 0$ because $x = 0$ is a characteristic line along which the PDEs reduce to an ODE that can be solved for v if u is known. (See Prob. 4.1.) For the same reason we do not need v at $y = \delta$. Almost invariably, initial conditions (i.e., entry boundary conditions) for parabolic equations are specified along a characteristic line, the y-axis in this case. Therefore the "obvious" boundary conditions outlined at the beginning of the paragraph *are* sufficient.

Note that we do not need to specify $\partial u/\partial y = 0$ or $\psi_{yy} = 0$ at $y = \infty$. From Eq. (3.1.16) it is clear that if $u = u_e$, the first term on the left equals the first term on the right, so a possible solution is $\partial u/\partial y = \partial^2 u/\partial y^2 = 0$. It is not clear that this is the only possible solution, but the choice between different solutions must be made on physical grounds and not by imposing boundary conditions in excess of the number equal to the order of the equation. In certain cases (Rosenhead, 1963, p. 247) several solutions of Eq. (4.2.6) do exist. All have $\partial u/\partial y = 0$ at $y = \delta$, but some approach it faster than others (i.e., have smaller δ). This, then, is an example of an ill-posed problem. However, physically, we argue that the meaningful solution is the one that approaches the free-stream conditions fastest because this is the solution that will appear first as vorticity propagates into an initially uniform free stream near the origin of the boundary layer. By invoking this additional requirement, our altered problem becomes well posed (i.e., there is only one solution with most rapid approach to the free stream).

In free shear layers (Fig. 4.5a and c) the external-stream velocity must,

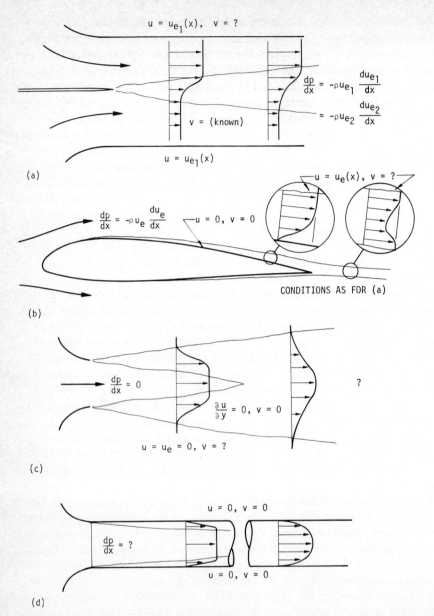

Fig. 4.5 Examples of shear layers: boundary conditions. A question mark denotes a result of the solution. (a) Mixing layer between parallel streams. (b) Boundary layer and wake of airfoil. (c) Merging mixing layers in jet. (d) Merging boundary layers in pipe flow.

nominally, be specified at both edges. Of course, the two values are connected by Bernoulli's equation if the static pressure is constant across the layer. Therefore if, say, the upper-stream velocity is specified as a function of x, the lower-stream velocity need be given only at $x = 0$. The difficulties associated with the v

boundary condition in free shear layers are less obvious. If the flow is symmetrical (Fig. 4.5c), no problem arises: we specify the initial (symmetrical) velocity profile and require $v = 0$ on the center line and $u = u_e$ at one edge. In practice, derivative boundary conditions are a little more difficult to incorporate in most numerical schemes. If the flow is not symmetrical (Fig. 4.5a), *we cannot find a boundary condition for v* from consideration of the shear layer and the TSL equations alone. In the real flow the behavior of v below the shear layer depends on the v-component equation of motion and the continuity equation, applied throughout the flow and not merely in the shear layer. Of course, a similar problem occurs in determining u_e either in the boundary layer or wake of Fig. 4.5b, but that problem is both more familiar and less perplexing. In a real external-flow problem, $u_e(x)$ must be obtained from a numerical solution for the inviscid flow around the body. If the external flow is significantly affected by the displacement effect of the shear layer, iteration between the external-flow and shear-layer calculations is necessary. The initial conditions in practical problems (such as the boundary layer on a round-nosed airfoil) can often be obtained from "similarity" solutions (Sec. 5.2), or, of course, from experimental data. In the case of a round-nosed airfoil, for instance, the calculation is started from the Hiemenz solution for laminar stagnation-line flow (Sec. 5.5). In research papers on shear-layer calculation methods and in comparisons of the predictions of such methods with experimental data, $u_e(x)$ is taken as known.

In so-called internal flows (pipes, ducts, or channels) the pressure gradient is not given explicitly by specifying $u_e(x)$. Instead, the mass flow is specified: it is constant along the duct unless there is transpiration through the walls. It is clear physically that this determines the pressure gradient, but the numerical solution usually involves an iterative procedure (see Secs. 5.7 and 7.3).

The three-dimensional TSL equations for a "boundary sheet" (Sec. 3.2) have a somewhat complicated region of influence, as sketched in Fig. 4.6. The (laminar) TSL equations permit the influence of viscous stresses to be transmitted "instantaneously" in the y-direction, whereas spanwise viscous diffusion is neglected, since the spanwise extent of a three-dimensional TSL is much larger than the thickness and spanwise viscous diffusion will only produce changes over a distance of the

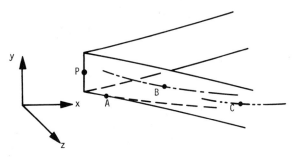

Fig. 4.6 Region of influence of point P in three-dimensional TSL. − − −, − · − , − · · − streamlines with largest w/u at A, B, and C, respectively. Nearer face of wedge touches streamlines at A, B, and C.

order of the thickness. Similar arguments apply in turbulent flow. Therefore the region of influence very close to P is approximately a wedge whose edge is a line in the y-direction passing through P and whose vertical faces are planes including the y-direction and the streamlines with the largest and smallest values of w/u. Further downstream, the faces of the wedge include the streamlines that locally have the largest and smallest values of w/u: in general, these are not the original streamlines that defined the faces near P. To calculate conditions at a given point, u and w profiles must be specified over the intersection of the domain of dependence with the initial plane $x = 0$, and u_e and w_e must be specified on the top face of the domain of dependence. Note that the substitution of the continuity equation in the x-component TSL momentum equation (for steady incompressible flow, say) gives

$$-u\frac{\partial v}{\partial y} - u\frac{\partial w}{\partial z} + v\frac{\partial u}{\partial y} + w\frac{\partial u}{\partial z} - \frac{1}{\rho}\frac{\partial p}{\partial x} + v\frac{\partial^2 u}{\partial y^2} - \frac{\overline{\partial u'v'}}{\partial y} \qquad (4.3.9)$$

which is an ODE for v as a function of y if u, w, and $\overline{u'v'}$ (and, of course, $\partial p/\partial x$) are known on the plane $x = 0$. Therefore as in two dimensions, initial v profiles need not be supplied. The conditions required on the initial plane $x = 0$ are profiles of u, w, and, in turbulent flow, the Reynolds stresses: if the initial plane is not at constant x, a combination of the x- and z-component momentum equations is needed to produce an equation for v containing u and w derivatives in the initial plane only. The behavior of characteristics in three dimensions is, in general, more complicated than in two. As can be seen from the above analysis, the special directions on which the order of the differential equations reduces by 1 are necessarily *surfaces* (usually curved) rather than lines. A further discussion is not necessary for present purposes but can be found in Wang (1971); see also Garabedian (1964). A final point to notice about boundary conditions for three-dimensional shear layers is that u_e and w_e are usually connected by the requirement that the y-component of vorticity, $\partial u/\partial z - \partial w/\partial x$, shall be zero in the free stream.

All time-dependent equations are parabolic or hyperbolic. The time-dependent incompressible (and, let us say, two-dimensional) NS equations seem parabolic because the region of influence in (x, y, t) space of a point $P(x_0, y_0, t_0)$ is the half space $t > t_0$. These equations can behave almost like hyperbolic equations in compressible flow because, except for viscous diffusion, disturbances originating at P are confined to the inside of a cone in (x, y, t) space, with its axis in the t-direction and a semiangle dx/dt or dy/dt equal to the speed of sound. The two-dimensional time-dependent TSL equations may be called degenerate hyperbolic, since the region of influence is the quarter space $x > x_0$, $t > t_0$, and all y. Three-dimensional time-dependent equations behave similarly, but a four-dimensional region of influence is less easy to visualize. The boundary conditions for a time-dependent incompressible NS problem comprise the same conditions as those in steady flow on the spatial boundaries of the domain, specified for each value of time, plus a complete specification of the variables within the domain at the initial time.

Examples of three-dimensional or time-dependent problems are discussed in Chap. 10.

4.3.4 Numerical Solution of Partial Differential Equations

Finite Difference Methods

In this class of methods a grid, usually rectangular, is placed on the flow field, and the flow quantities are to be calculated only at the nodes of the grid (the "net points"). Algebraic equations at each net point are obtained from the PDEs by approximating each derivative by an appropriate difference quotient. The boundary conditions must also be incorporated to ensure that there are as many equations as unknowns. Obviously, there are a tremendous number of different algebraic formulations that can be devised by the indicated procedures. The object, however, is to ensure that the algebraic equations (or difference equations) can be solved with reasonable efficiency and accuracy while representing acceptable approximations to the flow quantities. These questions have been studied in great detail and lead to the theory of consistency, stability, and convergence of difference schemes. In addition, much effort has been devoted to iterative and direct (noniterative) methods for solving various types of difference equations.

For parabolic and hyperbolic problems, difference schemes can be devised that are solved by "marching" through the net and determining the unknowns at each net point as the march proceeds. For flow problems the marching direction will naturally be in the predominant flow direction, x say. Some marching procedures will be implicit in that all the unknowns on some mesh line (generally orthogonal to the marching direction) must be simultaneously computed, but in "explicit" methods the equations for each point on the mesh line can be solved independently. The key requirement in all parabolic or hyperbolic problems is the well-known Courant condition, which simply says that the net points determining the values at a given new net point must at least include the correct domain of dependence of that point. Frequently, but not always, the Courant condition also ensures stability. Implicit schemes are usually unconditionally stable (i.e., no conditions are imposed on the distance between mesh points in the marching direction at least for linear equations).

For elliptic problems the difference equations are always implicit. In special cases, clever schemes have been devised for solving these large systems of linear algebraic equations (i.e., fast direct methods). More generally however, iterative methods are used to solve elliptic difference equations, and such methods must always be used in nonlinear problems. Some of these schemes are related to marching procedures for parabolic systems and may use alternate directions in sweeping the net.

Method of Lines

If in a PDE finite difference approximations are used for derivatives with respect to all independent variables but one, then there results a system of ODEs (in that one "continuous" variable). These equations are defined on lines through the net points, hence the term "method of lines." For example, the parabolic equation

$$\frac{\partial u(x,y)}{\partial x} = \sigma \frac{\partial^2 u(x,y)}{\partial y^2} \quad x > 0 \quad 0 \leqslant y \leqslant \delta(x) \tag{4.3.10}$$

on differencing over some net in the x-direction yields, say,

$$\frac{v_j(y) - v_{j-1}(y)}{\Delta x_j} = \sigma \frac{d^2 v_j(y)}{dy^2} \qquad 0 \leqslant y \leqslant \delta_j \qquad j = 1, 2, \ldots \tag{4.3.11}$$

Here $v_j(y)$ is to approximate $u(x_j, y)$. Many other alternatives to Eq. (4.3.11) could be suggested. The extension to three or four independent variables as well as to other equations is clear.

Usually, the resulting system of ODEs is solved numerically. This may be done by using special difference schemes for two-point-boundary value problems, or by using a shooting method with initial-value integrations of Runge-Kutta or predictor-corrector type (Sec. 4.2). In either case the method of lines has simply been used to develop some ultimate difference scheme or numerical method that may not have been suggested by employing standard finite difference techniques to the original problem. The stability, consistency, and convergence questions are still applicable but are frequently overlooked when applying the method of lines to derive the difference scheme.

Galerkin Method or Method of Integral Relations (MIR)

Here good approximations to the variation of the solution in one direction are guessed, as polynomials with unknown coefficients, say, and substituted into weighted integrals of the PDEs and their boundary conditions to give equations for the unknown coefficients. In a problem with two independent variables this yields ODEs for the coefficients. In higher dimensions, PDEs are obtained. There are many varieties of this method: apart from special names for particular varieties the general method is alternatively known as the "method of weighted residuals" or the "(generalized) Galerkin method." In discussion of shear layers it is often called simply the "integral" technique. It is applicable to all types of PDEs, being most efficient where a good approximation to the solution can be obtained from a function with only a few coefficients. Both this and the choice of suitable weighting functions for the integrals of the PDEs require knowledge of the physical problem and its solution. The method of integral relations is a particularly striking example of the use of accumulated physical knowledge to simplify computation: its range of applicability tends to be limited by that of the physical knowledge.

The ODEs or PDEs resulting from this method are solved by one of the techniques discussed above.

Finite-Element Methods

Here an approximation to the solution is devised separately for each of a number of elements (the finite elements) that together fill the domain of integration. Some conditions on the unknown coefficients are determined by continuity requirements between adjacent elements and by the boundary conditions. Other conditions are obtained by using the Galerkin method over sets of elements. Algebraic equations for the coefficients (or the variables at selected points) follow; these are the finite-element equations. The Galerkin method used over the elements can be similar to that used in the method of integral relations (MIR) technique or, for self-adjoint equations like those of structural mechanics, by using the variational principle for the PDEs and applying the Ritz procedure. The finite-element technique has been used extensively for elliptic problems and, to a more limited extent, for parabolic and hyperbolic problems.

Some typical numerical methods for TSL problems are described later in the book. For more details of particular methods see, for example, Cebeci and Smith (1974), Patankar and Spalding (1970), Mellor and Herring (1968), Birch et al. (1973), Roache (1972), and many papers in fluid mechanics journals, notably *Computers and Fluids*.

4.4 A CAUTION

The authors of this book are proud to have played a part in the application of digital computers to shear-layer calculation methods; having done so, we hope to be believed when we warn the reader about a grave shortcoming of computers. *They do not create information but only rearrange it*! This means that a calculation method, for shear layers or anything else, is at best only as reliable as the physical information used in formulating it. If a method based on the TSL equations is used where the TSL approximation is inaccurate, the answers it gives will be inaccurate. If a hypothesis about turbulent stresses is used in inappropriate conditions, the results will be misleading, even though the numerical statement of the hypothesis occupies only one or two lines in a thousand-line program. The developer of a calculation method has an obligation to the user to specify its limits of validity as accurately as possible; it is the user's responsibility to see that the method is kept within those limits. All too often, however, it is assumed that if a calculation can be completed without the computer's signaling a fatal error, then the results must be acceptable. The converse of this attitude is to agonize about numerical errors of 1 or 2% in a program embodying empirical constants that are in doubt by 10%. It appears that program developers are often less than responsible about providing adequate checks on numerical errors (any TSL calculation method should incorporate at least a momentum-integral check comparing the two sides of the momentum integral equation), but it is far more important that physical shortcomings should be documented. How many calculation methods for laminar boundary layers print out an error message when $d\delta/dx$ is no longer very small in comparison with unity as the TSL approximations require? How many methods for turbulent flow resist the efforts of engineers to run them at Reynolds numbers so low that the flow cannot be turbulent?

PROBLEMS

4.1. Find an equation to give v in a two-dimensional laminar TSL, given only $v(0)$, $u(y)$, and $\partial p/\partial x$.

4.2. Deduce the boundary conditions for a two-dimensional TSL by considering the single PDE for the stream function ψ.

4.3. A jet blows into a corner, as is shown in the sketch. Supposing that the width of the jet normal to the paper is very large and that the region at top left is unobstructed, state what boundary conditions are needed
   a. To predict the free mixing layer using the TSL equations.
   b. To predict the free mixing layer using the NS equations.
   c. To predict the whole flow shown in the sketch using the NS equations.

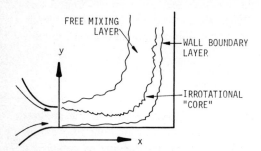

4.4. Sketch the likely behavior of successive shooting approximations to the solution of the TSL equations for a laminar boundary layer (x-component derivatives being inferred from a known solution upstream).

4.5. The solution of the Falkner-Skan equation (4.2.6) for negative wall shear is more difficult than the case for positive wall shear. For this reason a convenient procedure is to solve the system of Eqs. (4.2.6) and (4.2.7) as a "nonlinear eigenvalue" problem with m as the unknown parameter. That is, by Eq. (4.2.11c) the value of $v(0)$ is specified, and the appropriate value of m is sought.

Using Keller's shooting method with Newton's method, solve Eq. (4.2.6) by this procedure for $v(0) = -0.027$ and -0.066 and find the corresponding values of m for each case.

Hint: Solve the initial value problem, Eqs. (4.2.8)–(4.2.10), (4.2.11a), and (4.2.11c) with $v(0) = s$ fixed and seek to vary m so that Eq. (4.2.11b) is satisfied.

4.6. Using an efficient extrapolation scheme described below and Keller's shooting method, obtain the solutions of the Falkner-Skan equation for several values of m once a solution for one value of m is available. (For example, start with $m = 0$, and solve Eq. (4.2.6) for $m = -0.01, -0.05$, and -0.08.) Write ϕ in Eq. (4.2.12a) as $\phi(m, x)$ and expand it by Taylor's series to get

$$\left(\frac{\partial \phi}{\partial s}\right)_m \Delta s + \left(\frac{\partial \phi}{\partial m}\right)_s \Delta m = 0 \tag{P4.6.1}$$

where $\Delta s = s^{\nu+1} - s^{\nu}$ and $\Delta m = m_{new} - m_{old}$. Solving Eq. (P4.6.1) for Δs we get

$$s^{\nu+1} = s^{\nu} - \frac{(\partial \phi / \partial m)_s}{(\partial \phi / \partial s)_m} \Delta m \tag{P4.6.2}$$

To obtain $(\partial\phi/\partial m)_s$, solve a system of equations similar to those in Eqs. (4.2.13)–(4.2.16) obtained by differentiating Eqs. (4.2.8)–(4.2.11) with respect to m.

4.7. A long circular pipe with a smoothly faired entry leads from a large reservoir, containing a fluid of known properties and gage pressure p, to the atmosphere (gage pressure 0). What initial conditions and boundary conditions are needed to calculate laminar flow for the first few diameters of the pipe from the TSL equations?

Chapter 5

Analysis of Laminar Shear Layers

5.1 FORMULATION OF LAMINAR-FLOW PROBLEMS

In Chap. 3 we derived the TSL equations for laminar and turbulent flows. In this chapter we shall discuss the analysis and computation of two-dimensional and axisymmetric laminar shear layers. Note that the several coordinate transformations will require the reuse of some symbols with different meanings. Usage within a given section (Secs. 5.2, 5.3, etc.) is always consistent.

In this section we shall first discuss the various formulations of laminar-flow problems as special cases of the general problems discussed in Sec. 4.3. Most of the discussion applies also to turbulent-flow problems, which will be further treated in Chap. 6, but it is easier to discuss analytical and numerical problems in the laminar case where the fluid stresses are given by the simple formulas of Eq. (2.3.7).

For the sake of simplicity in this initial exposition we consider the TSL equations for *two-dimensional, steady, constant-property* laminar flows. According to Eqs. (2.2.3) and (3.1.16) they are

$$\frac{\partial u}{\partial x} + \frac{\partial v}{\partial y} = 0 \tag{5.1.1}$$

$$u\frac{\partial u}{\partial x} + v\frac{\partial u}{\partial y} = -\frac{1}{\rho}\frac{dp}{dx} + \nu\frac{\partial^2 u}{\partial y^2} \tag{5.1.2}$$

where p is a known function of x independent of y, so that $dp/dx = - \rho u_e \, du_e/dx$. Before proceeding we briefly discuss the restrictions given above. As was indicated in Sec. 3.5, no general treatment of fairly thin shear layers can be offered at present, and any specialized treatment (Chap. 11) will follow TSL practice qualitatively if not quantitatively. Three-dimensional or unsteady TSLs will be discussed in Chap. 10: unsteady flow (within the limits of the TSL equations) presents only numerical and not physical difficulties; and variation of fluid properties in high-speed (compressible) flow or low-speed flow with large temperature differences introduces only minor changes in the momentum equations and their solution. The main novelty in variable-property flow is the need to solve an equation for temperature or concentration simultaneously with those for the flow variables, and we postpone discussion to another work (Cebeci and Bradshaw, in preparation).

The boundary conditions for laminar shear layers (see Chap. 4) vary considerably from problem to problem, but there are two main types.

In *external flows* the shear layer, flowing in the x-direction, adjoins an effectively "inviscid" free stream extending to $y = \infty$. On the lower side may be either another inviscid stream extending to $y = - \infty$ (Fig. 1.1c), in which case the viscous region is called a "free shear layer," or a solid surface (usually taken as $y = 0$: Fig. 1.1b), in which case the viscous region is called a "wall shear layer." The variation of the static pressure with x within the shear layer depends on the shape of the solid body, if any, and on the displacement effect of the shear layer. The displacement effect may either be small enough to be neglected or represented by the displacement thickness as defined in Eq. (1.3.14), but in some cases of rapid streamwise change the pressure difference across the thickness of the shear layer may be large enough to invalidate Eqs. (5.1.2) and (1.3.14). The effect of a free shear layer on the external flow may be large if the shear layer has a large component of momentum normal to the undisturbed flow direction as in the cross-stream jet in Fig. 1.2. The simplest set of boundary conditions for a boundary layer with negligible $\partial p/\partial y$ is

$$y = 0 \quad u = 0 \quad v = v_w(x) \tag{5.1.3a}$$

$$y \to \infty \quad u = u_e(x) \tag{5.1.3b}$$

where v_w is the transpiration velocity, zero on a solid surface.

Internal flows (Figs. 1.1a and 1.1d) consist of a shear layer or layers filling part, or all, of the space between two solid boundaries. In this case the pressure distribution is set predominantly by the displacement effect of the shear layer; in the simplest case of slow change in the streamwise (x) direction the pressure is effectively independent of y. It is convenient to distinguish flows in which the shear layers fill the cross section, and flows such as that near the entrance to a duct (the left-hand part of Fig. 1.1d), where a region of effectively inviscid flow obeying Bernoulli's equation remains.

It is also convenient to distinguish the "entrance region," in which the flow field changes with x, and the "fully developed" region far downstream in a constant-area duct, in which the velocities do not. Note that properties do *not* cease to change with x as soon as the shear layers fill the duct, so the last two sentences make slightly different distinctions.

Therefore the basic difference between external and internal flows, which appears even in the simplest cases, is that the pressure is a given boundary condition for the TSL calculation in external flows such as boundary layers but is a *result* of the calculation in internal flows such as those in pipes. The relevant boundary condition in internal flows is the requirement of constant mass flow between the solid surfaces.

Strictly speaking, a TSL may affect the pressure distribution even in an external flow. In that case the effect is usually small but if it is to be determined, the inviscid flow must be calculated by solving an elliptic PDE. In contrast, simplified or even one-dimensional treatment of the inviscid flow suffices in ducts, but in this case the shear layer development almost wholly determines the pressure distribution.

In both cases a solution can be obtained either by a "differential" ("field") method, solving the PDEs (5.1.1) and (5.1.2) numerically, or by an "integral" method, solving ODEs already integrated in the y-direction. Integral methods may involve formal application of the MIR to the PDEs (Sec. 4.3) or, alternatively, may be based on empirical ODEs not explicitly related to the exact PDEs: Of the integral methods to be discussed below, the Pohlhausen method (Sec. 5.6.1) is a—nontypical—example of the MIR, while Thwaites' method (Sec. 5.6.2) uses an empirical ODE. Since an integral method is in some sense a simplification of the field method for the same problem, the general discussion below is based on the PDEs solved in a field method and followed by a treatment of integral methods. Numerical methods for solving the PDEs are discussed in Chap. 7. We concentrate on external flows in general and boundary layers in particular because these flows are the most common and calculation methods for them are the most highly developed. We return to the special problems of internal flows in Sec. 5.7, but much of the intervening material can also be applied to internal flows.

5.2 SIMILARITY CONCEPT IN LAMINAR SHEAR LAYERS

A very useful concept in shear layers is the concept of similarity. This concept can be conveniently explained by reference to the two-dimensional incompressible shear-layer equations given by Eqs. (5.1.1) and (5.1.2) when they are applied to an external boundary-layer flow subject to the boundary conditions (5.1.3) with $v_w = 0$. In general, the solution of Eqs. (5.1.1)–(5.1.3), for given v and $u_e(x)$, is

$$\frac{u}{u_e} = g(x,y) \tag{5.2.1}$$

where g is a general function of x and y. There are special cases in which

$$\frac{u}{u_e} = g(\eta) \tag{5.2.2}$$

where η, called a similarity variable, is a specific function of x and y. In such cases the number of independent variables is reduced from two (x and y) to one (η) so that Eqs. (5.1.1) and (5.1.2) become ODEs for u and v. In jet flows in still air we would replace u_e by the centerline velocity u_c. In wakes in a uniform stream we employ the so-called self-preserving form

$$\frac{u_c - u}{\Delta u} = g(\eta) \tag{5.2.3}$$

where in this case the velocity scale, Δu, is $u_e - u_c$, velocities being measured with respect to the local free stream instead of the faraway body that produced the wake. Similarity is seen to be a special case of self-preservation.

The simplest version of "separability of variables," treated in many elementary texts on DEs, corresponds to $u(x, y) = h(x)k(y)$, which is in the form of Eq. (5.2.2) with $u_e = h(x)$ and $\eta = y$. Below, we consider similarity only with reference to shear layers and not in full generality. In turbulent flows, complete similarity in the sense of Eq. (5.2.2) is rare because the viscous contribution to the shear stress necessarily depends directly on viscosity, whereas the Reynolds shear stress does not, so profile shape varies with Reynolds number. However, a more general form of self-preservation related to Eq. (5.2.3) applies in special cases (Sec. 6.2).

Similarity-type flows were widely studied in the precomputer era because of their comparative mathematical simplicity. In this book we use them to introduce the behavior of laminar shear layers because of their physical simplicity. We also show that their mathematical simplicity can be utilized as a basis for efficient calculation methods for general flows, in which the x-derivative terms resulting from nonsimilarity are either neglected or incorporated as "known" terms in a solution of the similarity equations at given x. Therefore although the original mathematical motivation for studying similar flows has been removed by the power of modern computing techniques, they still warrant a detailed discussion.

Sections 5.2–5.5 introduce similarity concepts with special reference to laminar shear layers. Numerical solutions for similar flows (Sec. 5.5) show the response of laminar boundary layers to different pressure gradients. Later, in Secs. 5.7 and 5.8, similar flows again provide simple examples of the behavior of internal flows and free shear layers, respectively. In Chap. 6, self-preserving flows are used as examples of turbulent shear layers. In Chap. 7 a similarity transformation is used to *reduce* (though not eliminate) dependence on x in various applications of a calculation method for nonsimilar flows.

The reader should note that, although similar flows are useful for exposition and similarity concepts are useful in calculation methods even for nonsimilar flows, only a few kinds of similar flows are found in practice. The most common are the laminar and (self-preserving) turbulent boundary layers in negligible pressure gradient, the trivial case of flow in a constant-section duct, and the circular jet in still air. Most of the other flows of interest to engineers correspond to general solutions, like Eq. (5.2.1), of the partial differential equations (5.1.1) and (5.1.2) or their turbulent counterparts or to three-dimensional solutions in which u/u_e is a function of x, y, and z.

To find the similarity variable η and the necessary conditions under which Eqs. (5.1.1) and (5.1.2) become ODEs, several methods can be used. Here we shall use the so-called group-theoretic method as discussed by Hansen (1964), which formalizes some of the other approaches found in fluid dynamics textbooks. To illustrate the method we shall consider Eqs. (5.1.1)–(5.1.3) and introduce the "linear group transformation" defined by

$$x = A^{\alpha_1}\bar{x} \qquad y = A^{\alpha_2}\bar{y} \qquad u = A^{\alpha_3}\bar{u} \qquad v = A^{\alpha_4}\bar{v} \qquad u_e = A^{\alpha_5}\bar{u}_e \tag{5.2.4}$$

Here $\alpha_1, \alpha_2, \ldots, \alpha_5$ are constants, and A is called the "parameter of transformation."

1. Under this group of transformations, Eqs. (5.1.1) and (5.1.2) can be written as

$$\frac{\partial u}{\partial x} + \frac{\partial v}{\partial y} = A^{\alpha_3 - \alpha_1}\frac{\partial \bar{u}}{\partial \bar{x}} + A^{\alpha_4 - \alpha_2}\frac{\partial \bar{v}}{\partial \bar{y}} = 0 \tag{5.2.5}$$

$$u\frac{\partial u}{\partial x} + v\frac{\partial u}{\partial y} - u_e\frac{du_e}{dx} - v\frac{\partial^2 u}{\partial y^2} = A^{2\alpha_3 - \alpha_1}\bar{u}\frac{\partial \bar{u}}{\partial \bar{x}} + A^{\alpha_3 + \alpha_4 - \alpha_2}\bar{v}\frac{\partial \bar{u}}{\partial \bar{y}}$$

$$- A^{2\alpha_5 - \alpha_1}\bar{u}_e\frac{d\bar{u}_e}{d\bar{x}} - A^{\alpha_3 - 2\alpha_2}v\frac{\partial^2 \bar{u}}{\partial \bar{y}^2} = 0 \tag{5.2.6}$$

2. We see that the equations will be invariant (unaltered by the transformation) if the powers of A in each term are the same. If this is true, then Eqs. (5.2.5) and (5.2.6) give, respectively,

$$\alpha_3 - \alpha_1 = \alpha_4 - \alpha_2 \tag{5.2.7}$$

$$2\alpha_3 - \alpha_1 = \alpha_3 + \alpha_4 - \alpha_2 = 2\alpha_5 - \alpha_1 = \alpha_3 - 2\alpha_2 \tag{5.2.8}$$

Partial solution of Eqs. (5.2.7) and (5.2.8) gives

$$\alpha_3 = \alpha_5 = \alpha_1 - 2\alpha_2 \qquad \alpha_4 = -\alpha_2 \tag{5.2.9}$$

Knowing the relations among the α's, we can find a relationship between the barred and unbarred quantities. This can be done by eliminating the parameter of transformation A. For example, from Eq. (5.2.4) we can write

$$A = \left(\frac{x}{\bar{x}}\right)^{1/\alpha_1} = \left(\frac{y}{\bar{y}}\right)^{1/\alpha_2}$$

or

$$\frac{y}{x^\alpha} = \frac{\bar{y}}{\bar{x}^\alpha} \tag{5.2.10a}$$

where $\alpha = \alpha_2/\alpha_1$. Similarly, from the rest of Eqs. (5.2.4) and from (5.2.9) we can write

$$\frac{u}{x^{1-2\alpha}} = \frac{\bar{u}}{\bar{x}^{1-2\alpha}} \qquad \frac{v}{x^{-\alpha}} = \frac{\bar{v}}{\bar{x}^{-\alpha}} \qquad \frac{u_e}{x^{1-2\alpha}} = \frac{\bar{u}_e}{\bar{x}^{1-2\alpha}} \tag{5.2.10b}$$

These combinations of variables in Eq. (5.2.10) are seen to be invariant under the linear group of transformations [Eq. (5.2.4)] and are called absolute variables. According to Morgan's theorem (see Hansen, 1964) these invariants are similarity variables if the boundary conditions [Eq. (5.1.3)] can be transformed and expressed independent of x. We now put

$$\eta = \frac{y}{x^\alpha} \quad \hat{f}(\eta) = \frac{u}{x^{(1-2\alpha)}} \quad \hat{g}(\eta) = \frac{v}{x^{(-\alpha)}} \tag{5.2.11}$$

and

$$\hat{h}(\eta) = \frac{u_e}{x^{(1-2\alpha)}} = C \tag{5.2.12}$$

where $\hat{f}$ and $\hat{g}$ are functions of η. In Eq. (5.2.12), $\hat{h}(\eta) = C$ is a constant, since the mainstream velocity u_e is a function of x only and thus cannot be a nonconstant function of η because this would introduce dependence on y.

We now transform the boundary conditions [Eq. (5.1.3)] and express them in terms of the possible similarity variables defined in Eqs. (5.2.11) and (5.2.12). Using Eqs. (5.2.11) and (5.2.12) in Eq. (5.1.3), we get

$$\eta = 0 \quad \hat{f} = \hat{g} = 0 \tag{5.2.13a}$$
$$\eta \to \infty \quad \hat{f} = C \tag{5.2.13b}$$

Thus for constant values of η, constant values of $\hat{f}$ and $\hat{g}$ result. The boundary conditions [Eq. (5.2.13)] are independent of x, and therefore the absolute invariants in Eqs. (5.2.11) and (5.2.12) are similarity variables.

It is now a simple matter to transform the DEs (5.1.1) and (5.1.2) to the following ODEs, in which we have written m for $1 - 2\alpha$ to conform with the usual notation so that $u_e = cx^m$:

$$m\hat{f} - \frac{1-m}{2}\eta\hat{f}' + \hat{g}' = 0 \tag{5.2.14}$$

and

$$m\hat{f}^2 - \frac{1-m}{2}\eta\hat{f}\hat{f}' + \hat{g}\hat{f}' = mC^2 + \nu\hat{f}'' \tag{5.2.15}$$

The above example illustrates the group-theoretic method of obtaining the similarity variables. Recall the steps in the argument:

1. Define a group of transformations and substitute them into the DEs.
2. Require that each DE be invariant, which means that the DE is identical in form in terms of either (x, y, u, v) or $(\bar{x}, \bar{y}, \bar{u}, \bar{v})$. Relations among the constants in the transformation are obtained.

3. Eliminate the parameter of transformation to give the absolute invariants, which will become the similarity variables.
4. Check the boundary conditions to see if they can be transformed into constant values. If they can, then we can proceed to step 5. If not, similarity solutions do not exist.
5. Transform the DEs.

We shall now consider another example of the procedure of obtaining similarity variables. This time we consider a two-dimensional jet coming out of an orifice into still air. In this case, as we shall see in Sec. 5.8, the governing equations for laminar flow are given by Eq. (5.1.1) and by Eq. (5.1.2) with zero pressure gradient, that is, Eq. (5.1.2) becomes

$$u \frac{\partial u}{\partial x} + v \frac{\partial u}{\partial y} = v \frac{\partial^2 u}{\partial y^2} \tag{5.2.16}$$

These equations are subject to the following boundary conditions (see Sec. 4.3.3):

$$y = 0 \quad \frac{\partial u}{\partial y} = 0 \quad v = 0 \tag{5.2.17a}$$

which express symmetry about $y = 0$, and

$$y = \infty \quad u = 0 \tag{5.2.17b}$$

In addition to the continuity and momentum equations the total momentum in the x-direction, denoted by J, remains constant and is independent of the distance x from the orifice. Hence

$$J = \rho \int_{-\infty}^{\infty} u^2 \, dy = \text{const} \tag{5.2.18}$$

For convenience let us express u and v in terms of the stream function ψ, that is,

$$u = \frac{\partial \psi}{\partial y} \quad v = -\frac{\partial \psi}{\partial x} \tag{5.2.19}$$

automatically satisfying Eq. (5.1.1). With the definition of stream function we can write Eqs. (5.2.16), (5.2.18), and (5.2.17) as

$$\frac{\partial \psi}{\partial y} \frac{\partial^2 \psi}{\partial x \, \partial y} - \frac{\partial \psi}{\partial x} \frac{\partial^2 \psi}{\partial y^2} = v \frac{\partial^3 \psi}{\partial y^3} \tag{5.2.20}$$

$$\frac{J}{\rho} = \int_{-\infty}^{\infty} \left(\frac{\partial \psi}{\partial y} \right)^2 dy \tag{5.2.21}$$

$$y = 0 \qquad \frac{\partial^2 \psi}{\partial y^2} = 0 \qquad \frac{\partial \psi}{\partial x} = 0 \tag{5.2.22a}$$

$$y = \infty \qquad \frac{\partial \psi}{\partial y} = 0 \tag{5.2.22b}$$

According to the above procedure we introduce a one-parameter linear transformation,

$$x = A^{\alpha_1} \bar{x} \qquad y = A^{\alpha_2} \bar{y} \qquad \psi = A^{\alpha_3} \bar{\psi} \tag{5.2.23}$$

and substitute it into Eqs. (5.2.20)–(5.2.22) to get

$$A^{2\alpha_3 - \alpha_1 - 2\alpha_2} \left(\frac{\partial \bar{\psi}}{\partial \bar{y}} \frac{\partial^2 \bar{\psi}}{\partial \bar{x} \, \partial \bar{y}} - \frac{\partial \bar{\psi}}{\partial \bar{x}} \frac{\partial^2 \bar{\psi}}{\partial \bar{y}^2} \right) = A^{\alpha_3 - 3\alpha_2} \nu \frac{\partial^3 \bar{\psi}}{\partial \bar{y}^3} \tag{5.2.24}$$

$$\frac{J}{\rho} = A^{2\alpha_3 - \alpha_2} \int_{-\infty}^{\infty} \left(\frac{\partial \bar{\psi}}{\partial \bar{y}} \right)^2 d\bar{y} \tag{5.2.25}$$

$$\bar{y} = 0 \qquad \frac{\partial^2 \bar{\psi}}{\partial \bar{y}^2} = 0 \qquad \frac{\partial \bar{\psi}}{\partial \bar{x}} = 0 \tag{5.2.26a}$$

$$\bar{y} = \infty \qquad \frac{\partial \bar{\psi}}{\partial \bar{y}} = 0 \tag{5.2.26b}$$

Equations (5.2.24) and (5.2.25) are invariant if

$$2\alpha_3 - \alpha_1 - 2\alpha_2 = \alpha_3 - 3\alpha_2 \tag{5.2.27}$$

and

$$2\alpha_3 - \alpha_2 = 0 \tag{5.2.28}$$

From Eqs. (5.2.27) and (5.2.28) we get

$$\alpha_2 = \tfrac{2}{3}\alpha_1 \qquad \alpha_3 = \tfrac{1}{3}\alpha_1 \tag{5.2.29}$$

The absolute invariants of the transformation are therefore

$$\frac{y}{x^{2/3}} = \frac{\bar{y}}{\bar{x}^{2/3}} \qquad \frac{\psi}{x^{1/3}} = \frac{\bar{\psi}}{\bar{x}^{1/3}} \tag{5.2.30}$$

from which we can write

$$\eta = \frac{y}{3\sqrt{\nu}\,x^{2/3}} \qquad f(\eta) = \frac{\psi}{\sqrt{\nu}\,x^{1/3}} \tag{5.2.31}$$

Here $3\sqrt{\nu}$ and $\sqrt{\nu}$ in the above equations are introduced for dimensional purposes.

By introducing the above transformations into Eqs. (5.2.20) and (5.2.21) it can be shown that

$$f''' + ff'' + (f')^2 = 0 \qquad (5.2.32)$$

and

$$J = \rho \frac{\sqrt{\nu}}{3} \int_{-\infty}^{\infty} (f')^2 \, d\eta \qquad (5.2.33)$$

subject to the following transformed boundary conditions:

$$\eta = 0 \quad f'' = 0 \quad f = 0 \qquad (5.2.34a)$$
$$\eta = \infty \quad f' = 0 \qquad (5.2.34b)$$

It is worth noting that similarity can be achieved only far downstream of the orifice (very roughly, $x > 20$ orifice heights). In general, similarity is attained only some distance downstream from where the boundary conditions permit it because the initial profiles are not usually the similarity functions.

5.3 FALKNER-SKAN TRANSFORMATION FOR EXTERNAL BOUNDARY LAYERS

As was mentioned in Sec. 5.2, similarity transformations are useful even in non-similar flows. In this section we introduce the best known of these, the Falkner-Skan transformation for two-dimensional and axisymmetric laminar flows. The differences between this and the transformation of Sec. 5.2 should be noted. Basically, the Falkner-Skan procedure assumes that $\alpha = \frac{1}{2}$ in the coordinate transformation whatever the value implied by the local variation of u_e. In the analysis below, the $\bar{x}\bar{y}$-plane is the plane of a two-dimensional or Mangler-transformed flow.

We define the Falkner-Skan transformation by

$$\eta = \left(\frac{u_e}{\nu \bar{x}} \right)^{1/2} \bar{y} \qquad (5.3.1)$$

where $u_e = u_e(x)$, and introduce a dimensionless stream function $f(\bar{x}, \eta)$ by

$$\bar{\psi}(\bar{x}, \bar{y}) = (u_e \nu \bar{x})^{1/2} f(\bar{x}, \eta) \qquad (5.3.2)$$

From Eq. (5.3.1) we can write

$$\left(\frac{\partial}{\partial \bar{x}} \right)_{\bar{y}} = \left(\frac{\partial}{\partial \bar{x}} \right)_\eta + \frac{\partial \eta}{\partial \bar{x}} \left(\frac{\partial}{\partial \eta} \right)_{\bar{x}} = \left(\frac{\partial}{\partial \bar{x}} \right)_\eta - \frac{\eta}{2\bar{x}} \left(\frac{\partial}{\partial \eta} \right)_{\bar{x}} \qquad (5.3.3a)$$

$$\left(\frac{\partial}{\partial \bar{y}}\right)_{\bar{x}} = \left(\frac{u_e}{\nu \bar{x}}\right)^{1/2} \left(\frac{\partial}{\partial \eta}\right)_{\bar{x}}$$ (5.3.3b)

Introducing Eqs. (5.3.2) and (5.3.3) into Eq. (3.3.7), we get

$$\bar{u} = u_e f' \qquad \bar{v} = -\frac{\partial}{\partial \bar{x}}[(u_e \nu \bar{x})^{1/2} f] + \frac{\eta}{2}\left(\frac{u_e \nu}{\bar{x}}\right)^{1/2} f'$$ (5.3.4)

where a prime denotes differentiation with respect to η. Substituting Eqs. (5.3.3) and (5.3.4) into the Mangler-transformed momentum equation (3.3.9), with $\rho \overline{u'v'} = 0$ and $dp/dx = -\rho u_e du_e/dx$, we get the transformed momentum equation for two-dimensional and axisymmetric laminar flows ($k = 0$ and $k = 1$, respectively):

$$[(1 + t)^{2k} f'']' + \frac{m + 1}{2} ff'' + m[1 - (f')^2] = \bar{x}\left(f' \frac{\partial f'}{\partial \bar{x}} - f'' \frac{\partial f}{\partial \bar{x}}\right)$$ (5.3.5)

Here

$$t = -1 + \left[1 + \left(\frac{L}{r_0}\right)^2 \frac{2 \cos \phi}{L} \left(\frac{\nu \bar{x}}{u_e}\right)^{1/2} \eta\right]^{1/2}$$ (5.3.6)

The quantity m is a dimensionless pressure-gradient parameter defined by

$$m = \frac{\bar{x}}{u_e} \frac{du_e}{d\bar{x}}$$ (5.3.7)

and is therefore consistent with its use in Sec. 5.2, where $u_e \propto x^m$. The boundary conditions for Eq. (5.3.5), including mass transfer at the wall, that is, $v = v_w$ when $y = 0$, are

$$\eta = 0 \qquad f' = 0 \qquad f(\bar{x}, 0) \equiv f_w = -\frac{1}{(u_e \nu \bar{x})^{1/2}} \int_0^{\bar{x}} \bar{v}_w d\bar{x}$$ (5.3.8a)

$$\eta = \eta_\infty \qquad f' = 1$$ (5.3.8b)

In (5.3.8b), η_∞ corresponds to the transformed boundary-layer thickness. In physical coordinates the boundary-layer thickness $\delta(x)$ usually increases with increasing downstream distance for both laminar and turbulent flows. In transformed coordinates, η_∞ is nearly constant for most laminar flows and increases with increasing streamwise Reynolds number for turbulent flows, roughly as $(u_e x/\nu)^{0.3}$ in small pressure gradients. The slow change of η_∞ with x, implying fairly small x-derivatives at constant η within the shear layer, means that in a numerical solution, large streamwise steps can be taken even near the front of the body when the boundary layer is very thin. It yields large savings in computer time, especially for high-Reynolds-number problems. Note that once the velocity profile is obtained, it is necessary to get the physical distance y by inverting the coordinate transformation (see Prob. 5.4). In axisymmetric flows with transverse curvature this is a nontrivial operation.

For two-dimensional laminar flows the transformed momentum equation (5.3.5) and its boundary condition (5.3.8) become

$$f''' + \frac{m+1}{2} ff'' + m\,[1 - (f')^2] = x\left(f'\,\frac{\partial f'}{\partial x} - f''\,\frac{\partial f}{\partial x}\right) \tag{5.3.9}$$

$$\eta = 0 \quad f' = 0 \quad f_{\text{w}} = -\frac{1}{(u_e \nu x)^{1/2}} \int_0^x v_{\text{w}}\,dx \tag{5.3.10a}$$

$$\eta = \eta_\infty \quad f' = 1 \tag{5.3.10b}$$

We note that for axisymmetric laminar flows with zero transverse curvature effect, Eq. (5.3.5) and its boundary conditions are of the same form as Eqs. (5.3.9) and (5.3.10). We note also that if the x-wise derivatives are expressed as upstream finite differences, e.g., $\partial f/\partial x = (f^n - f^{n-1})/(x_n - x_{n-1})$ where f^{n-1} is the known value of f at x^{n-1}, Eq. (5.3.9) is equivalent to an ODE for f as a function of η at the given value of x denoted by x_n. In similar flows (Sec. 5.2) the x-derivatives are zero by definition.

We now express some of the boundary-layer parameters in terms of transformed coordinates to pave the way to the calculations to be discussed later. Using the transformation [Eq. (5.3.1)], we can write the local skin-friction coefficient, $c_{\text{f}} \equiv \tau_{\text{w}}/\tfrac{1}{2}\rho u_e^2$, the displacement thickness, δ^*, and the momentum thickness, θ [defined by Eqs. (3.4.10) and (3.4.11), respectively], as

$$c_{\text{f}} = 2\left(\frac{r_0}{L}\right)^k \frac{f_{\text{w}}''}{\sqrt{\text{Re}_{\bar{x}}}} \tag{5.3.11}$$

$$\delta^* = \left(\frac{L}{r_0}\right)^k \bar{x}\,\frac{\delta_1^*}{\sqrt{\text{Re}_{\bar{x}}}} \tag{5.3.12}$$

$$\theta = \left(\frac{L}{r_0}\right)^k \bar{x}\,\frac{\theta_1}{\sqrt{\text{Re}_{\bar{x}}}} \tag{5.3.13}$$

where

$$\text{Re}_{\bar{x}} = \frac{u_e \bar{x}}{\nu} \tag{5.3.14}$$

$$\delta_1^* = \int_0^{\eta_\infty} (1 - f')\,d\eta = \eta_\infty + f(\bar{x},0) - f(\bar{x},\eta_\infty) \tag{5.3.15}$$

$$\theta_1 = \int_0^{\eta_\infty} f'(1 - f')\,d\eta \tag{5.3.16}$$

5.4 SIMILAR FLOWS FOR LAMINAR BOUNDARY LAYERS

To discuss the requirements for similarity in two-dimensional laminar flows, we begin with the assumption that f is a function of η only, where η is given by Eq. (5.3.1), and write Eq. (5.3.9) as

$$f''' + \frac{m+1}{2} ff'' + m[1 - (f')^2] = 0 \qquad (5.4.1)$$

which is the same as Eq. (4.2.6). The solutions of Eq. (5.4.1) are independent of x provided that the boundary conditions [Eq. (5.3.10)] and the pressure-gradient parameter m are independent of x. These requirements are satisfied if f_w and m are constants.

The requirement that f_w is constant leads to the following boundary conditions for Eq. (5.4.1):

$$\eta = 0 \quad f = f_w = \text{const} \quad f' = f'_w = 0 \qquad (5.4.2a)$$

$$\eta = \infty \quad f' = 1 \qquad (5.4.2b)$$

Note that nonzero f_w corresponds to surface mass transfer. For flows with suction, f_w is negative, and for flows with blowing, f_w is positive.

The requirement that m is constant leads to

$$u_e = Cx^m \qquad (5.4.3)$$

where C is a constant. Thus, as was found in Sec. 5.2, we get similar boundary-layer flows *only when the external velocity varies with the surface distance x as prescribed by Eq. (5.4.3) and when the boundary conditions are independent of x.* The importance of similar flows is quite evident; we solve an ODE (5.4.1) rather than a PDE (5.3.9). For various values of m and for a set of specified boundary conditions [Eq. (5.4.2)], the solutions of Eq. (5.4.1) can be obtained with great accuracy.

EXAMPLE 5.1

For a two-dimensional similar laminar flow, what is the value of m for which the wall shear, τ_w, is independent of x?

Solution By using the Falkner-Skan transformation we can write the wall shear as

$$\tau_w = \mu \left(\frac{\partial u}{\partial y}\right)_w = \mu u_e f''_w \left(\frac{\partial \eta}{\partial y}\right) = \mu u_e f''_w \sqrt{\frac{u_e}{\nu x}} \qquad (E5.1.1)$$

Since μ, ν, and f''_w are independent of x,

$$\tau_w \sim \frac{u_e^{3/2}}{x^{1/2}} \qquad (E5.1.2)$$

But for a similar flow,

$$u_e \sim x^m \tag{E5.1.3}$$

Substituting Eq. (E5.1.3) into Eq. (E5.1.2) yields

$$\tau_w \sim x^{(3m-1)/2} \tag{E5.1.4}$$

Therefore for τ_w to be independent of x, m must be equal to $\frac{1}{3}$.

EXAMPLE 5.2

For a two-dimensional laminar flow over a flat plate ($m = 0$) with surface mass transfer, find the wall mass transfer velocity, v_w, as a function of x required for similarity. Assume that $v_w \sim x^n$, where n is a constant.

Solution   From Eq. (5.3.8a), for two-dimensional flow, we can write

$$f_w = -\frac{1}{(u_e \nu x)^{1/2}} \int_0^x v_w \, dx \tag{E5.2.1}$$

Integration of Eq. (E5.2.1) with $v_w \sim x^n$ yields

$$f_w \sim x^{n+1/2} \tag{E5.2.2}$$

since u_e is constant. Thus for similarity, n must be $-\frac{1}{2}$. Then the variation of mass transfer velocity, v_w, with x is

$$v_w \sim \frac{1}{\sqrt{x}} \tag{E5.2.3}$$

Equation (5.4.1) was first given by Falkner and Skan (1930), who also produced solutions for a range of values of m. Later, Hartree (1937) studied Eq. (5.4.1) in greater detail, first eliminating m by making the linear transformation

$$Y = \left(\frac{m+1}{2}\right)^{1/2} \eta \qquad F = \left(\frac{m+1}{2}\right)^{1/2} f \tag{5.4.4}$$

so that Eq. (5.4.1) becomes

$$F''' + FF'' + \beta[1 - (F')^2] = 0 \tag{5.4.5}$$

subject to the boundary conditions

$$Y = 0 \quad F' = 0 \quad F_{\mathrm{w}} = \mathrm{const} \tag{5.4.6a}$$

$$Y = Y_{\infty} \quad F' = 1 \tag{5.4.6b}$$

In Eqs. (5.4.5) and (5.4.6), primes denote differentiation with respect to Y, and β is a new dimensionless pressure-gradient parameter related to m by

$$\beta = \frac{2m}{m+1} \tag{5.4.7}$$

and to the general parameter of Sec. 5.2 by

$$\beta = \frac{1 - 2\alpha}{1 - \alpha} \tag{5.4.8}$$

For positive values of β, Eq. (5.4.5) has unique solutions, but more than one solution exists for each negative value of β. Hartree calculated a range of solutions for positive and negative values of m (or β). For $\beta = -0.199$ it is found that

$$F''(0) \equiv 0$$

so that $\tau_{\mathrm{w}} \equiv 0$ for all x. This solution therefore represents a flow whose external velocity forces the boundary layer to incipient separation at all streamwise stations.

A second family of solutions of Eq. (5.4.5), the so-called "lower-branch" solutions, was given later by Stewartson (1954) and recently by Cebeci and Keller (1971) and by others, again for negative values of β in the range $-0.199 \leqslant \beta \leqslant 0$. (See Prob. 4.5.) These solutions all have the property that $F'' \leqslant 0$, so that at $y = 0$, $\partial u / \partial y \leqslant 0$ for all x. Accordingly, these solutions can only correspond physically to flows in a laminar boundary layer beyond separation.

The solutions of Eq. (5.4.5) are sometimes referred to as "wedge-flow" solutions. The flow is diverted through an angle of magnitude $\beta \pi / 2$ (see Fig. 5.1). Thus, physically, for $0 \leqslant \beta \leqslant 2$ the flow is that past an infinite wedge whose vertex angle is $\beta \pi$.

5.5 SIMILAR SOLUTIONS OF THE FALKNER–SKAN EQUATION FOR POSITIVE WALL SHEAR

If we integrate Eq. (5.4.1) across the boundary layer, use the definitions of transformed displacement and momentum thickness, δ_1^* and θ_1 [see Eqs. (5.3.15) and (5.3.16), respectively], and note that f'' is zero at the edge of the boundary layer, we get ($f_0'' \equiv f_{\mathrm{w}}''$)

$$-f_0'' + \frac{m+1}{2} \int_0^{\eta_{\infty}} ff'' \, d\eta + m(\delta_1^* + \theta_1) = 0 \tag{5.5.1}$$

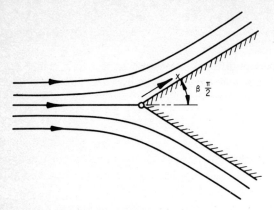

Fig. 5.1 Flow past a wedge. In the neighborhood of the leading edge, the external velocity distribution is $u_e(x) = Cx^{\beta/(2-\beta)}$.

The integral in this equation is simply the momentum thickness θ_1, as can be shown by integrating Eq. (5.3.16) and noting that $f'(\eta_\infty) = 1$. Equation (5.5.1) can therefore be written as

$$-f_0'' + m\delta_1^* + \left(\frac{3m+1}{2}\right)\theta_1 = 0 \qquad\qquad (5.5.2)$$

and shows the relation between the three boundary-layer parameters, f_0'', θ_1, and δ_1^* for a given m. It is, of course, the momentum integral equation (3.4.9). This equation can be used to get θ_1 without evaluating Eq. (5.3.16) numerically.

Before we discuss the numerical solutions of Eqs. (5.4.1) and (5.4.2) for various values of m, it is useful to single out two flows that are quite common in practice. The first one corresponds to $m = 0$ ($\beta = 0$) and is the flow over a flat plate at zero incidence. In this case the external velocity, u_e, is constant, and Eq. (5.4.1) reduces to the Blasius equation,

$$f''' + \tfrac{1}{2}ff'' = 0 \qquad\qquad (5.5.3)$$

The second flow corresponds to $m = 1$ ($\beta = 1$, wedge half angle 90°), which is the case of a two-dimensional stagnation flow. The external velocity is $u_e = Cx$, and Eq. (5.5.2) reduces to the Hiemenz equation,

$$f''' + ff'' - (f')^2 + 1 = 0 \qquad\qquad (5.5.4)$$

The Falkner-Skan equation has been solved for different boundary conditions and for various values of m. Here we present the results for $f(0) = 0$, $f'(0) = 0$, and $f''(0) \geqslant 0$.

We first discuss the solution of the Blasius equation (5.5.3). We set $m = 0$, specify a generous estimate of η_∞ (it will turn out that $\eta = 8$ is ample), and by the numerical method described in Sec. 4.2.2 calculate the functions f, f', and f'' as a function of η.

The accuracy of the solutions depends on the number of η-points taken across the boundary layer. To get high accuracy (say five digits) with the Runge-Kutta method, we choose $\Delta\eta = 0.10$. For a value of $\eta_\infty = 8$ this choice corresponds to 81 η-points. In general, approximately 40 points should be sufficient to get four-digit accuracy. Figure 5.2 shows the f and f' profiles across the boundary layer. Tabulations are given in many textbooks; Schlichting (1968) gives results to five-digit accuracy (mostly obtained on desk calculators) that agree almost everywhere with the present results. We note from these results that for most of the boundary layer, f' varies almost linearly with η. As η approaches η_∞, f' asymptotically approaches 1 and f'' approaches 0.

To obtain the normal velocity component, v, from the solution of the Falkner-Skan equation, we use Eqs. (5.3.4), (5.3.14), and (5.4.3) and write

$$\frac{v}{u_e} = \frac{1}{\sqrt{Re_x}} \left(\frac{\eta}{2} f' - \frac{m+1}{2} f \right) \tag{5.5.5}$$

Figure 5.3 shows the variation of $(v/u_e)\sqrt{Re_x}$ across the boundary layer for a flat-plate flow. We note that as η approaches η_∞, $(v/u_e)\sqrt{Re_x}$ approaches a constant value (0.8604). This means that at the outer edge there is a flow outward because the increasing boundary-layer thickness causes the fluid to be displaced from the

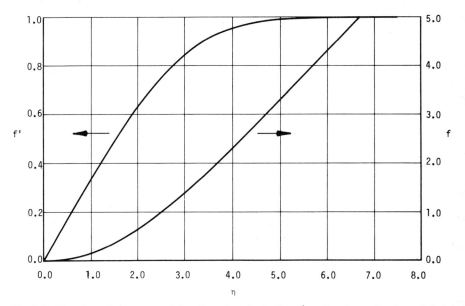

Fig. 5.2 The dimensionless stream function, f, and velocity, f', as functions of η for a flat-plate flow ($m = 0$).

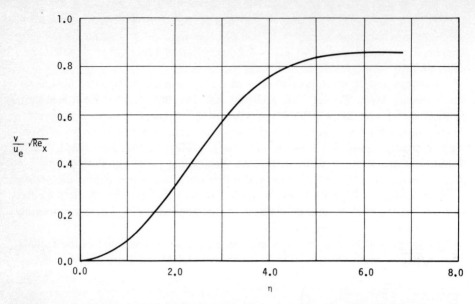

Fig. 5.3 The variation of dimensionless normal velocity, v, with η for a flat-plate flow ($m = 0$). Note that at the boundary-layer edge, v reaches a constant value ($= 0.8604$).

wall as it flows along it: in the present case, Eq. (3.4.15) reduces to $v_e/u_e = d\delta^*/dx$.

The local skin-friction coefficient and displacement thickness for a flat-plate flow can be obtained by using the definitions of Eqs. (5.3.11), (5.3.12), and (5.3.15) and the results in Table 5.1.

$$c_f = \frac{0.664}{\sqrt{\mathrm{Re}_x}} \qquad (5.5.6)$$

$$\frac{\delta^*}{x} = \frac{1.721}{\sqrt{\mathrm{Re}_x}} \qquad (5.5.7)$$

The constant in the latter expression is, of course, just twice the 0.8604 just quoted

Table 5.1 Solutions of the Falkner-Skan Equation for Positive Wall Shear

m	$f''(0) \equiv \frac{1}{2} c_f \sqrt{\mathrm{Re}_x}$	$\delta_1^* \equiv \delta^* \sqrt{u_e/\nu x}$	$\theta_1 = \theta \sqrt{u_e/\nu x}$	H
1	1.23259	0.64791	0.29234	2.216
1/3	0.75745	0.98536	0.42900	2.297
0.1	0.49657	1.34782	0.55660	2.422
0	0.33206	1.72074	0.66412	2.591
−0.01	0.31148	1.78000	0.67892	2.622
−0.05	0.21351	2.1174	0.75148	2.818
−0.0904	0.0	3.4277	0.86797	3.949

for v/u_e. We note that Eq. (5.5.2) reduces to $\theta_1 = 2f_0''$, while the numerical calculations give $f_0'' = 0.33206$; hence

$$\frac{\theta}{x} = \frac{0.664}{\sqrt{Re_x}} \tag{5.5.8}$$

The ratio of δ^* to θ, the shape factor defined in Sec. 1.3, is

$$H = \frac{\delta^*}{\theta} = 2.591 \tag{5.5.9}$$

If δ is defined to be the distance from the surface where $f' = 0.995$ (Sec. 1.3), this corresponds to an η of approximately 5.3. Then with Eq. (5.3.1) we can write δ as

$$\frac{\delta}{x} = \frac{5.3}{\sqrt{Re_x}} \tag{5.5.10}$$

EXAMPLE 5.3

A thin flat plate is immersed in a stream of atmospheric pressure air at $25°C$ moving at a velocity of 15 m s^{-1}. (a) At a distance 20 cm from the leading edge determine the distance from the plate at which the local velocity is half the mainstream velocity. At this point calculate v. (b) Find the boundary-layer thickness and the local skin-friction coefficient at a distance 1 m from the leading edge. Assume that $v = 1.5 \times 10^{-5}$ m^2 s^{-1}.

Solution   (a) From Fig. 5.2, when $u/u_e = 0.5$,

$$\eta = \left(\frac{u_e}{vx}\right)^{1/2} y = 1.5 = \sqrt{Re_x}\,\frac{y}{x}$$

but

$$Re_x = \frac{u_e x}{v} = (15)\left(\frac{20}{100}\right)\frac{1}{1.5 \times 10^{-5}} = 2 \times 10^5$$

Therefore

$$y = 0.067 \text{ cm}$$

From Fig. 5.2 at $\eta = 1.5$, $f = 0.37$. By using Eq. (5.5.5),

$$v = \frac{15}{\sqrt{2 \times 10^5}}\left[\frac{1.5}{2}\left(\frac{1}{2}\right) - \frac{1}{2}(0.37)\right] = 0.00637 \text{ m s}^{-1}$$

(b) Since $Re_x = 1 \times 10^6$, from Eq. (5.5.10),

$$\delta = \frac{5.3}{\sqrt{1 \times 10^6}} = 5.3 \times 10^{-3} \text{ m} = 0.53 \text{ cm}$$

From Eq. (5.5.6),

$$c_f = \frac{0.664}{\sqrt{1 \times 10^6}} = 0.664 \times 10^{-3}$$

Table 5.1 shows the computed values of $f''(0)$, δ_1^*, and θ_1, and Fig. 5.4 shows the velocity profiles for several values of m, including the Hiemenz value $m = 1$ and the "separation" value $m = -0.0904$, where $\tau_w = 0$. The effect of the pressure gradient on the profiles is clearly seen. With increasing adverse pressure gradient (decelerating flow) the slope of the velocity profile decreases (even when it is plotted as u/u_e versus y/δ), and at $m = -0.0904$ it becomes zero. The velocity profile corresponding to this value of m is called the *separating velocity* profile. On the other hand, with increasing favorable pressure gradient (accelerating flow) the slope of the velocity profile increases.

The velocity profile corresponding to an infinite value of m ($\beta = 2$) cannot be obtained from Eq. (5.4.1). It can, however, be obtained from Hartree's equation (5.4.5) and gives $H = 2.155$. That equation in turn has problems when β becomes infinite, which corresponds to an accelerated stream with the external velocity given by

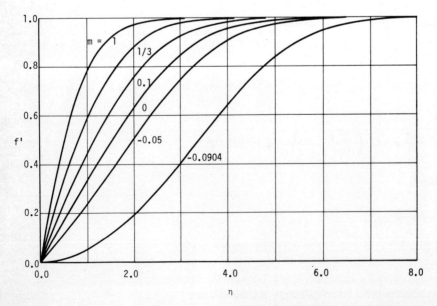

Fig. 5.4 The dimensionless velocity, f', as a function of η for various values of m.

$$u_e = \frac{u_1}{-x} \tag{5.5.11}$$

For $u_1 > 0$ it represents a two-dimensional flow (sometimes referred to as the sink flow) in a convergent channel. Equation (5.4.5) can be placed in a suitable form when β becomes infinite by defining new independent and dependent variables:

$$\zeta = \beta^{1/2} Y \tag{5.5.12}$$

$$\phi(\zeta) = \beta^{1/2} F(Y) \tag{5.5.13}$$

With these relations, Eq. (5.4.5) can be written as

$$\phi''' + \frac{1}{\beta}\phi\phi'' + 1 - (\phi')^2 = 0 \tag{5.5.14}$$

subject to the boundary conditions

$$\phi(0) = \phi'(0) = 0 \qquad \phi'(\zeta_\infty) = 1 \tag{5.5.15}$$

Here the primes denote differentiation with respect to ζ.

When β is infinite, Eq. (5.5.14) becomes

$$\phi''' + 1 - (\phi')^2 = 0 \tag{5.5.16}$$

A closed form solution to Eqs. (5.5.15) and (5.5.16) was first given by Pohlhausen (1921; see Schlichting, 1968) as

$$\phi(\zeta) = \zeta + 2\sqrt{3} - 3\sqrt{2} \, \tanh\left[\frac{\zeta}{\sqrt{2}} + \tanh^{-1}\left(\frac{2}{3}\right)^{1/2}\right] \tag{5.5.17}$$

Noting that $[\tanh^{-1}(\frac{2}{3})^{1/2} = 1.146]$, we can write the dimensionless velocity u/u_e as

$$\frac{u}{u_e} = \phi' = 3 \tanh^2\left(\frac{\zeta}{\sqrt{2}} + 1.146\right) - 2 \tag{5.5.18}$$

This leads to

$$H = 2.07 \qquad \phi''(0) = \frac{2}{\sqrt{3}} = 1.1547 \tag{5.5.19}$$

The only common closed-form analytical solution for the external-flow boundary layer equations is that for the *asymptotic suction profile*, a boundary layer with v_w negative and independent of x. Sufficiently far downstream from the start of suction,

$$u(y) = u_e \left[1 - \exp\left(\frac{v_w y}{\nu} \right) \right] \tag{5.5.20}$$

independent of x (see Sec. 8.2). Here $H = 2$ (compared with 2.6 for $v_w = 0$) and $\frac{1}{2} c_f = -v_w/u_e$.

The main lessons to be learned from this section are the power of the Falkner-Skan transformation in reducing similar flows to simple numerical problems and the quantitative effect of pressure gradient and mass transfer on the profile shape and skin-friction coefficient, described qualitatively in Sec. 1.3.

5.6 CALCULATION METHODS FOR NONSIMILAR FLOWS

Nonsimilar flows are those for which u/u_e is a function both of x and η; non-self-preserving flows are those for which even $(u - u_e)/u_0$ cannot be reduced to a function of η alone by a suitable choice of the velocity scale u_0 (not necessarily u_e) as a function of x, $u_0(x)$. However, in this chapter we will let "nonsimilar" include "non-self-preserving." In practice nonsimilar flows are more important than similar flows because $u_e(x)$ rarely varies according to Eq. (5.4.3), as is required for similar flows, and because the surface boundary conditions may not fulfill the requirements for similarity even if u_e does.

In Sec. 4.3.4 we very briefly introduced the main methods of solving PDEs. Finite-element methods have not been used much in shear-layer problems, but several methods for laminar or turbulent shear layers exist in both of the following categories.

1. Finite-difference solutions of PDEs, sometimes called the field or differential methods.
2. Streamwise integrations of ODEs (obtained by using the MIR on shear layer PDEs), sometimes called the integral method. A special case of MIR is the Pohlhausen method to be described in Sec. 5.6.1. In some cases, like the Thwaites method of Sec. 5.6.2 and the Head method of Sec. 6.8, some or all of the ODEs are derived from fits to experimental data or "exact" solutions without formal application of MIR: we will call these "type 2a" methods. Nearly all precomputer methods of calculating turbulent flow were of this type.

Methods of types 1 and 2 can in principle give exact results if the number of integration steps, or the number of coefficients in the MIR profile families, is large enough: in practice the number of steps in type 1 is chosen to give adequate accuracy in a given run, while the number of coefficients in type 2 is chosen once for all. Methods of type 2a are necessarily approximate, although the momentum integral equation is usually satisfied exactly. It is important to note that the reduction of PDEs to ODEs in similar flows has no connection with the reduction of PDEs to ODEs by MIR: in one case the remaining independent variable is η (in effect, y), and in the other it is x.

The more advanced methods of type 1 are discussed in Chaps. 7 and 8, laminar and turbulent flows being treated together as far as possible. Methods of type 2 and

2a are dedicated either to laminar or turbulent flow; the profile families or empirical functions necessarily differ in the two cases. Here and in Chap. 6 we present some simple methods of types 2 and 2a: the Pohlhausen method is a simple example of MIR, while the Thwaites method for laminar boundary layers and the Head method for turbulent boundary layers are among the most useful methods for obtaining quick rough answers. In practice, Thwaites' method is often used for calculating the small initial regions of laminar flow on bodies at high Reynolds number before switching over to a type 1 method for the turbulent boundary layer. Head's method is used for comparative calculations or for external-velocity distributions not too different from the cases of monotonic adverse pressure gradient used to specify the empirical functions in the method.

5.6.1 Pohlhausen's Method

In this method (Schlichting, 1968) we assume a velocity profile $u(x, y)$ that satisfies the momentum integral equation (3.4.8) and a set of boundary conditions

$$y = 0 \quad u = 0 \quad \text{and} \quad y \to \infty \quad u = u_e(x) \qquad (5.6.1a)$$

We also use additional "boundary conditions" obtained by evaluating the momentum equation (5.1.2) at the wall with $v_w = 0$, that is,

$$\nu \frac{\partial^2 u}{\partial y^2} = \frac{1}{\rho} \frac{dp}{dx} = -u_e \frac{du_e}{dx} \qquad (5.6.1b)$$

and also some additional boundary conditions obtained from differentiating the edge boundary condition with respect to y, namely,

$$y \to \infty \quad \frac{\partial u}{\partial y}, \frac{\partial^2 u}{\partial y^2}, \frac{\partial^3 u}{\partial y^3}, \cdots \to 0 \qquad (5.6.1c)$$

Note that Eq. (5.6.1b) or Eq. (5.6.1c) are *properties of the solution* of the PDEs and *not* boundary conditions for the PDEs (see the discussion in Sec. 4.3.3).

We assume a fourth-order polynomial for the velocity profile, u/u_e, and write

$$\frac{u}{u_e} = a_0 + a_1\eta + a_2\eta^2 + a_3\eta^3 + a_4\eta^4 \qquad (5.6.2)$$

where η is now used for y/δ. This polynomial contains five coefficients that can be determined from the boundary conditions in Eq. (5.6.1). Using Eqs. (5.6.1a) and (5.6.1b) and the first two conditions in Eq. (5.6.1c), we get

$$a_0 = 0 \quad a_1 = 2 + \frac{\Lambda}{6} \quad a_2 = -\frac{\Lambda}{2} \quad a_3 = -2 + \frac{\Lambda}{2} \quad a_4 = 1 - \frac{\Lambda}{6} \qquad (5.6.3)$$

where Λ [see Eq. (1.3.11)] is a pressure-gradient parameter defined by

$$\Lambda = \frac{\delta^2}{\nu} \frac{du_e}{dx} \tag{5.6.4}$$

and can be interpreted physically as a typical ratio of pressure forces to viscous forces. With Eq. (5.6.3) we can write Eq. (5.6.2) as

$$\frac{u}{u_e} = (2\eta - 2\eta^3 + \eta^4) + \tfrac{1}{6}\Lambda\eta(1 - \eta)^3 \tag{5.6.5}$$

It is important to note that this is a very special case of the MIR. It uses only one integrated equation obtained by using Eq. (5.6.5) to evaluate the quantities δ^*, H, and c_f appearing in the momentum integral equation (3.4.8). More usually, several such ODEs are needed to define the profile parameters, and the Pohlhausen technique of constraining the parameters by boundary conditions deduced from the expected behavior of the *solution* of the PDEs, in addition to the true boundary conditions on the PDEs, is a rather hazardous one. It works only if the effect of a given boundary condition on the profile shape is strong: in a turbulent shear layer the velocity profile is virtually independent of the local pressure gradient, except very close to the wall, so that Eq. (5.6.1b) would be a highly unsuitable choice.

Figure 5.5 shows the velocity profiles for various values of Λ. Here $\Lambda = 0$ corresponds to a flat-plate flow, a negative value of Λ corresponds to a decelerating

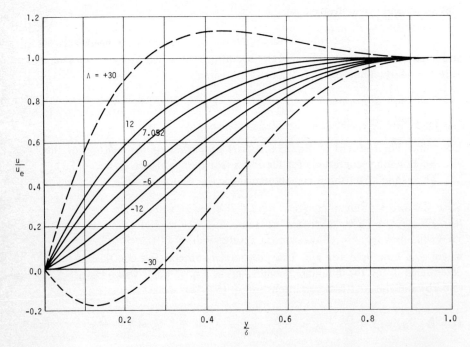

Fig. 5.5 The variation of dimensionless velocity, u/u_e, across the boundary layer according to Eq. (5.6.5).

flow, and a postive value of Λ corresponds to an accelerating flow. The profile at separation corresponds to $\Lambda = -12$, giving $\partial u/\partial y = 0$ at $\eta = 0$ from Eq. (5.6.5). The positive values of Λ are restricted to 12, since for incompressible flows the velocity profiles cannot have overshoots within the boundary layer. Thus Λ is restricted to the range of $-12 \leqslant \Lambda \leqslant 12$.

Once the velocity profile is known, the boundary layer parameters θ, δ^*, and τ_w or c_f appearing in the momentum integral equation (3.4.8) can be determined. From the definitions of τ_w, δ^*, and θ it follows that

$$\tau_w = \frac{\mu u_e}{\delta}(2 + \tfrac{1}{6}\Lambda) \tag{5.6.6}$$

$$\delta^* = \delta(\tfrac{3}{10} - \tfrac{1}{120}\Lambda) \tag{5.6.7}$$

$$\theta = \frac{\delta}{315}(37 - \frac{1}{3}\Lambda - \frac{5}{144}\Lambda^2) \tag{5.6.8}$$

Substituting Eqs. (5.6.6)–(5.6.8) into Eq. (3.4.8) leads to an equation of the form

$$\frac{dZ}{dx} = \frac{g(\Lambda)}{u_e} + h(\Lambda)Z^2\frac{d^2 u_e}{dx^2} \tag{5.6.9}$$

where $Z = \delta^2/v = \Lambda/(du_e/dx)$ and $g(\Lambda)$ and $h(\Lambda)$ are known functions of Λ. As usual, u_e and its derivatives are known. We note that the auxiliary relations for c_f and H that follow from Eqs. (5.6.6)–(5.6.8) are *not* differential equations: only the initial value of δ (or θ) and the distribution of $u_e(x)$ are needed to start the calculation.

EXAMPLE 5.4

Assume that the velocity profile for a laminar flow on a flat plate can be approximated by

$$u = a + by \tag{E5.4.1}$$

Use the momentum integral equation to obtain expressions for the dependence of local boundary-layer thickness, δ, and skin-friction coefficient, c_f, on the local Reynolds number, Re_x. Compare the results with exact solutions.

Solution Using the boundary conditions $u = 0$ at $y = 0$ and $u = u_e$ at $y = \delta$, we find that Eq. (E5.4.1) can be written as

$$\frac{u}{u_e} = \frac{y}{\delta} \equiv \eta \tag{E5.4.2}$$

With Eq. (E5.4.2) and with the definition of θ and c_f we can write

$$\theta = \delta \int_0^1 \eta(1 - \eta)\, d\eta = \frac{\delta}{6} \tag{E5.4.3}$$

$$\frac{c_f}{2} = \frac{\nu}{\delta u_e} \tag{E5.4.4}$$

Substituting Eqs. (E5.4.3) and (E5.4.4) into the momentum integral equation for zero pressure-gradient flow, that is,

$$\frac{d\theta}{dx} = \frac{c_f}{2} \tag{E5.4.5}$$

we get

$$\frac{d\delta}{dx} = \frac{6\nu}{\delta u_e} \tag{E5.4.6}$$

Integrating and noting that $\delta = 0$ at $x = 0$, we get

$$\frac{\delta}{x} = \frac{\sqrt{12}}{\sqrt{\mathrm{Re}_x}} = \frac{3.46}{\sqrt{\mathrm{Re}_x}} \tag{E5.4.7}$$

Substituting Eq. (E5.4.7) into Eq. (E5.4.4) for δ, we obtain

$$c_f = \frac{2}{3.46} \frac{1}{\sqrt{\mathrm{Re}_x}} = \frac{0.578}{\sqrt{\mathrm{Re}_x}} \tag{E5.4.8}$$

The exact results are Eqs. (5.5.10) and (5.5.6), respectively. The reason for the underestimation of δ is that the linear profile we have chosen is in effect the tangent to the real profile at $y = 0$ (see Fig. 1.9 for the best fit linear profile).

Before electronic digital computers became generally available, the Pohlhausen method was the most sophisticated one in general use because solution of the PDEs by finite-difference methods was impracticable. Now, it is less commonly used: for accurate work, differential solutions are preferred, while for the many cases where only the initial part of the shear layer is laminar, so that high accuracy is not needed, engineers prefer the very simple method due to Thwaites (1949; see also Rosenhead 1963).

5.6.2 Thwaites' Method

For convenience let us rewrite the momentum integral equation (3.4.8) as

$$\frac{d\theta}{dx} + \frac{\theta}{u_e}(H + 2)\frac{du_e}{dx} = \frac{c_f}{2} \tag{5.6.10}$$

If H and c_f are known as functions of θ or of some suitable combination of θ and u_e, Eq. (5.6.10) can be integrated, at least by a numerical process. Such functions were found in Thwaites' method by writing the following boundary conditions for Eq. (5.1.2) from which Eq. (5.6.10) can be derived by integrating with respect to y:

$$y = 0 \qquad \frac{\partial^2 u}{\partial y^2} = -\frac{u_e}{\theta^2}\lambda \qquad \frac{\partial u}{\partial y} = \frac{u_e}{\theta}l \tag{5.6.11}$$

These equations define λ and l. The variable l may be calculated by any particular solution of the boundary-layer equations, and it is found in all known cases to adhere reasonably closely to a universal function of λ, which Thwaites denoted by $l(\lambda)$. In the same way, if H is regarded as depending only on λ, a reasonably valid universal function for H can also be found, namely, $H(\lambda)$.

By putting $y = 0$ in Eq. (5.1.2) and using Eq. (5.6.11), we find

$$\lambda = \frac{\theta^2}{\nu}\frac{du_e}{dx} \tag{5.6.12}$$

Also,

$$\frac{c_f}{2} = \frac{\tau_w}{\rho u_e^{\,2}} = \frac{\nu}{u_e^{\,2}}\left(\frac{\partial u}{\partial y}\right)_w = \frac{\nu l(\lambda)}{u_e \theta} \tag{5.6.13}$$

The assumptions that l or c_f and H are functions of λ only are quasi-similarity assumptions. The Falkner-Skan solutions of Sec. 5.5 could be used to give $l(\lambda)$ and $H(\lambda)$. With these two results, Eq. (5.6.10) may be rewritten in the form

$$\frac{u_e}{\nu}\frac{d\theta^2}{dx} = 2\{-[H(\lambda) + 2]\lambda + l(\lambda)\} \equiv F(\lambda) \tag{5.6.14}$$

Here $F(\lambda)$ is another universal function. Thwaites writes an expression for $F(\lambda)$ chosen to fit known solutions of Eq. (5.1.2) as well as possible [see Rosenhead (1963), Fig. 6.7, for a demonstration of the accuracy of fit]:

$$F(\lambda) = 0.45 - 6\lambda = 0.45 - 6\frac{\theta^2}{\nu}\frac{du_e}{dx} \tag{5.6.15}$$

Substituting Eq. (5.6.15) into Eq. (5.6.14) and multiplying the resulting equation by $u_e^{\,5}$, we can write, after some rearranging,

$$\frac{1}{\nu}\frac{d}{dx}(\theta^2 u_e^{\,6}) = 0.45 u_e^{\,5}$$

which, upon integration, leads to

$$\frac{\theta^2 u_e^{\,6}}{\nu} = 0.45 \int_0^x u_e^5 \, dx + \left(\theta^2 \, \frac{u_e^{\,6}}{\nu}\right)_0 \tag{5.6.16}$$

In terms of dimensionless quantities defined by

$$x^* \equiv \frac{x}{L} \qquad u^* \equiv \frac{u}{u_{\text{ref}}} \qquad u_e^* \equiv \frac{u_e}{u_{\text{ref}}} \qquad R_L \equiv \frac{u_{\text{ref}} L}{\nu} \tag{5.6.17}$$

Eq. (5.6.16) can be written as

$$\left(\frac{\theta}{L}\right)^2 R_L = \frac{0.45}{(u_e^*)^6} \int_0^{x^*} (u_e^*)^5 \, dx^* + \left(\frac{\theta}{L}\right)_0^2 R_L \left(\frac{u_{e0}^*}{u_e^*}\right)^6 \tag{5.6.18}$$

For a stagnation-point flow ($m = 1$), Eq. (5.6.18) gives

$$\left(\frac{\theta}{L}\right)_0^2 R_L = \frac{0.075}{(du_e^*/dx^*)_0} \tag{5.6.19}$$

where (du^*/dx^*) denotes the slope of the external-velocity distribution for stagnation-point flow. Note that the last term in Eq. (5.6.18) is zero in calculations starting from a stagnation point, because $u_{e0}^* = 0$.

Once θ is calculated for a given external-velocity distribution, the other boundary-layer parameters H and c_f can be determined from the relations given below.

For $0 \leqslant \lambda \leqslant 0.1$,

$$\begin{aligned} l &= 0.22 + 1.57\lambda - 1.8\lambda^2 \\ H &= 2.61 - 3.75\lambda + 5.24\lambda^2 \end{aligned} \tag{5.6.20}$$

For $-0.1 \leqslant \lambda \leqslant 0$,

$$l = 0.22 + 1.402\lambda + \frac{0.018\lambda}{0.107 + \lambda}$$

$$H = \frac{0.0731}{0.14 + \lambda} + 2.088 \tag{5.6.20}$$

EXAMPLE 5.5

The inviscid flow solution for the velocity along the surface of a circular cylinder that is situated in a cross flow with a velocity u_∞ is

$$u_e = 2u_\infty \sin \phi \tag{E5.5.1}$$

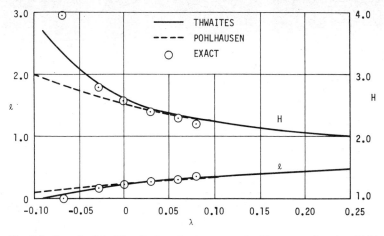

Fig. 5.6 Functions in Thwaites' method compared with exact values (see Table 5.1) and those given in Pohlhausen's method (see Schlichting, 1968).

where ϕ ($\equiv x/r_0$) is measured from the stagnation point. Using this velocity distribution, compute $\theta/L \sqrt{R_L}$, $\delta^*/L \sqrt{R_L}$, and $[\tau_w/(\frac{1}{2}) \rho u_\infty^2] \sqrt{R_L}$ by Thwaites' method at an angle of $\phi = 40°$. Take L equal to the cylinder radius.

Solution From Eq. (5.6.18),

$$\frac{\theta}{L} \sqrt{R_L} = \left(\frac{0.45}{2 \sin^6 40} \int_0^{40°} \sin^5 \phi \, d\phi \right)^{1/2} = 0.213$$

From the definition of H,

$$\frac{\delta^*}{L} \sqrt{R_L} = \frac{\theta}{L} \sqrt{R_L} H(\lambda) = 0.213 H(\lambda)$$

but

$$\lambda = \frac{\theta^2}{\nu} \frac{du_e}{dx} = \left(\frac{\theta}{L} \right)^2 R_L \frac{du_e^*}{d\theta} = (0.213)^2 \, 2(\cos 40) = 0.0695$$

Thus from Fig. 5.6,

$$H \simeq 2.35 \qquad l(\lambda) \simeq 0.32$$

Then

$$\frac{\delta^*}{L} \sqrt{R_L} = (0.213)(2.35) = 0.501$$

From Eq. (5.6.13),

$$\frac{\tau_w}{\rho u_\infty{}^2} = \frac{\tau_w}{\rho u_e{}^2}\left(\frac{u_e}{u_\infty}\right)^2 = \frac{\nu l}{u_e \theta}\left(\frac{u_e}{u_\infty}\right)^2 = \frac{l u_e^*}{\theta/L R_L}$$

Therefore

$$\frac{\tau_w}{1/2\rho u_\infty{}^2}\sqrt{R_L} = \frac{2l u_e^*}{\theta/L\sqrt{R_L}} = \frac{2(0.32)2\sin 40}{0.213} = 3.86$$

Thwaites' method can also be used for axisymmetric boundary layers by using the Mangler transformation. If we denote two-dimensional variables by the subscript 2 and the axisymmetric ones by 3 [these correspond, respectively, to barred and unbarred variables in Eqs. (3.3.1) and (3.3.2)], then according to the Mangler transformation for cases where $\delta \ll r_0$ (no transverse curvature effect), we write

$$\theta_2 = \left(\frac{r_0}{L}\right)^k \theta_3 \qquad\qquad (5.6.21a)$$

$$dx_2 = \left(\frac{r_0}{L}\right)^{2k} dx_3 \qquad\qquad (5.6.21b)$$

By using Eqs. (5.6.21), Eq. (5.6.18) can be written as

$$\left(\frac{\theta_3}{L}\right)^2 R_L = \frac{0.45}{(u_e^*)^6(r_0^*)^{2k}}\int_0^{x_3^*}(u_e^*)^5(r_0^*)^{2k}dx_3^* + \left(\frac{\theta_3}{L}\right)_0^2 R_L\left(\frac{u_{e0}^*}{u_e^*}\right)^6 \qquad (5.6.22)$$

For a three-dimensional stagnation-point flow ($m = \frac{1}{3}$),

$$\left(\frac{\theta_3}{L}\right)^2 R_L = \frac{0.056}{(du_e^*/dx^*)_0} \qquad\qquad (5.6.23)$$

In Eqs. (5.6.21)

$$r_0^* = \frac{r_0}{L} \qquad x_3^* = \frac{x_3}{L}$$

Once θ_3 is calculated from Eq. (5.6.22), then the variables δ^*, H, and c_f can be calculated from Eq. (5.6.20).

5.6.3 Fortran Program for Thwaites' Method

In this section we present a Fortran program for predicting the laminar boundary-layer development on two-dimensional and axisymmetric bodies by Thwaites' method. It is programed in Fortran IV for the IBM 370/165. The integration of Eq.

(5.6.22) is performed by using the trapezoidal rule. The distance along the surface, $s(x^*$ or $x_3^*)$, at a given surface location, x_i, y_i $(i = 1, 2, \ldots, I)$, is calculated from the simple formula

$$S_i = S_{i-1} + \sqrt{(x_i - x_{i-1})^2 + (y_i - y_{i-1})^2}$$

For axisymmetric bodies, y_i is replaced by $(r_0^*)_i$. For surface distance calculation it is assumed that the calculation starts from a stagnation point. The output of the program consists of Re_x, δ^*, θ, H, c_f, and R_θ as functions of S. The input consists of

Card 1 The first three variables are punched as integers with field length of 3; the remaining variables are punched in F10.0 format.

NXT   Total number of x-stations, not to exceed 30.
KASE Flow index, 0 for two-dimensional flow, 1 for two-dimensional flow that starts as stagnation-point flow, and 2 for axisymmetric flow.
KDIS Index for surface distance; 1 when surface distance is input, 0 when surface distance is calculated.
UREF Reference velocity, u_{ref}, feet per second or meters per second.
BIGL Reference length, L, feet or meters.
CNU   Kinematic viscosity, ν, square feet per second or square meters per second.

1	2	3	4	5	6	7	8	9	10	11	12	13	14	15	16	17	18	19	20	21	22	23	24	25	26	27	28	29	30	31	32	33	34	35	36	37	38	39
NXT		KASE		KDIS			UREF								BIGL									CNU														

Load sheet for card 1.

Cards 2 to NXT + 1 Variables are punched in 3F10.0 format, one set of three per card. Number of cards equals NXT of card 1.

X   Dimensionless chordwise or axial distance, x/L. If KDIS = 1, then X is the surface distance.
UE   Dimensionless velocity, u_e/u_{ref}.
R   Dimensionless two-dimensional body ordinate or body of revolution radius; r/L.

1	2	3	4	5	6	7	8	9	10	11	12	13	14	15	16	17	18	19	20	21	22	23	24	25	26	27	28	29	30
		X									UE									R									

Load sheet for cards 2 to NXT + 1.

```
      DIMENSION S(30),X(30),R(30),UE(30)
C ** INPUT
      READ(5,8000) NXT,KASE,KD1S,UREF,BIGL, CNU
      WRITE(6,9000) NXT,KASE,UREF,BIGL,CNU
      IF(KD1S .EQ. 1) GO TO 100
      READ(5,8100) (X(I),UE(I),R(I),I=1,NXT)
C ** CALCULATION OF SURFACE DISTANCE
      S(1)  = 0.0
      DO 50 I=2,NXT
   50 S(I) = S(I-1)+SQRT((X(I)-X(I-1))**2+(R(I)-R(I-1))**2)
      GO TO 400
  100 READ(5,8100) (S(I),UE(I),R(I),I=1,NXT)
C ** THWAITES METHOD
  400 WRITE(6,9200)
      CF    = 0.0
      URSUM = 0.0
      RL    = UREF*BIGL/CNU
      F2    = 0.0
      DO 500 I=1,NXT
      R2    = 1.
      IF(KASE .EQ. 2) R2 = R(I)**2
      F2    = UE(I)**5*R2
      IF(I.EQ.1) GOTO 480
      IF(I.EQ.2 .AND. KASE.GE.1) GOTO 480
      URSUM = URSUM + 0.5*(F1+F2)*(S(I)-S(I-1))
      DUEDS = (UE(I)-UE(I-1))/(S(I)-S(I-1))
      CONST = 0.45/(F2*UE(I))
      THTATM= CONST*URSUM
      GO TO 495
  480 DUEDS = (UE(2)-UE(1))/S(2)
      THTATM=0.
      IF(KASE .EQ. 1) THTATM = 0.075/DUEDS
      IF(KASE .EQ. 2) THTATM = 0.056/DUEDS
  495 THETA = SQRT(THTATM/RL)*BIGL
      RTHETA= THETA*UE(I)/CNU*UREF
      RS    = UE(I)*S(I)/CNU*UREF*BIGL
      CLMBDA= THTATM*DUEDS
      IF(CLMBDA .LT. 0.0) GO TO 496
      H     = 2.61-3.75*CLMBDA+5.24*CLMBDA**2
      CL    = 0.22+1.57*CLMBDA-1.8*CLMBDA**2
      GO TO 497
  496 H     = 0.0731/(0.14+CLMBDA) + 2.088
      CL    = 0.22+1.402*CLMBDA+0.018*CLMBDA/(CLMBDA+0.107)
  497 DELS  = THETA*H
      IF(I .GT. 1) CF = 2.0*CL/(UE(I)*THETA/BIGL*RL)
      WRITE(6,9300) I,S(I),RS,DELS,THETA,H,CF,RTHETA
      F1    = F2
  500 CONTINUE
      STOP
C - - - - - - - - - - - - - - - - - - - - - - - - - - - - -
 8000 FORMAT(3I3,3F10.0)
 8100 FORMAT(3F10.0)
 9000 FORMAT(1H0,5HNXT =,I3,14X,5HKASE=,I3/1H ,5HUREF=,E14.6,3X,
     1       5HBIGL=,E14.6,3X,5HNU  =,E14.6/)
 9200 FORMAT(1H0,2X,1HI,5X,1HS,11X,2HRS,9X,4HDELS,7X,5HTHETA,9X,1HH,11X,
     1       2HCF,9X,6HRTHETA)
 9300 FORMAT(1H ,I3,7E12.4)
      END
```

EXAMPLE 5.6

Consider an incompressible laminar flow past a prolate spheroid at zero angle of attack, $\alpha = 0$ (see Fig. 5.7). Using the Fortran program for Thwaites' method, calculate the local skin-friction coefficient, c_f, and the displacement thickness, δ^*, at $x/a = -\frac{1}{2}$ and 0 for $u_{ref} = 12$ m s^{-1}, $\nu = 1.5 \times 10^{-5}$ m^2 s^{-1}, and $b/a = \frac{1}{2}$.

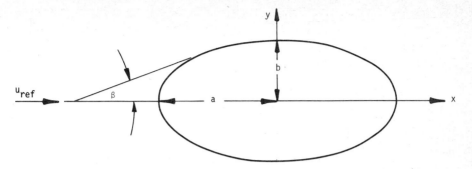

Fig. 5.7 Flow past a prolate spheroid at zero angle of attack.

Solution A prolate spheroid is an ellipsoid of revolution whose length along its symmetry axis is greater than the diameter of its largest circular cross section. According to inviscid flow theory, for zero angle of attack the external-velocity distribution around the prolate spheriod is given by

$$u_e(s) = u_{ref} A \cos \beta \qquad (E5.6.1)$$

Here s represents the surface distance, and β denotes the angle between the line tangent to the elliptic profile and the positive x-axis. The parameter A is a function of the thickness ratio t ($\equiv b/a$) of the elliptic profile. It is given by (see also Fig. 5.8)

$$A = \frac{(1 - t^2)^{3/2}}{\sqrt{1 - t^2} - 1/2t^2 \ln [(1 + \sqrt{1 - t^2})/(1 - \sqrt{1 - t^2})]} \qquad (E5.6.2)$$

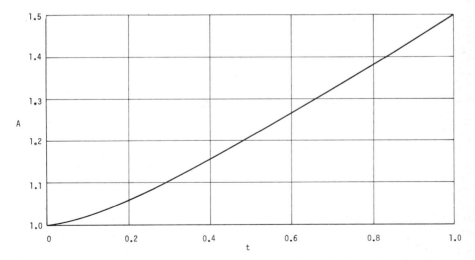

Fig. 5.8 Maximum velocity ratio for flow parallel to the major axis of a prolate spheroid.

Before we can use the Fortran program, we need to specify the external velocity, u_e^*, and the body, r_0^* and R_L. To compute the external velocity, we must compute $\cos \beta$. It can be calculated as follows.

The equation of a prolate spheroid is

$$\frac{x^2}{a^2} + \frac{r_0^2}{b^2} = 1 \tag{E5.6.3}$$

The surface distance, which is defined by $(ds)^2 = (dx)^2 + (dr_0)^2$, can be shown to be given by

$$\frac{ds}{dx} = \left[\frac{1 + \xi^2(t^2 - 1)}{1 - \xi^2}\right]^{1/2} \tag{E5.6.4}$$

where $\xi = x/a$. Since $\cos \beta = dx/ds$,

$$\cos \beta = \left[\frac{1 - \xi^2}{1 + \xi^2(t^2 - 1)}\right]^{1/2} \tag{E5.6.5}$$

To compute the dimensionless body radius r_0^*, we use Eq. (E5.6.3) to get

$$r_0^* = t\sqrt{1 - \xi^2} \tag{E5.6.6}$$

From Fig. 5.8 we find that for $t = 0.5$, $A = 1.22$. Using this value and the expression given by Eq. (E5.6.5), we can now compute u_e^*, and from Eq. (E5.6.6), r_0^* for various values of ξ, which in our case correspond to x_f. The data generated in this manner are transferred to cards 2 to NXT + 1. All together we take 20 x-stations. With KASE = 2, u_{ref}, $L \equiv a \,(\equiv 1)$, and ν given, KDIS = 0; we can now compute the boundary-layer development using the computer program. We find that at $\xi = -\frac{1}{2}$, $c_f = 1.80 \times 10^{-3}$ and $\delta^* = 0.777 \times 10^{-3}$ m and at $\xi = 0$, $c_f = 0.886 \times 10^{-3}$ and $\delta^* = 1.34 \times 10^{-3}$ m.

5.7 LAMINAR INTERNAL FLOWS

As was pointed out in Secs. 4.3.3 and 5.1, the prediction of the velocity field in the entrance region of internal flows requires the solution of the continuity and the momentum equations together with a relationship obtained from the conservation of mass. As in external flows the governing system of equations can be solved either in physical or in transformed coordinates. Each coordinate system has its advantages. To pave the way for a numerical method that will be discussed later in Chap. 7 for internal entrance-region problems, we consider both coordinate systems in Sec. 5.7.1. Fully developed flow equations (applicable in constant-area ducts far from the entry, so that the flow is no longer changing with x) are ordinary DEs that do not require the use of transformed coordinates (strictly speaking, the similarity coordinates are the physical coordinates, in which the equations can be easily solved, as is shown in Sec. 5.7.2).

5.7.1 Entrance-Region Flow in Symmetrical Ducts

The governing equations for the entrance region are given by Eqs. (3.3.6) and (3.3.9) in Mangler transformed coordinates, so entrance region refers to a two-dimensional duct or a circular pipe. Note that for a pipe of constant cross section, $\bar{x} = x$, the assumption being that we choose the reference length L equal to the pipe radius r_0. To remove the singularity at the entrance of the duct ($\bar{x} = 0$), we use the similarity variable η defined in Eq. (5.3.1) with the external (or edge) velocity $u_e(x)$ replaced by a constant reference velocity u_0. We could continue to use u_e in the entrance region but would in any case have to abandon it at downstream stations because when the two shear layers merge, the potential core disappears and the edge velocity has no meaning. We also use the same definition of the dimensionless stream function $f(\bar{x}, \eta)$ defined by Eq. (5.3.2), replacing u_e again by a constant reference velocity u_0. Thus with the transformation

$$\eta = \left(\frac{u_0}{\nu\bar{x}}\right)^{1/2} \bar{y} \quad \text{with} \quad d\bar{y} = \left(\frac{r}{L}\right)^k dy \tag{5.7.1}$$

and the dimensionless stream function $f(\bar{x}, \eta)$ defined by

$$f(\bar{x}, \eta) = \frac{\psi(\bar{x}, \bar{y})}{(u_0 \nu\bar{x})^{1/2}} \tag{5.7.2}$$

it can be shown that Eq. (3.3.9) can be written as

$$[(1 - t)^{2k} f'']' + \tfrac{1}{2} ff'' = \bar{x} \frac{dp^*}{d\bar{x}} + \bar{x}\left(f' \frac{\partial f'}{\partial \bar{x}} - f'' \frac{\partial f}{\partial \bar{x}}\right) \tag{5.7.3}$$

Here primes denote differentiation with respect to η and

$$p^* = \frac{p}{\rho u_0^2} \quad t = 1 - \left(1 - 2\left(\frac{L}{r_0}\right)^2 \sqrt{\frac{\bar{x}}{L}} \frac{\eta}{\sqrt{R_L}}\right)^{1/2} \quad R_L = \frac{u_0 L}{\nu} \tag{5.7.4}$$

where L is a reference length that would normally be taken as the pipe radius r_0. The wall boundary conditions for Eq. (5.7.3) are

$$f(\bar{x}, 0) = f_w(\bar{x}) \quad f'(\bar{x}, 0) = 0 \tag{5.7.5a}$$

At the boundary-layer edge, $f''(x, \eta_\infty) = 0$. Substituting this in Eq. (5.7.3) yields a boundary condition for f', namely, the x-derivative of Bernoulli's equation for incompressible flow:

$$\frac{d}{d\bar{x}} \frac{[f'(\eta_\infty)]^2}{2} = -\frac{dp^*}{d\bar{x}} \tag{5.7.5b}$$

The core velocity u_e, proportional to $f'(\eta_\infty)$, is deduced from the condition of constant mass flow, $\int \rho u \, dA = \text{const}$, where the integration is over the duct cross section. In a plane duct of height $2h$, for instance, the mass flow is $2\rho(h - \delta^*)u_e$. We postpone discussion of numerical procedures until Chap. 7.

Equation (5.7.3) applies for both two-dimensional and axisymmetric flows. By setting $k = 0$ or $k = 1$ we get the transformed momentum equation for either two-dimensional flows or axisymmetric flows. The relation between the physical variables x and y and the Mangler variables $\bar{x}$ and $\bar{y}$ are given by Eqs. (3.3.1) and (3.3.2). Obviously, for two-dimensional flows, $\bar{y} = y$ as well as $\bar{x} = x$.

Although the similarity variable η removes the singularity at the leading edge, stretches the coordinate normal to the flow, maintains the transformed boundary-layer thickness nearly constant ($\eta_\infty \approx 8$), and thus provides a very useful computational advantage, it is necessary to abandon the similarity variable in favor of physical variables at some $x = x_0$ before the calculated boundary-layer thickness δ exceeds the duct half width or pipe radius. The value of x_0 can be determined from Eq. (5.7.1). Using the definition of R_L, we can write Eq. (5.7.1) as

$$\left(\frac{\bar{y}}{L}\right) = \left(\frac{\bar{x}}{L}\right)^{1/2} \frac{\eta}{(R_L)^{1/2}}$$

Since η_∞ is approximately constant and equal to 8, the governing equations can be solved in transformed variables for values of $\bar{x}/L$ less than

$$\frac{\bar{x}}{L} < \left(\frac{R_L}{64}\right)\left(\frac{\bar{y}_c}{L}\right)^2 \tag{5.7.6}$$

where $\bar{y}_c$ is the distance between the duct wall and the centerline.

When we use physical coordinates, we can again write the momentum equation as a third-order equation by using the definition of the stream function. By the introduction of a dimensionless distance, Y, defined by

$$Y = \left(\frac{u_0}{\nu L}\right)^{1/2} \bar{y} \tag{5.7.7}$$

and a dimensionless stream function $F(\bar{x}, Y)$ defined by

$$F(\bar{x}, Y) = \frac{\psi(\bar{x}, \bar{y})}{(u_0 \nu L)^{1/2}} \tag{5.7.8}$$

together with the definition of $r(\equiv r_0 - y)$ for internal axisymmetric flows, we can write Eq. (3.3.9) as

$$[(1 - t)^{2k}F'']' = \frac{dp^*}{d\bar{x}} + F'\frac{\partial F'}{\partial \bar{x}} - F''\frac{\partial F}{\partial \bar{x}} \tag{5.7.9}$$

Here primes denote differentiation with respect to Y, and t is [see Eq. (3.3.10)]

$$t = 1 - \left[1 - \frac{2}{(r_0/L)^2} \frac{Y}{\sqrt{R_L}} \right]^{1/2} \tag{5.7.10}$$

The wall boundary conditions for Eq. (5.7.9) are

$$F(\bar{x},0) = F_w(\bar{x}) \qquad F'(\bar{x},0) = 0 \tag{5.7.11a}$$

At the centerline, symmetry requires $\partial u/\partial y = 0$ in untransformed coordinates. In transformed coordinates, $\partial u/\partial y$ at the centerline, where $\bar{y} = \bar{y}_c$ or $Y_c = \sqrt{R_L}\,\bar{y}_c/L$, is

$$\frac{\partial u}{\partial Y}(\bar{x}, Y_c) = \frac{0}{0}$$

which is indeterminate. To determine the centerline boundary condition, we consider Eq. (5.7.9) and find that

$$F''(\bar{x}, Y_c) = -\frac{1}{2}\left(\frac{r_0}{L}\right)^{2k} \sqrt{R_L}\left[\frac{dp^*}{d\bar{x}} + \frac{1}{2}\frac{d}{d\bar{x}}(F')^2\right] \tag{5.7.11b}$$

As will be discussed in Chap. 7, we first solve Eqs. (5.7.3) and (5.7.5) for values of $\bar{x}$ less than those given by Eq. (5.7.6). Then we switch to physical coordinates and solve Eqs. (5.7.9) and (5.7.11), again with the constraint of constant mass flow. We now present results obtained by this procedure for the entrance-region laminar flow in a pipe and in parallel plates.

Entrance-Region Laminar Flow in a Pipe

Figures 5.9 and 5.10 show the velocity profiles, u/u_0, and dimensionless centerline (maximum) velocity, u_c/u_0, as functions of $2x^*/R_d$ in the entrance region of a pipe. Here $x^* = x/r_0$ and $R_d = u_0 d/\nu$. Figure 5.10 also shows the measured values of the centerline velocity obtained by Pfenninger (1951). According to the results of Fig. 5.10, the centerline velocity has almost reached its asymptotic value of 2 at $2x^*/R_d = 0.20$. Thus the entrance length for a laminar flow in a circular pipe is

$$\frac{l_e}{d} = \frac{R_d}{20} \tag{5.7.12}$$

Figure 5.11 shows the pressure drop Δp^* in the entrance region together with the pressure drop for a fully developed flow [see Eq. (5.7.28)] as a function of $2x^*/R_d$. The figure also shows the experimental data of Shapiro et al. (1954).

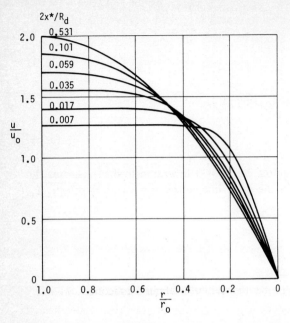

Fig. 5.9 Laminar velocity profiles in the entrance region of a pipe. Note that far downstream, where the flow becomes fully developed, the centerline velocity ratio becomes 2.0.

EXAMPLE 5.7

Calculate the pressure drop in the entrance region of a circular pipe. Compare this value with that for fully developed flow with the same mass-flow conditions and determine the maximum entrance length obtainable for laminar flow.

Solution The entrance length is given by Eq. (5.7.12). From Fig. 5.11 the pressure drop in the entrance region is $\Delta p^*_{\text{entrance}} = 2.2$. For fully developed flow the corresponding pressure drop is given by Eq. (5.7.28) or $\Delta p^* = 1.6$. Because the fully developed flow remains laminar for $R_d < 2000$, we see from the entrance-

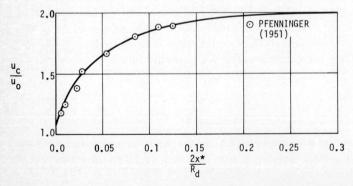

Fig. 5.10 Variation of the centerline velocity ratio in the entrance region of a pipe for laminar flow.

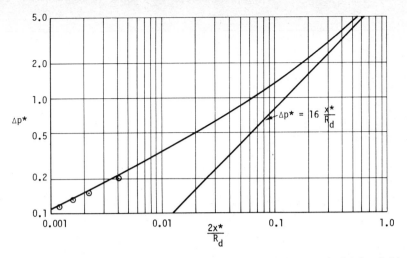

Fig. 5.11 Pressure drop for a laminar flow in a pipe. The line on the left-hand side is for the entrance region, and the one on the right-hand side [indicated by Eq. (5.7.28)] is for fully developed flow. ⊙ Shapiro et al. (1954).

length expression that laminar flow will persist for at least 100 tube diameters under smooth conditions.

Entrance-Region Laminar Flow between Parallel Plates

Figure 5.12 shows the velocity profile u/u_0, and Fig. 5.13 shows the dimensionless centerline (maximum) velocity u_c/u_0 as a function of x^*/R_h, in the entrance region of two parallel plates. Here $x^* = x/h$, $R_h = u_0 h/\nu$. According to Fig. 5.13, for all

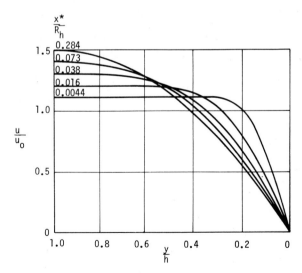

Fig. 5.12 Laminar velocity profiles in the entrance region of a two-dimensional duct. Note that for downstream of the duct, flow becomes fully developed, and the centerline velocity ratio becomes 1.5.

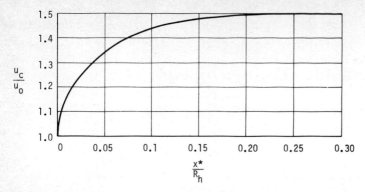

Fig. 5.13 Variation of the centerline velocity ratio in the entrance region of a two-dimensional duct for laminar flow.

practical purposes, the centerline velocity has reached its asymptotic value 1.50 at $x^*/R_h = 0.20$. If we take this value to be the beginning of fully developed flow, then the entrance length for Poiseuille flow is

$$\frac{l_e}{h} = \frac{R_h}{5} \tag{5.7.13}$$

Figure 5.14 shows the pressure drop (Δp^*) in the entrance region together with the pressure drop for a fully developed flow [see Eq. (5.7.36)] as a function of x^*/R_h.

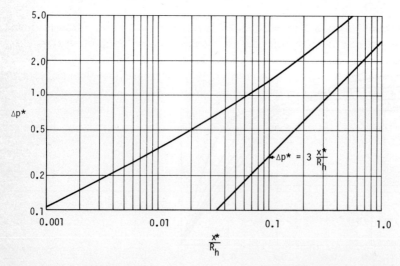

Fig. 5.14 Pressure drop for a laminar flow in a two-dimensional duct. The line on the left-hand side is for the entrance region, and the one on the right-hand side (indicated by the equation) is for fully developed flow.

EXAMPLE 5.8

Consider a laminar flow between two stationary flat plates with $h = 2$ cm, length 5 m, and $R_h = 10^3$. Compute the dimensionless pressure drop, Δp^*, with and without consideration of the entrance region.

Solution From Eq. (5.7.13),

$$l_e = h \frac{R_h}{5} = 2 \frac{1000}{5} = 400 \text{ cm} = 4 \text{ m}$$

Therefore according to Fig. 5.14, Δp^* in the entrance region is 2.5. Fully developed flow takes place in the last meter of the duct. Thus according to Fig. 5.14 or Eq. (5.7.36), Δp^* in the fully developed portion is 0.15. As a result, the total dimensionless pressure drop is

$$\Delta p^*_{total} = 2.5 + 0.15 = 2.65$$

If we assume fully developed flow everywhere, then from Eq. (5.7.36),

$$\Delta p^*_{total} = 0.75$$

The reason why the pressure drop is greater in the entrance region is that mass flow conservation requires the central part of the flow to accelerate (see Fig. 5.12).

5.7.2 Fully Developed Flows

We now present some well-known results for simple duct flows. Readers should regard these as recapitulations of what they already know, using the methodology developed in earlier chapters for use in the more general problems discussed here.

For a two-dimensional or axisymmetric fully developed incompressible steady laminar flow the momentum equation can be written as

$$\frac{dp}{dx} = \frac{1}{r^k} \frac{d}{dy}(r^k \tau) = \frac{\mu}{r^k} \frac{d}{dy}\left(r^k \frac{du}{dy}\right) \tag{5.7.14}$$

Laminar Flow in a Pipe

Let us first consider flow in a *circular pipe*, $k = 1$. Noting that $r = r_0 - y$ so that $dr = -dy$, we can write Eq. (5.7.14) as

$$\frac{dp}{dx} = \frac{\mu}{r} \frac{d}{dr}\left(r \frac{du}{dr}\right) \tag{5.7.15}$$

The boundary conditions for Eq. (5.7.15) are

$$x = x_i \quad p = p_i; \quad x = x_0 \quad p = p_0 \tag{5.7.16a}$$

$$r = r_0 \quad u = 0; \quad r = 0 \quad du/dr = 0 \tag{5.7.16b}$$

Integrating Eq. (5.7.15) with respect to x and applying Eq. (5.7.16) to the resulting expression, we get

$$\frac{p_0 - p_i}{L} = \frac{\mu}{r} \frac{d}{dr} \left(r \frac{du}{dr} \right) \tag{5.7.17}$$

where $L = x_0 - x_i$. Integrating Eq. (5.7.17) twice with respect to r and applying Eq. (5.7.16b) to the resulting expression, we get the velocity distribution,

$$u = \frac{(p_i - p_0){r_0}^2}{4\mu L} \left[1 - \left(\frac{r}{r_0} \right)^2 \right] \tag{5.7.18}$$

Thus u is a parabolic function of r, necessarily zero at the wall and a maximum at the pipe axis. Its maximum value is

$$u_{max} = \frac{(p_i - p_0){r_0}^2}{4\mu L} \tag{5.7.19}$$

Substituting u_{max} into Eq. (5.7.18), we get

$$u = u_{max} \left[1 - \left(\frac{r}{r_0} \right)^2 \right] \tag{5.7.20}$$

The volume flow rate in a pipe is

$$Q = \int_0^{r_0} 2\pi r u \, dr \tag{5.7.21}$$

In terms of an average velocity, u_{av} (equal to the uniform velocity at inlet), defined by

$$u_{av} = \frac{\displaystyle\int_0^{r_0} 2\pi r u \, dr}{\pi {r_0}^2} \tag{5.7.22}$$

the volume flow rate becomes

$$Q = u_{av} \pi {r_0}^2 \tag{5.7.23}$$

The average velocity can be obtained by substituting Eq. (5.7.20) into (5.7.22) and integrating the resulting expression:

$$u_{av} = \frac{u_{max}}{2}$$ (5.7.24)

With Eqs. (5.7.24) and (5.7.19), Eq. (5.7.23) can be written as

$$Q = \frac{\pi r_0^4}{8\mu L}(p_i - p_0)$$ (5.7.25)

Equation (5.7.25) provides a relationship between flow rate and pressure loss over a given length of pipe for fully developed laminar flows. It is known as the *Hagen-Poiseuille* equation.

A useful parameter in duct flows is a dimensionless wall shear stress called the friction factor. It is very similar to the local skin-friction coefficient c_f defined for external boundary-layer flows except for $\frac{1}{8}$ rather than $\frac{1}{2}$. It is defined by

$$f = \frac{\tau_w}{1/8\rho u_{av}^2} = \frac{-2r_0}{1/2\rho u_{av}^2}\frac{dp}{dx}$$ (5.7.26)

the pressure drop per diameter divided by the dynamic pressure at entry. The wall shear stress τ_w is

$$\tau_w = \mu\left(\frac{du}{dy}\right)_w = -\mu\left(\frac{du}{dr}\right)_w = \frac{p_i - p_0}{2L}r_0$$

Substituting the definition of τ_w into Eq. (5.7.26), we can express the friction factor as a function of Reynolds number R_d based on pipe diameter:

$$f = \frac{64}{R_d}$$ (5.7.27)

Sometimes it is useful to write Eq. (5.7.25) in a slightly different form. Substituting the expression of Q given by Eq. (5.7.23) into Eq. (5.7.25) and using the definition of R_d, we can write Eq. (5.7.25) as

$$\Delta p^* = \frac{32}{R_d}\frac{L}{d}$$ (5.7.28)

where $\Delta p^* = (p_i - p_0)/\rho u_{av}^2$.

EXAMPLE 5.9

A water pipe is 5 m long and 5 cm in diameter. If the entry flow is uniform, find the highest speed for which fully developed laminar flow is attained at the exit.

Solution   From Eq. (5.7.12),

$$R_d = \frac{u_{av}d}{\nu} = 20\frac{l_e}{d} = \frac{20 \times 5}{0.05} = 2000$$

Taking $\nu = 1.1 \times 10^{-6}$ m^2 s^{-1} and $u_{av} = 0.044$ m s^{-1}, $R_d = 2000$, which happens to be approximately the critical Reynolds number at which transition occurs with disturbed entry conditions. Typical speeds in laminar flow (except in thin boundary layers) are quite low.

Laminar Flow between Parallel Plates

Another fully developed laminar-flow case of importance is flow between parallel plates. In this case, Eq. (5.7.14) reduces to

$$\frac{dp}{dx} = \mu\frac{d^2u}{dy^2} \tag{5.7.29}$$

The boundary conditions are

$$x = x_i \quad p = p_i; \quad x = x_0 \quad p = p_0 \tag{5.7.30a}$$
$$y = 0 \quad u = 0; \quad y = 2h \quad u = u_0 = \text{const} \tag{5.7.30b}$$

Integrating Eq. (5.7.29) first with respect to x and then with respect to y and using the boundary conditions (5.7.30), we get the velocity distribution between parallel plates with one plate moving with a constant velocity, u_0:

$$u = u_0\frac{y}{2h} - \frac{1}{2\mu}\frac{p_0 - p_i}{L}y(2h - y) \tag{5.7.31}$$

The type of flow represented by Eq. (5.7.31) is known as *Couette* flow, and two special cases can be examined. First, for both plates stationary ($u_0 \equiv 0$), Eq. (5.7.31) becomes

$$u = -\frac{1}{2\mu}\frac{p_0 - p_i}{L}y(2h - y) \tag{5.7.32}$$

which is a parabolic velocity distribution with u_{max} at $y = h$. This is called *plane Poiseuille* flow.

Second, for zero pressure gradient, that is, with the whole fluid flow caused by the motion of the top plate,

$$u = u_0\frac{y}{2h} \tag{5.7.33}$$

This is a linear velocity distribution, and the flow is known as *simple Couette* flow.

The volume flow rate between two parallel stationary plates per unit depth is given by

$$u_0 2h = \int_0^{2h} u \, dy \tag{5.7.34}$$

For the velocity distribution [Eq. (5.7.32)],

$$\frac{u_{max}}{u_{av}} = \frac{3}{2} \tag{5.7.35}$$

An expression similar to Eq. (5.7.28) can also be obtained for plane Poiseuille flow. It is given by

$$\Delta p^* = \frac{3}{R_h} \frac{L}{h} \tag{5.7.36}$$

Laminar Flow in Tubes of Other Cross-Sectional Shapes

In steady, incompressible fully developed flow the equation of motion in x-, y-, z-coordinates, Eq. (2.3.9), always reduces to

$$0 = -\frac{1}{\rho} \frac{dp}{dx} + \nu \left(\frac{\partial^2 u}{\partial y^2} + \frac{\partial^2 u}{\partial z^2} \right) \tag{5.7.37}$$

which for a given pressure gradient is a Poisson equation (see Prob. 2.5) for u in the xz-plane. This elliptic PDE can be solved by standard methods for any boundary in the xz-plane—that is, any cross-sectional shape—under the simple boundary condition $u = 0$. The same equation describes the (small) deflection of a soap bubble on the end of a pipe, and the semiquantitative results provided by blowing such a bubble on a complicated cross section can be instructive.

In the inlet region the only simplification of the equations of motion that can be made with confidence is the neglect of the second derivatives of u, v, and w with respect to x; $\partial p/\partial y$ and $\partial p/\partial z$ cannot be neglected in general, and matching of the shear-layer solution to the potential flow is quite complicated (see Chap. 11). However, the terms appearing in Eq. (5.7.37) still dominate, and it is possible to combine a marching solution for the equations as a whole with an elliptic solution for each cross section (see Pratap and Spalding, 1975, for example).

In order to compute flows through ducts of cross section other than circular, we define the Reynolds number by

$$R_{d_e} = \frac{u_{av} d_e}{\nu} \tag{5.7.38}$$

where

$$d_e = \frac{4A}{s} \tag{5.7.39}$$

We note that Eq. (5.7.39) replaces the diameter, d, and is called the equivalent diameter. A/s is the ratio of *cross-sectional area of flow* to the *wetted perimeter* and is called the hydraulic radius, r_h. Thus $d_e = 4r_h$. For an annulus of inner and outer diameters d_1 and d_2, the equivalent diameter is $\pi(d_2{}^2 - d_1{}^2)/\pi(d_2 + d_1) = d_2 - d_1$. For ducts of not-too-complicated cross-sectional shape, quite good estimates of the friction factor can be obtained by using R_{d_e} in the circular-pipe formula (5.7.27) and replacing r_0 by $d_e/2$ in the definition of f, Eq. (5.7.26).

5.8 FREE SHEAR LAYERS

As in the case of some external flows the TSL equations admit similarity solutions for some laminar and turbulent free shear flows. The similarity variables for different free shear flows can be found by a number of methods. In Sec. 5.2 we discussed the group-theoretic method. Here we shall use a different approach to find the similarity variables of some special free shear laminar flows, shown in the sketches of Table 5.2. These flows go turbulent at very low Reynolds numbers, and one of the few practical cases where they are important is in the operation of fluidic devices. Later in Chap. 6, we shall use the same approach for similar turbulent free shear flows. It should be noted that the shear layer thickness δ used to scale the cross-stream coordinate is chosen for convenience in each case. It is not the distance from the axis at which the velocity or velocity defect is 0.005 of the maximum, analogous to δ_{995} in boundary layers. It should also be noted that the similarity solutions become valid only at large distances from the origin, say 20 nozzle diameters in jets or 100 body diameters in wakes, for reasons stated at the end of Sec. 5.2.

5.8.1 Axisymmetric Jet

We consider an axisymmetric jet emerging from a small circular hole and mixing with the surrounding fluid at rest (Fig. 4.5c). Let the x-direction coincide with the jet axis and the origin lie in the hole. Since the streamlines are nearly parallel within the jet, although the streamlines in the entraining flow are more nearly normal to the axis, the pressure variation in the jet is small and can be neglected. The governing equations subject to TSL approximations follow from Eqs. (3.2.4) and (3.2.5) and can be written as

$$\frac{\partial u}{\partial x} + \frac{1}{r}\frac{\partial}{\partial r}(rv) = 0 \tag{5.8.1}$$

$$u\frac{\partial u}{\partial x} + v\frac{\partial u}{\partial r} = \frac{1}{r}\frac{\partial}{\partial r}(r\tau) \tag{5.8.2}$$

Here

$$= \frac{1}{\rho r}\frac{\partial}{\partial r}(r\tau)$$

$$\tau = \mu \frac{\partial u}{\partial r} - \rho \overline{u'v'} \qquad\qquad (5.8.3)$$

They are subject to the boundary conditions

$$r = 0 \quad v = 0 \quad \frac{\partial u}{\partial r} = 0 \qquad\qquad (5.8.4a)$$

$$r = \infty \quad u = 0 \qquad\qquad (5.8.4b)$$

We note that Eq. (5.8.4a) follows from symmetry conditions.

Since the pressure is constant and the motion is steady, the total momentum in the x-direction is constant, that is,

$$J = 2\pi \int_0^\infty \rho u^2 r \, dr = \text{const} \qquad\qquad (5.8.5)$$

To find the similarity solution for the system given by Eqs. (5.8.1)–(5.8.5), valid for large distances from the orifice, we start with Eq. (5.2.2) and write it as

$$\frac{u(x,r)}{u_c(x)} = F(\eta) \qquad\qquad (5.8.6)$$

where u_c is the velocity along the axis $r = 0$. Here η denotes the similarity variable defined by

$$\eta = \frac{r}{\delta(x)} \qquad\qquad (5.8.7)$$

We assume that the stream function $\psi(x, r)$ is related to a dimensionless stream function $f(\eta)$ by

$$\psi(x,r) = u_c(x)\delta^2(x)f(\eta) \qquad\qquad (5.8.8)$$

With Eqs. (5.8.6) and (5.8.7) we can write Eq. (5.8.5) as

$$J = 2\pi \rho M^2 \int_0^\infty F^2 \eta \, d\eta \qquad\qquad (5.8.9)$$

where

$$M = u_c \delta \qquad\qquad (5.8.10)$$

We note that since the total momentum J is constant, then M must be constant also, since the integral in Eq. (5.8.9) is a pure number.

By using Eqs. (5.8.6), (5.8.7), and (5.8.10) and noting that $F = f'/\eta$, we can write Eqs. (5.8.1) and (5.8.2) as

$$\rho u_c^2 \frac{d\delta}{dx}\left[\frac{(f')^2}{\eta} + f\left(\frac{f'}{\eta}\right)'\right] + (\tau\eta)' = 0 \tag{5.8.11}$$

In terms of similarity variables the boundary conditions [Eq. (5.8.4)] become

$$\eta = 0 \quad f = 0 \quad f' = 0 \tag{5.8.12a}$$
$$\eta = \infty \quad f' = 0 \tag{5.8.12b}$$

Equation (5.8.11) with τ defined by Eq. (5.8.3) applies to both laminar and turbulent axisymmetric jets. For laminar flows, using the definition of τ, we can write Eq. (5.8.11) as

$$\frac{M}{\nu}\frac{d\delta}{dx}\left[\frac{(f')^2}{\eta} + f\left(\frac{f'}{\eta}\right)'\right] + \left[\left(\frac{f'}{\eta}\right)'\eta\right]' = 0 \tag{5.8.13}$$

We see that for a circular laminar jet the system of Eqs. (5.8.1)–(5.8.5) will have a similarity solution if the coefficients of Eq. (5.8.13) are independent of x, that is, if

$$\frac{M}{\nu}\frac{d\delta}{dx} = \text{const} = c_1 \tag{5.8.14}$$

where we expect $c_1 > 0$. With this restriction and with $c_1 = 1$, Eq. (5.8.13) becomes

$$\frac{(f')^2}{\eta} + f\left(\frac{f'}{\eta}\right)' + \left[\left(\frac{f'}{\eta}\right)'\eta\right]' = 0 \tag{5.8.15a}$$

and can be written as

$$\left(\frac{ff'}{\eta}\right)' + \left[\left(\frac{f'}{\eta}\right)'\eta\right]' = 0 \tag{5.8.15b}$$

Integrating Eq. (5.8.15b), we get

$$\frac{ff'}{\eta} + f'' - \frac{f'}{\eta} = \text{const} \tag{5.8.16a}$$

To perform the integration, we note that

$$\lim_{\eta \to \infty} \frac{f'(\eta)}{\eta} = 0$$

and that as $\eta \to \infty$, $f''(\eta) \to 0$. Therefore the integration constant in Eq. (5.8.16a) is zero. Rewriting Eq. (5.8.16a) as

$$ff' + \eta f'' - f' = 0 \tag{5.8.16b}$$

we find that its solution, subject to $f(0) = 0$, is

$$f(\eta) = \frac{\frac{1}{2}\eta^2}{1 + \frac{1}{8}\eta^2} \tag{5.8.17}$$

Then

$$\frac{u}{u_c} = \frac{f'(\eta)}{\eta} = \frac{1}{(1 + \frac{1}{8}\eta^2)^2} \tag{5.8.18}$$

Figure 5.15 shows the velocity profile for a circular jet according to Eq. (5.8.18) as well as the velocity profile for a two-dimensional jet (see Prob. 5.10).

Inserting Eq. (5.8.18) into Eq. (5.8.9), we find

$$M = \frac{\sqrt{3}}{2} \sqrt{\frac{J/\rho}{2\pi}} \tag{5.8.19}$$

With $c_1 = 1$ (which defines δ) in Eq. (5.8.14) we find, after integration,

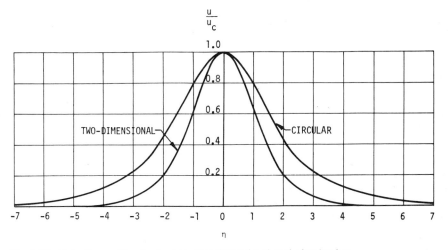

Fig. 5.15 Velocity profiles for laminar two-dimensional and circular jets.

$$\delta = \frac{vx}{M} = 2.894\left(\frac{J}{\rho}\right)^{-1/2} vx \qquad\qquad\qquad (5.8.20)$$

where u/u_c is approximately 0.79 at $r = \delta$. Therefore

$$u_c = \frac{3}{8\pi}\frac{J}{\rho}(vx)^{-1} = 0.119\frac{J}{\rho}(vx)^{-1} \qquad\qquad\qquad (5.8.21)$$

The mass flow rate, $\dot{m} = 2\pi \displaystyle\int_0^\infty \rho u r\, dr$, is

$$\dot{m} = 8\pi\mu x \qquad\qquad\qquad\qquad\qquad\qquad (5.8.22)$$

It is interesting to note that $\dot{m}$ is independent of J in this case.

EXAMPLE 5.10

A fluidic device called a "turbulence amplifier" consists of a 0.1 mm diameter jet orifice, fed by a long pipe of the same diameter, in one wall of a chamber, with a pressure tapping on the opposite wall, 1 cm away and on the jet axis. 1.5×10^{-7} kg s^{-1} of air is supplied through the pipe. Calculate the difference between the pressure in the chamber (assumed atmospheric pressure) and at the pressure tapping.

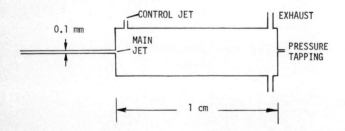

Solution Taking $\rho = 1.19$ kg m^{-3}, we find that at the orifice

$$u_{av} = \frac{1.5 \times 10^{-7}}{(\pi/4) \times 10^{-8} \times 1.19} = 16.05 \text{ m s}^{-1}$$

Now $u_{max} = 2u_{av}$ and $u = u_{max}(1 - r/r_0^2)$, where r_0 is the pipe radius. Hence $J \equiv \int \rho u^2\, dA = 2\pi\rho u_{av}^2 r_0^2 = 4.81 \times 10^{-6}$ N (Newtons). From Eq. (5.8.21) we find that the velocity on the centerline 1 cm from the effective origin of the jet is

$$\frac{0.119 \times 4.81 \times 10^{-6}}{1.19} \frac{1}{10^{-2} \times 1.5 \times 10^{-5}} = 3.21 \text{ m s}^{-1}$$

so that the dynamic pressure is $\frac{1}{2}$ X 1.19 X $(3.21)^2$ = 6.13 N m^{-2}. The net impingement region will be quite short, and so the change of total pressure on the centerline as the fluid decelerates will be negligible. Therefore the static-pressure rise will be about 0.13 N (equal to about a 0.064 cm water gauge in standard conditions).

Will the jet remain laminar? The Reynolds number, $u_{av}(2r_0)/v$, is about 100, and although this is well above the stability limit for a jet, it is probable that the flow will still be effectively laminar 1 cm from the orifice.

5.8.2 Two–Dimensional Wake

The TSL equations also admit a similarity solution for laminar and turbulent flows in the wake of a two-dimensional body. The solutions are, at best, valid only far enough downstream for the pressure disturbances introduced by the body to be negligible. In addition, the solutions are subject to the restriction that the velocity defect in the wake

$$u_1(x,y) = u_\infty - u(x,y) \tag{5.8.23}$$

is small in comparison with the velocity of the free stream u_∞, so that higher-order terms can be dropped. The governing equations for laminar or turbulent flows in the wake of a two-dimensional body, subject to Eq. (5.8.23), are

$$-\frac{\partial u_1}{\partial x} + \frac{\partial v}{\partial y} = 0 \tag{5.8.24}$$

$$-u_\infty \frac{\partial u_1}{\partial x} = \frac{1}{\rho} \frac{\partial \tau}{\partial y} \tag{5.8.25}$$

Here τ is

$$\tau = -\mu \frac{\partial u_1}{\partial y} - \rho \overline{u'v'} \tag{5.8.26}$$

The boundary conditions are

$$y = 0 \quad v = 0 \quad \frac{\partial u_1}{\partial y} = 0 \tag{5.8.27a}$$

$$y = \infty \quad u_1 = 0 \tag{5.8.27b}$$

For simplicity let us consider flow over a flat plate. By Newton's second law, the drag of the plate, F, say, is equal to the flux of momentum defect in the wake; therefore

$$F = \rho b \int_{-\infty}^{\infty} u(u_\infty - u) \, dy \tag{5.8.28}$$

where b is the width of the plate. Substituting the definition of u_1 into Eq. (5.8.28) and neglecting the second-order terms, we get

$$F = \rho b \int_{-\infty}^{\infty} u_1 u_\infty \, dy = \text{const} \tag{5.8.29}$$

To find the similarity solution for the system of Eqs. (5.8.24)–(5.8.27) and (5.8.29), we start with Eq. (5.2.3) and write

$$u = u_\infty + u_c(x) f(\eta) \tag{5.8.30}$$

Here $u_c(x)$ denotes the wake centerline velocity *defect* and f is negative. The definition of η is similar to Eq. (5.8.7),

$$\eta = \frac{y}{\delta(x)} \tag{5.8.31}$$

With the definition of u_1, Eq. (5.8.30) can be written as

$$f(\eta) = -\frac{u_1}{u_c(x)} \tag{5.8.32}$$

Using Eqs. (5.8.31) and (5.8.32), we can write Eq. (5.8.25) as

$$u_\infty \left(f \frac{du_c}{dx} - u_c f' \frac{\eta}{\delta} \frac{d\delta}{dx} \right) = \frac{1}{\rho} \frac{\partial \tau}{\partial y} \tag{5.8.33}$$

The boundary conditions become

$$\eta = 0 \quad f' = 0 \quad \eta = \infty \quad f = 0 \tag{5.8.34}$$

[We shall not need the boundary condition $v = 0$, since Eq. (5.8.25) does not contain v, which is in fact negligibly small.]

Equation (5.8.33) applies to both laminar and turbulent flows. For laminar flows, substituting for τ from Eq. (5.8.26) with the Reynolds stress neglected allows us to write Eq. (5.8.33) as

$$\frac{\delta^2 u_\infty}{\nu u_c} \frac{du_c}{dx} f - \frac{\delta u_\infty}{\nu} \frac{d\delta}{dx} \eta f' = f'' \tag{5.8.35}$$

We see that for a flow in the wake of a flat plate we will have a similarity solution if

$$\frac{\delta^2 u_\infty}{\nu u_c} \frac{du_c}{dx} = \text{const} \qquad \frac{\delta u_\infty}{\nu} \frac{d\delta}{dx} = \text{const} = c_2 \tag{5.8.36}$$

Assuming that $c_2 = 1$ (which defines δ), we can integrate the second relation in Eq. (5.8.36) to obtain

$$\delta = \left(\frac{2\nu x}{u_\infty}\right)^{1/2} \tag{5.8.37}$$

In terms of similarity variables f and η, Eq. (5.8.29) can be written as

$$u_c \delta = -\frac{F}{\rho b u_\infty \displaystyle\int_{-\infty}^{\infty} f \, d\eta} = \text{const} \tag{5.8.38}$$

With the definition of δ given by Eq. (5.8.37) we find that

$$u_c = A x^{-1/2} \tag{5.8.39}$$

where A is constant.

Using Eqs. (5.8.37) and (5.8.39) we find that the constant in the first expression of Eq. (5.8.36) is -1. This, together with $c_2 = 1$, enables us to write Eq. (5.8.35) as

$$f'' + (\eta f)' = 0 \tag{5.8.40}$$

Integrating twice and using the boundary condition $f'(0) = 0$, we get

$$f(\eta) = B e^{-\eta^2/2} \tag{5.8.41}$$

Substituting Eqs. (5.8.37), (5.8.39), and (5.8.41) into Eq. (5.8.38), we find that

$$AB = -\frac{F}{\rho b}(2\nu u_\infty)^{-1/2}(2\pi)^{-1/2} \tag{5.8.42}$$

With Eqs. (5.8.39) and (5.8.42), Eq. (5.8.30) becomes

$$u = u_\infty - \frac{F}{2\rho b}\left(\frac{1}{\pi \nu u_\infty}\right)^{1/2} x^{-1/2} e^{-\eta^2/2} \tag{5.8.43}$$

For the special case of a flat plate at zero incidence F (the drag on both sides!) is, from Eq. (5.5.8) written at the trailing edge,

$$F \equiv 2\theta_{TE} b \rho u_\infty{}^2 = 1.328 b \rho u_\infty{}^2 \sqrt{\frac{\nu l}{u_\infty}}$$

where l denotes the length of the plate. Substituting the above expression into Eq.

(5.8.43) and rearranging, we get

$$\frac{u}{u_\infty} = 1 - \frac{0.664}{\sqrt{\pi}}\left(\frac{x}{l}\right)^{-1/2} e^{-\eta^2/2} \tag{5.8.44}$$

EXAMPLE 5.11

The excitation unit of an electrostatic dust precipitator consists of a set of thin parallel wires in a plane normal to an airflow. The spacing between the wires is 1 cm, and the airspeed is 1 m s^{-1}. Calculate the distance downstream at which the wakes of adjacent wires meet.

Solution We neglect the blockage of the flow caused by the presence of the wires and consider the wake of an isolated cylinder. The blockage (displacement thickness) of the wake will be of the same order as the solid blockage of the wire and can also be neglected. Take the edge of the wake as the point where $y/\delta \equiv \eta = 3$ and $\exp(-\eta^2/2) \approx 0.01$. Then using Eq. (5.8.37) and requiring $3\delta = 0.5$ cm, we get

$$x = \left(\frac{0.5}{3 \times 10^2}\right)^2 \frac{1}{2 \times 1.5 \times 10^{-5}} = 9.26 \text{ cm}$$

Note that the wire diameter does not appear. The low speed increases the usually very low growth rate of laminar shear layers.

5.8.3 Mixing Layer between Two Uniform Streams

The boundary-layer equations also admit similarity solutions for laminar and turbulent flows in which mixing takes place downstream of a splitter plate ending at $x = 0$ between two uniform streams that move with velocities u_1 and u_2. One of the velocities may be zero; in that case it is sometimes referred to as a *half jet*.

The procedure by which we can obtain the similarity solutions for this case is very similar to the procedure for the case of the two-dimensional jet (see Prob. 5.31). The governing equations are given by Eqs. (5.8.1) and (5.8.2) with the restriction that $r = 1$ and $\partial/\partial r = \partial/\partial y$. The boundary conditions [Eq. (5.4.4b)] are replaced by new ones given by

$$y = \infty \quad u = u_1; \quad y = -\infty \quad u = u_2 \tag{5.8.45a}$$

As was pointed out in Sec. 4.3.3, the v boundary condition in free shear layers depends on conditions outside the shear layer. The choice of $v = 0$ on the centerline of symmetrical shear layers is effectively a definition of the position of the centerline in xy-space. In the asymmetrical mixing layer we define the x-axis, $y = 0$, as the line on which $v = 0$: it is in general inclined at an angle of less than $5°$

to the plane of the splitter plate. Thus

$$y = \infty \quad u = u_1; \quad y = -\infty \quad u = u_2 \tag{5.8.45a}$$
$$y = 0 \quad v = 0 \tag{5.8.45b}$$

To find the similarity solution, we define

$$f'(\eta) = \frac{u(x,y)}{u_1} \tag{5.8.46}$$

$$\psi(x,y) = u_1 \delta(x) f(\eta) \tag{5.8.47}$$

with η defined by Eq. (5.8.31). By using Eqs. (5.8.31), (5.8.46), and (5.8.47), we can write the two-dimensional TSL equations (subject to the restrictions discussed above) as

$$-\frac{u_1^2}{\delta} \frac{d\delta}{dx} ff'' = \frac{1}{\rho} \frac{\partial \tau}{\partial y} \tag{5.8.48}$$

In terms of similarity variables, Eq. (5.8.45) becomes

$$\eta = \infty \quad f' = 1; \quad \eta = -\infty \quad f' = \frac{u_2}{u_1} \equiv \lambda; \quad \eta = 0 \quad f = 0 \tag{5.8.49}$$

Equation (5.8.48) with τ defined by Eq. (5.8.3) ($\partial/\partial r = \partial/\partial y$) applies to both laminar and turbulent flows: for laminar flows, Eq. (5.8.48) becomes

$$f''' + \frac{u_1 \delta}{\nu} \frac{d\delta}{dx} ff'' = 0 \tag{5.8.50}$$

For similarity we must have

$$\frac{u_1 \delta}{\nu} \frac{d\delta}{dx} = \text{const} \quad = \frac{1}{2} \tag{5.8.51}$$

Taking the constant equal to $\frac{1}{2}$, we see that

$$\delta = \left(\frac{\nu x}{u_1}\right)^{1/2} \tag{5.8.52}$$

and

$$f''' + \frac{1}{2} ff'' = 0 \tag{5.8.53}$$

No closed-form solutions are known for the system given by Eqs. (5.8.49) and (5.8.53). Several solutions obtained by numerical methods were provided by Lock (1954). The solution of Eq. (5.8.53) by using Keller's shooting method (Sec. 4.2) can be obtained as follows. We again write Eq. (5.8.53) as a first-order system in a form similar to Eqs. (4.2.8)-(4.2.10) with $m = 0$. The boundary conditions [Eq. (5.8.49)] become

$$f(0) = 0 \qquad u(\eta_\infty) = 1 \qquad u(-\eta_\infty) = \lambda \tag{5.8.54}$$

The modified shooting method solves the initial-value problem for Eqs. (4.2.8)-(4.2.10) with $m = 0$, subject to the initial conditions

$$f(0) = 0 \qquad u(0) = a \qquad v(0) = b \tag{5.8.55}$$

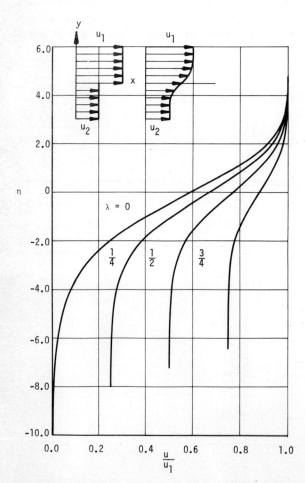

Fig. 5.16 Velocity profiles for the mixing of two uniform laminar streams at different velocities, $\lambda = u_2/u_1$.

Then a and b are sought to satisfy the two-edge boundary conditions

$$u(\eta_\infty, a, b) - 1 = 0 \qquad u(-\eta_\infty, a, b) - \lambda = 0 \qquad\qquad (5.8.56)$$

These equations are solved by Newton's method described in Sec. 4.2. Note that in this case two systems of variational equations must be solved to get the partial derivatives with respect to a and b.

Figure 5.16 shows solutions obtained by this method for $\lambda = 0, \frac{1}{4}, \frac{1}{2}$, and $\frac{3}{4}$.

5.8.4   Power Laws for Width and for Centerline Velocity of Similar Free Shear Layers

In the previous sections we have discussed the similarity solutions of several free shear layers for laminar flows. On the basis of similarity considerations we obtained the variations of the width (δ) and the centerline velocity (u_c or u_1) of several shear layers. Table 5.2 summarizes the results.

Table 5.2   Power Laws for Width and for Centerline Velocity of Laminar Similar Free Shear Layers

Flow	Sketch	Width, δ	Centerline velocity, $u_c(x)$ or u_1
Two-dimensional jet[a]		$x^{2/3}$	$x^{-1/3}$
Axisymmetric jet		x	x^{-1}
Two-dimensional wake		$x^{1/2}$	$x^{-1/2}$
Axisymmetric wake[a]		$x^{1/2}$	x^{-1}
Two uniform streams		$x^{1/2}$	x^0

[a]See Probs. 5.31 and 5.32.

PROBLEMS

5.1. For a two-dimensional laminar wake, the governing TSL equations and their boundary conditions subject to the restriction that the velocity defect in the wake u_1 is small in comparison with the velocity of the free stream u_∞ are given by Eqs. (5.8.50)–(5.8.53). Using these equations together with the equation for the drag of a flat plate (see p. 135), show by the group-theoretic method that the governing equations and their boundary conditions can be written as

$$\nu f'' + \tfrac{1}{2} u_\infty (f + \eta f') = 0$$
$$\eta = 0 \quad f' = 0; \quad \eta = \eta_\infty \quad f = 0$$

Here

$$f(\eta) = \frac{u_1}{x^{-1/2}} \quad \eta = \frac{y}{\sqrt{x}}$$

and a prime denotes differentiation with respect to η.

5.2. By using the transformation

$$\eta = \frac{y}{\sqrt{\nu t}} \quad u = f(\eta, t)$$

express the nonsteady momentum equation

$$\frac{\partial u}{\partial t} = \nu \left(\frac{\partial^2 u}{\partial y^2} \right)$$

in transformed coordinates.

5.3. A useful transformation often used in boundary-layer problems is von Mises' transformation, which uses new independent variables (x, ψ), where ψ is the stream function. Show that with this transformation the momentum equation

$$u \frac{\partial u}{\partial x} + v \frac{\partial u}{\partial y} = u_e \frac{du_e}{dx} + \nu \frac{\partial^2 u}{\partial y^2}$$

subject to the boundary conditions

$$y = 0 \quad u = v = 0$$
$$y \to \infty \quad u \to u_e$$

can be written as

$$u \frac{\partial u}{\partial x} - u_e \frac{du_e}{dx} = \nu u \frac{\partial}{\partial \psi} \left(u \frac{\partial u}{\partial \psi} \right) \tag{P5.3.1}$$

subject to the new boundary conditions

$$\psi = 0 \quad u = 0$$
$$\psi \to \infty \quad u \to u_e$$

(P5.3.2)

Note that when u is determined from Eqs. (P5.3.1) and (P5.3.2), then v follows from the continuity equation.

5.4. Show that the relation between y and η for an axisymmetric flow with significant transverse curvature and local surface angle ϕ is

$$y = \frac{r_0}{\cos \phi} \left[-1 + \sqrt{1 + \frac{2 \cos \phi}{L} \left(\frac{L}{r_0}\right)^2 \sqrt{\frac{v \bar{x}}{u_e}} \, \eta} \right]$$

5.5. For a two-dimensional similar laminar flow, what is the value of m for which the wall shear τ_w is proportional to x?

5.6. For a two-dimensional stagnation-point laminar flow with mass transfer, find the variation of wall mass transfer velocity v_w with x required for similarity.

5.7. Derive Eq. (5.5.2).

5.8. Show that for similar flows by differentiating

$$\delta^* = \delta_1^* \sqrt{\frac{v x}{u_e}}$$

we get

$$dR_{\delta*} = \frac{(\delta_1^*)^2}{\delta^*} \frac{m + 1}{2} \, dx$$

Here $R_{\delta*} = u_e \delta^* / v$.

5.9. Water at $20°C$ flows at a velocity of 3 m s^{-1} past a flat plate. Plot the velocity profiles, u/u_e versus y, at stations 0.5, 1, and 2 m from the leading edge. Also plot the variation of the local skin-friction coefficient over the first 2 m of the plate and determine the average skin-friction drag.

5.10. Air at $25°C$ and 1 atm pressure flows normal to a 1 cm diameter circular cylinder at a velocity of 10 m s^{-1}. According to the inviscid-flow theory the external velocity in the vicinity of the stagnation point, which can be obtained by expanding Eq. (E5.5.1) in Ex. 5.5 by Taylor series and approximating it close to the stagnation point, is

$$u_e = \frac{2 u_\infty x}{r_0}$$

Calculate the displacement thickness of the boundary layer at the stagnation point and discuss the significance of the results.

5.11. Using the definition of δ_1^* and Eq. (5.5.2), show that for the sink flow

$$\delta_1^* = 3\sqrt{2} - 2\sqrt{3}$$
$$\theta_1 = \frac{8}{\sqrt{3}} - 3\sqrt{2}$$

5.12. a. Assume that the velocity profile for a laminar flow in a flat plate can be approximated by a second-order polynomial given by

$$u = a + by + cy^2$$

Use the momentum integral equation to obtain expressions for the dependence of local boundary-layer thickness and skin-friction coefficient on the local Reynolds number. Compare the results with exact solutions.

b. Repeat (a) for the velocity profile approximated by

$$u = b \sin ay$$

5.13. Repeat Prob. 5.12 for a velocity profile approximated by

$$\frac{u}{u_e} = a + b\frac{y}{\delta} + c\left(\frac{y}{\delta}\right)^2 + d\left(\frac{y}{\delta}\right)^3$$

and find the coefficients for a constant-pressure boundary layer. Show that the value of H is 2.69 compared with the "exact" value of 2.60. Sketch a typical profile in an *adverse* pressure gradient. What is the sign of c?

5.14. a. Using the linear velocity profile discussed in Example 5.4 and the continuity equation (5.1.1), show that for the general case of suction and blowing the normal velocity component v can be written as

$$v = v_w + \left(u_e\frac{d\delta}{dx} - \delta\frac{du_e}{dx}\right)\frac{1}{2}\left(\frac{y}{\delta}\right)^2 \tag{P5.14.1}$$

b. Using Eq. (P5.14.1), the linear velocity profile and the momentum equation (5.1.2) subject to the boundary condition $\partial u/\partial y = 0$ at $y = \delta$, show that for the general case of suction and blowing

$$vu = \left(\delta^2 u_e\frac{du_e}{dx} - u_e^2\delta\frac{d\delta}{dx}\right)\left[\frac{1}{24}\left(\frac{y}{\delta}\right)^4 - \frac{1}{6}\left(\frac{y}{\delta}\right)\right] + v_w u_e \delta\left[\frac{1}{2}\left(\frac{y}{\delta}\right)^2 - \frac{y}{\delta}\right]$$

$$- \delta^2 u_e\frac{du_e}{dx}\left[\frac{1}{2}\left(\frac{y}{\delta}\right)^2 - \left(\frac{y}{\delta}\right)\right] \tag{P5.14.2}$$

c. Using the boundary condition that $u = u_e$ at $y = \delta$, show that Eq. (P5.14.2) can be written

$$u_e^2\delta\frac{d\delta}{dx} = -3\delta^2 u_e\frac{du_e}{dx} + 8vu_e + 4v_w u_e\delta \tag{P5.14.3}$$

d. Substituting Eq. (P5.14.3) into Eq. (P5.14.2), show that we get an iterated expression for the x-component of the velocity profile,

$$\frac{u}{u_e} = \frac{1}{3}\left[4\left(\frac{y}{\delta}\right) - \left(\frac{y}{\delta}\right)^4\right] + \left(\frac{\delta^2}{\nu}\frac{du_e}{dx} - \frac{v_w\delta}{\nu}\right)\left[\frac{1}{3}\frac{y}{\delta} - \frac{1}{2}\left(\frac{y}{\delta}\right)^2 + \frac{1}{6}\left(\frac{y}{\delta}\right)^4\right]$$

(P5.14.4)

which is like that of Pohlhausen for $v_w = 0$.

e. Show that after substituting Eq. (P5.14.3) into Eq. (P5.14.1), the variation of normal velocity v across the boundary layer is given by

$$\frac{v\delta}{\nu} = 4\left(\frac{y}{\delta}\right)^2 - 2\frac{\delta^2}{\nu}\frac{du_e}{dx}\left(\frac{y}{\delta}\right)^2 + \frac{v_w\delta}{\nu}\left[1 + 2\left(\frac{y}{\delta}\right)^2\right]$$

(P5.14.5)

5.15. Using the definition of momentum thickness and the results given in Table 5.1, show that for a two-dimensional stagnation-point flow a more accurate expression for Eq. (5.6.19) is

$$\left(\frac{\theta}{L}\right)^2 R_L = \frac{0.085}{(du_e^*/dx)_0}$$

5.16. The transformed boundary-layer equation (5.3.9) is a PDE. With some approximations it can be reduced to a system of ODEs. If we let

$$g = \frac{\partial f}{\partial x} \quad \text{and} \quad g' = \frac{\partial f'}{\partial x}$$

and consider the zero mass transfer case, then Eqs. (5.3.9) and (5.3.10) can be written as

$$f''' + \frac{m+1}{2}ff'' + m[1 - (f')^2] = x(f'g' - f''g)$$

(P5.16.1)

$$\eta = 0 \quad f = f' = 0; \quad \eta = \eta_\infty \quad f' = 1$$

(P5.16.2)

Show that if we differentiate (P5.16.1) and (P5.16.2) with respect to x, and neglect the terms $\partial/\partial x\ (f'f' - f''g)$, then Eqs. (P5.16.1) and (P5.16.2) can be written as

$$g''' + n[\tfrac{1}{2}f'' + 1 - (f')^2] + \frac{m+1}{2}(fg'' + gf'') - (2m+1)fg' + f''g = 0$$

(P5.16.3)

$$\eta = 0 \quad g = g' = 0; \quad \eta = \eta_\infty \quad g' = 0$$

(P5.16.4)

Here

$$n = \frac{dm}{dx}$$

(P5.16.5)

The system of equations (P5.16.1)–(P5.16.5), sometimes referred to as the *local nonsimilarity equations*, was first proposed by Sparrow et al. (1970). Although their solutions are not as accurate as Eq. (5.3.9), they are quite useful, since they avoid the solution of a PDE and, for flows away from separation, give satisfactory results.

5.17. Using the shooting method discussed in Sec. 4.2, obtain the solutions of the system of equations given in Prob. 5.16 for an external-velocity distribution

$$u_e = u_{ref}[1 - \tfrac{1}{8}x]$$

in the region $0 \leqslant x \leqslant 0.3$.

5.18. Show that $(\delta^*/\tau_w) \, dp/dx$ represents the ratio of net pressure force to wall shear force acting on a two-dimensional boundary layer. Is it uniquely related to the pressure-gradient parameter $(\theta^2/v) \, du_e/dx$ used in Thwaites' method?

5.19. Calculate the nondimensional displacement thickness, momentum thickness, and skin-friction coefficient on a semi-infinite circular cone whose half angle is $120°$. Note that this flow is "inside" the cone and away from the apex. The external velocity varies as

$$u_e \sim s^{2.2}$$

5.20. The coordinates of the symmetrical NACA 0012 airfoil and its external-velocity distribution (in potential flow) for zero angle of attack are given below. Compute the laminar boundary-layer development for a chord Reynolds number of $R_c \equiv u_\infty c/v = 6 \times 10^6$ in the region $0 \leqslant x/c \leqslant 0.20$. Find

a. Displacement thickness distribution.
b. Local skin-friction distribution.

$x, \% \, c$	$y, \% \, c$	u_e/u_∞	$x, \% \, c$	$y, \% \, c$	u_e/u_∞
0	0	0	25	5.941	1.174
0.5	—	0.800	30	6.002	1.162
1.25	1.894	1.005	40	5.803	1.135
2.5	2.615	1.114	50	5.294	1.108
5.0	3.555	1.174	60	4.563	1.080
7.5	4.200	1.184	70	3.664	1.053
10	4.683	1.188	80	2.623	1.022
15	5.345	1.188	90	1.448	0.978
20	5.737	1.183	95	0.807	0.952
			100	0.126	0.915

5.21. Consider a semi-infinite ogival body of revolution as shown in the sketch. The equation of the nose is ($t \equiv b/a$):

$$\left(\frac{x}{a} - 1\right)^2 + \left(\frac{r}{a} + \frac{1 - t^2}{2t}\right)^2 = \left(\frac{1 + t^2}{2t}\right)^2 \qquad 0 \leqslant \frac{x}{a} \leqslant 1$$

The inviscid-velocity distribution for this region is given by the following tabular values. The velocity distribution in the region very close to the nose can be approximated by

$$\frac{u_e}{u_\infty} = 1.1135\left(\frac{s}{a}\right)^{1/3} \quad \text{for} \quad 0 \leqslant \frac{s}{a} \leqslant 0.05$$

where s is the surface distance.

x/a	u_e/u_∞
0	0
0.05	0.485
0.1	0.623
0.2	0.802
0.3	0.930
0.4	1.025
0.5	1.090
0.6	1.140
0.7	1.168
0.8	1.183
0.9	1.177
1.0	1.130

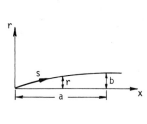

For $u_\infty = 50$ m s^{-1}, $v = 1.5 \times 10^{-5}$ m^2 s^{-1}, $a = 1$ m, and $b = 0.2$, calculate the displacement thickness, momentum thickness, and the local skin-friction coefficient at $x/a = 0.75$.

5.22. For plane Poiseuille flow between stationary parallel plates the velocity profile is given by Eq. (5.7.32). Show that when the plates are located at $\bar{y} = y/h = 1$ and $\bar{y} = -1$, so that the maximum dimensionless velocity $\bar{u}(\equiv u/u_0)$ at $\bar{y} = 0$ is unity, Eq. (5.7.32) can be written as

$$\bar{u} = 1 - (\bar{y})^2$$

5.23. Derive Eq. (5.7.11b).

5.24. Develop Eq. (5.7.15) by using a force balance on an element of fluid.

5.25. Show that the velocity distribution for fully developed laminar flow in a horizontal annulus of outer radius r_o and inner radius r_i is given by:

$$u = -\frac{1}{4\mu}\frac{dp}{dx}\left[\left(r_o^2 - r^2\right) + \frac{r_o^2 - r_i^2}{\ln(r_o/r_i)}\ln\frac{r_o}{r}\right]$$

Here x is the flow direction, and r is any radius.

5.26. Consider a laminar flow through a circular pipe with internal diameter d equal to 2 cm, length equal to 3 m, and $R_d = 10^3$. Compute the dimensionless pressure drop Δp^* with and without consideration of the entrance region.

5.27. Derive Eq. (5.7.35).

5.28. For water at $15°C$ and for air at the same temperature and atmospheric pressure flowing through a given smooth pipe, what must be the ratio of volume rates of flow in order for the friction factor as given by Eq. (5.7.27) to be the same?

5.29. Assuming fully developed flow, calculate the pressure drop for the following cases:

    a. 0.005 kg s^{-1} of air at 25°C and at atmospheric pressure flowing in a smooth 3 m long pipe with a diameter of 10 cm.

    b. 5 × 10^3 cm^3 s^{-1} of air at 30°C and at atmospheric pressure flowing in a smooth-walled rectangular duct 30 cm × 10 cm × 200 cm long.

5.30. Find the inclination to the horizontal of a long 1 cm diameter drain pipe required to run full of water at $R_d = 2000$.

5.31. The governing TSL equations for a two-dimensional jet coming out from a slot mixing with the surrounding fluid at rest are

$$\frac{\partial u}{\partial x} + \frac{\partial v}{\partial y} = 0 \tag{P5.31.1}$$

$$u\frac{\partial u}{\partial x} + v\frac{\partial u}{\partial y} = \frac{1}{\rho}\frac{\partial \tau}{\partial y} \tag{P5.31.2}$$

They are subject to the boundary conditions

$$y = 0 \quad v = 0 \quad \frac{\partial u}{\partial y} = 0; \quad y = \infty \quad u = 0 \tag{P5.31.3}$$

Here τ is given by Eq. (5.8.3) with $\partial/\partial r = \partial/\partial y$. As in the circular jet, the total momentum J in the x-direction is constant:

$$J = \rho \int_{-\infty}^{\infty} u^2 \, dy = \text{const} \tag{P5.31.4}$$

To find the similarity solution for the system of Eqs. (P5.31.3)–(P5.31.4), assume

$$\frac{u(x,y)}{u_c(x)} = f'(\eta)$$

$$\psi(x,y) = u_c(x)\delta(x)f(\eta)$$

and carry out a procedure similar to the laminar circular jet case. Show that for similarity we must have

$$\frac{\delta^2}{\nu}\frac{du_c}{dx} = \text{const} = c_1$$

Take $c_1 = -1$ and show that

$$u_c = M^{2/3}(3\nu x)^{-1/3}$$

$$f(\eta) = \sqrt{2} \tanh\left(\frac{\eta}{\sqrt{2}}\right)$$

$$f'(\eta) = \operatorname{sech}^2\left(\frac{\eta}{\sqrt{2}}\right)$$

$$u_c = \left(\frac{3}{32}\right)^{1/3}\left(\frac{J}{\rho}\right)^{2/3}(\nu x)^{-1/3} = 0.454\left(\frac{J}{\rho}\right)^{2/3}(\nu x)^{-1/3}$$

$$\delta = \left(\frac{3}{4\sqrt{2}}\right)^{-1/3}\left(\frac{J}{\rho}\right)^{-1/3}(3\nu x)^{2/3} = 2.57\left(\frac{J}{\rho}\right)^{-1/3}(\nu x)^{2/3}$$

$$\dot{m} = \rho\frac{3}{2}\left(\frac{32}{3}\right)^{1/3}\left(\frac{J}{\rho}\right)^{1/3}(\nu x)^{1/3} = 3.302\rho\left(\frac{J}{\rho}\right)^{1/3}(\nu x)^{1/3}$$

5.32. With the restriction that the velocity defect in the wake is small, the momentum equation for an axisymmetric wake can be written as

$$u_\infty\frac{\partial u_1}{\partial x} = \frac{1}{r\rho}\frac{\partial}{\partial r}(r\tau) \tag{P5.32.1}$$

Here τ is

$$\tau = \mu\frac{\partial u_1}{\partial r} - \rho\overline{u'v'} \tag{P5.32.2}$$

The boundary conditions are

$$y = 0 \quad \frac{\partial u_1}{\partial y} = 0; \quad y = \infty \quad u_1 = 0 \tag{P5.32.3}$$

The drag of the body subject to Eq. (5.8.23) is

$$F = 2\pi\rho u_\infty\int_0^\infty u_1 r\, dr = \text{const} \tag{P5.32.4}$$

To find the similarity solution for the system of Eqs. (P5.32.1)–(P5.32.4), carry out a procedure similar to the two-dimensional laminar wake. For example, use Eqs. (5.8.30) and (5.8.32) and the definition of η given by Eq. (5.8.7) and express Eq. (P5.32.1) in terms of similarity variables f (negative again) and η. Show that for similarity we must have

$$\frac{u_\infty}{u_c}\frac{du_c}{dx}\frac{\delta^2}{\nu} = \text{const} \qquad \frac{u_\infty\delta}{\nu}\frac{d\delta}{dx} = \text{const} = c_1 \tag{P5.32.5}$$

Assume that $c_1 = 1$ and integrate the second relation in Eq. (P5.32.5) and show that

$$u_c = Ax^{-1} \tag{P5.32.6}$$

where A is constant. Show that the constant in the first expression in Eq. (P5.32.5) is -2. This together with $c_1 = 1$ allows the momentum equation expressed in similarity variables to be written as

$$(\eta f')' + (\eta^2 f)' = 0 \tag{P5.32.7}$$

so that its solution is the same as the velocity profile for a two-dimensional wake, and

$$u = u_\infty - \frac{F}{\rho} \frac{e^{-\eta^2/2}}{2\pi \nu x} \tag{P5.32.8}$$

5.33. For laminar mixing of two streams of nearly equal velocity ($u_1 - u_2 \ll u_1$), linearize the momentum equation. Discuss the solution, the velocity profile, the relation of the free shear layer to the external flow, and the shear stress on the dividing streamline. Compare it with Lock's solutions for the cases $u_2/u_1 = 0$ and $u_2/u_1 = 0.501$.

5.34. Find $\tau_w/(\frac{1}{2}\rho u_{ref}^2) \sqrt{R_L}$ in Example 5.5 by Görtler series, and compare the result obtained by Thwaites' method. Take $j = 1$ and 2.

APPENDIX 5A: Görtler Series Method

Series methods were developed for solving nonsimilar problems on desk calculators. They represent an interesting historical milestone in the development of boundary-layer theory. Their advantage over other methods is that many of the coefficients can be worked out once and for all and looked up by the desk calculator operator. Because electronic computers are not very efficient at looking up tables, series methods have been superseded for most purposes by the methods described elsewhere in this book but may have some uses.

Of the several series methods for two-dimensional flows, we shall consider the one due to Görtler (1957). Unlike the Howarth-Blasius series method (Schlichting, 1968), which applies only to flows that start as a stagnation point flow ($m = 1$ at $x = 0$), Görtler series method applies to flows that start either as a flat-plate flow ($m = 0$ at $x = 0$) or as a stagnation-point flow. The method can be described briefly as follows.

In terms of the dimensionless variables

$$x^* = \frac{x}{L} \qquad y^* = \frac{y}{L}\sqrt{R_L} \qquad u^* = \frac{u}{u_{ref}} \qquad v^* = \frac{v}{u_{ref}}\sqrt{R_L} \qquad u_e^* = \frac{u_e}{u_{ref}}$$

$$R_L = \frac{u_{ref}L}{\nu} \tag{5A.1}$$

Eq. (5.1.2) can be written as

$$u^* \frac{\partial u^*}{\partial x^*} + v^* \frac{\partial u^*}{\partial y^*} = u_e^* \frac{du_e^*}{dx^*} + \frac{\partial^2 u^*}{\partial y^{*2}} \tag{5A.2}$$

We now introduce the transformation

$$\xi = \int_0^{x^*} u_e^* \, dx^* \qquad \eta = \frac{u_e^*}{(2\xi)^{1/2}} y^* \tag{5A.3}$$

together with a dimensionless stream function, $F(\xi, \eta)$, defined by

$$\psi^*(x^*, y^*) = (2\xi)^{1/2} F(\xi, \eta) \tag{5A.4}$$

into Eq. (5A.2) to get

$$F''' + FF'' + \beta[1 - (F')^2] = 2\xi \left(F' \frac{\partial F'}{\partial \xi} - F'' \frac{\partial F}{\partial \xi} \right) \tag{5A.5}$$

Here the primes denote differentiation with respect to η and

$$\beta = \frac{2\xi}{u_e^*} \frac{du_e^*}{d\xi} \qquad F' = \frac{u^*}{u_e^*} \tag{5A.6}$$

For the case of no mass transfer the boundary conditions in Eq. (5.1.3) become

$$\eta = 0 \quad F = F' = 0; \quad \eta = \eta_\infty \quad F' = 1 \tag{5A.7}$$

The solution of Eqs. (5A.5) and (5A.7) is obtained by expanding the stream function, $F(\eta, \xi)$, by power series in ξ:

$$F(\xi, \eta) = \sum_{k=0}^{\infty} F_k(\eta) \xi^k \tag{5A.8a}$$

and writing $\psi^*(x^*, y^*)$ as

$$\psi^*(x^*, y^*) = (2\xi)^{1/2} \sum_{k=0}^{\infty} F_k(\eta) \xi^k \tag{5A.8b}$$

The dimensionless stream function, $F_k(\eta)$, is related to the universal functions f_k for values of k up to 5, starting from $k = 1$ by the following relations:

$$
\begin{aligned}
F_1 &= \beta_1 f_1 \\
F_2 &= \beta_1{}^2 f_{11} + \beta_2 f_2 \\
F_3 &= \beta_1{}^3 f_{111} + \beta_1 \beta_2 f_{12} + \beta_3 f_3 \\
F_4 &= \beta_1{}^4 f_{1111} + \beta_1{}^2 \beta_2 f_{112} + \beta_1 \beta_3 f_{13} + \beta_3{}^2 f_{22} + \beta_4 f_4 \\
F_5 &= \beta_1{}^5 f_{11111} + \beta_1{}^3 \beta_2 f_{1112} + \beta_1{}^2 \beta_3 f_{113} + \beta_1 \beta_2{}^2 f_{122} + \beta_1 \beta_4 f_{14} + \beta_2 \beta_3 f_{23} + \beta_5 f_5
\end{aligned}
\tag{5A.9}
$$

The relations [Eq. (5A.9)] apply to the derivatives of F_k and f_k also, for example, $F_1' = \beta_1 f_1'$, $F_1'' = \beta_1 f_1''$, etc., and some of their derivatives. Table 5A.1 presents the

Table 5A.1 Universal Wall Shear Functions for the Görtler Series for $\beta = 0, 1$

$\beta(0)$	F_0''	f_1''	f_{11}''	f_2''	f_{111}''	f_{12}''	f_3''	f_{1111}''	f_{112}''	f_{13}''
0	0.469600	1.032361	−0.714746	0.908119	1.103512	−1.191046	0.829995	−2.313327	2.775762	−1.047926
1	1.232587	0.493840	−0.077205	0.464540	0.022415	−0.136636	0.442383	−0.008272	0.058722	−0.124239

$\beta(0)$	f_{22}''	f_4''	f_{11111}''	f_{1112}''	f_{113}''	f_{122}''	f_{14}''	f_{23}''	f_5''
0	−0.505493	0.774210	5.600941	−7.803477	2.450686	2.355199	−0.949480	−0.899401	0.731424
1	−0.061283	0.0424639	0.003560	−0.029057	0.052941	0.051995	−0.114928	−0.112469	0.409895

universal wall shear functions for $\beta(0) = 0$ and 1. Other tables reported by Görtler (1957) present the universal functions in Eq. (5A.9) and their derivatives.

The coefficients β_j ($j = 1$-5) are obtained either by expanding the pressure-gradient parameter β as

$$\beta(\xi) = \sum_j^5 \beta_j \xi^j \tag{5A.10}$$

or by expanding the given velocity distribution u_e^* as

$$u_e^*(x^*) = \sum_k^5 u_k(x^*)^k \tag{5A.11}$$

with k starting from 0 for $\beta(0) = 0$ and k starting from 1 for $\beta(0) = 1$, and determining the coefficients β_j from the relations given below:

$\beta(0) = 0$:

$$\beta_1 = 2u_1 \qquad \beta_2 = 4(-u_1^2 + u_2) \qquad \beta_3 = 2(4u_1^3 - 7u_1 u_2 + 3u_3)$$
$$\beta_4 = \tfrac{1}{3}(-48u_1^4 + 118u_1^2 u_2 - 66u_1 u_3 - 28u_2^2 + 24u_4) \tag{5A.12}$$
$$\beta_5 = \tfrac{1}{6}(192u_1^5 - 605u_1^3 u_2 + 375u_1^2 u_3 + 320u_1 u_2^2 - 192u_1 u_4$$
$$- 150u_2 u_3 + 60u_5)$$

$\beta(0) = 1$:

$$\beta_1 = 3u_3 \qquad \beta_2 = -13u_3^2 + \tfrac{40}{3}u_5 \qquad \beta_3 = 54u_3^3 - 96u_3 u_5 + 42u_7$$
$$\beta_4 = -221u_3^4 + \tfrac{1648}{3}u_3^2 u_5 - 312u_3 u_7 - \tfrac{1184}{9}u_5^2 + \tfrac{576}{5}u_9 \tag{5A.13}$$
$$\beta_5 = 898u_3^5 - \tfrac{8560}{3}u_3^3 u_5 + 1790u_3^2 u_7 + 1520u_3 u_5^2 - 928u_3 u_9$$
$$- 720u_5 u_7 + \tfrac{880}{3}u_{11}$$

In Eq. (5A.11), u_0 always must start as unity. It is apparent that once $\psi^*(x^*, y^*)$ is known, then the velocity profiles as well as the boundary-layer parameters such as θ, δ^*, and c_f can be computed by the Görtler series method.

EXAMPLE 5A.1

For a two-dimensional incompressible laminar flow with external-velocity distribution given by

$$u_e^*(x) = 1 - \tfrac{1}{8}x^* \tag{E5A.1.1}$$

find c_f by using the Görtler series method at $x^* = 0.8$.

Solution From Eq. (5A.3),

$$\xi = x^* - \frac{(x^*)^2}{16} \tag{E5A.1.2}$$

Thus at $x^* = 0.8$, $\xi = 0.76$. From Eqs. (5A.11) and (E5A.1.1) we see that $u_1 = -\frac{1}{8}$, $u_k = 0$, $k = 2, \ldots, 5$. Therefore from Eq. (5A.12),

$$\beta_1 = -\frac{1}{4} \qquad \beta_2 = -\frac{1}{16} \qquad \beta_3 = -\frac{1}{64} \qquad \beta_4 = -\frac{1}{256} \qquad \beta_5 = -\frac{1}{1024} \qquad \text{(E5A.1.3)}$$

The local skin friction in dimensionless physical coordinates is

$$c_f = \frac{\mu(\partial u/\partial y)_w}{1/2\rho u_e^2} = \frac{2}{\sqrt{R_L}\,(u_e^*)^2}\left(\frac{\partial u^*}{\partial y^*}\right)_w = \frac{2}{\sqrt{R_L}\,(u_e^*)^2}\left(\frac{\partial^2 \psi^*}{\partial y^{*2}}\right)_w$$

Thus

$$c_f\sqrt{R_L} = \frac{2}{(u_e^*)^2}\left(\frac{\partial^2 \psi^*}{\partial y^{*2}}\right)_w = \left(\frac{2}{\xi}\right)^{1/2} F_w'' \qquad \text{(E5A.1.4)}$$

Using the relations in Eq. (5A.9) for the second derivatives of F_k and the given values of the universal functions $f_1''(0)$, $f_2''(0)$, etc., in Table 5A.1, we find the wall values of F_k'' in Eq. (5A.8a) to be

$$F_0''(0) = 0.469600 \qquad F_1''(0) = -0.25809025 \qquad F_2''(0) = -0.10142905$$
$$F_3''(0) = -0.04882114 \qquad F_4''(0) = -0.02897155 \qquad \text{(E5A.1.5)}$$
$$F_5''(0) = -0.01958904$$

With $\xi = 0.76$ we find from Eq. (E5A.1.4) that $c_f\sqrt{R_L} = 0.305735$.

Chapter 6

Analysis of Turbulent Shear Layers

6.1 INTRODUCTION

With modern computers the calculation of laminar shear layers obeying the TSL equations is routine: general methods were discussed in Chap. 5, and some particular examples of PDE solution methods are given in Chap. 7. Problems in which a laminar shear layer interacts with the external flow (Chap. 11) present more difficulties, but again the problems are purely numerical because the stresses are known to be simply related to the rate of strain. Transition from laminar to turbulent flow is not usually calculated in detail, although some promising attempts have been made and will be discussed in Chap. 9. For the relatively simple case of two-dimensional constant-property boundary layers the transition position can be satisfactorily calculated by using several empirical correlations. We briefly present some of these correlations here and use them later in the chapter when we consider the complete development of boundary layers at high Reynolds number where the final results are not too severely affected by assumptions about transition. One such useful expression is based on a combination of Michel's method (1951) and Smith and Gamberoni's e^9 correlation curve (1956). It is given by Cebeci and Smith (1974) as a connection between R_θ ($\equiv u_e\theta/\nu$) and Re_x ($\equiv u_e x/\nu$) at transition (see Fig. 6.1)

$$R_{\theta\,\mathrm{tr}} = 1.174\left(1 + \frac{22{,}400}{\mathrm{Re}_{x\,\mathrm{tr}}}\right)\mathrm{Re}_{x\,\mathrm{tr}}^{0.46} \tag{6.1.1}$$

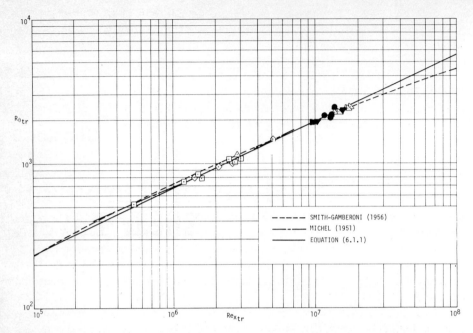

Fig. 6.1 Empirical transition correlation curves for two-dimensional incompressible flows. The symbols denote various experimental data taken from Smith and Gamberoni (1956).

EXAMPLE 6.1

Determine the transition Reynolds number, $\text{Re}_{x_{tr}}$, on a flat plate by using Eq. (6.1.1).

Solution According to Eq. (6.1.1) we need to know the development of R_θ as a function of Re_x. The values of R_θ and Re_x that satisfy Eq. (6.1.1) are the transitional values. In our problem the functional relationship between R_θ and Re_x is given by Eq. (5.5.8), which can also be written as

$$R_\theta = 0.664\sqrt{\text{Re}_x} \tag{E6.1.1}$$

One obvious way to find $\text{Re}_{x_{tr}}$ is to set Eq. (E6.1.1) equal to Eq. (6.1.1) and solve for $\text{Re}_{x_{tr}}$, that is,

$$0.664\sqrt{\text{Re}_{x_{tr}}} = 1.174\left(1 + \frac{22{,}400}{\text{Re}_{x_{tr}}}\right)\text{Re}_x^{0.46} \tag{E6.1.2}$$

The other way is to use a graphical procedure in which we plot R_θ as a function of Re_x and find the point where that curve intersects the curve represented by Eq. (6.1.1) (see Fig. 6.2). According to Fig. 6.2, $\text{Re}_{x_{tr}} = 3 \times 10^6$. Note that before transition the (R_θ, Re_x) values are under the curve. This is also true for (R_θ, Re_x) values obtained in flows with pressure gradient.

Note also that the equations are rather ill-conditioned. In an adverse pressure gradient, θ grows more rapidly with x, and the angle of intersection on the graph is larger.

Another method that can be used to predict transition is due to Granville (1953). This method uses as a transition criterion a single curve of $(R_{\theta_{tr}} - R_{\theta_i})$ as a function of the mean Pohlhausen parameter $\bar{\lambda}$ (see Fig. 6.3). The parameter $\bar{\lambda}$ is an average λ [see Eq. (5.6.12)] that may be computed from

$$\bar{\lambda} = \frac{4}{45} - \frac{1}{5}\left[\frac{R_{\theta_{tr}}^2 - (u_e/u_{e_i})R_{\theta_i}^2}{Re_x - (u_e/u_{e_i})Re_{x_i}}\right] \tag{6.1.2}$$

where the subscript i denotes the point of instability to Tollmien-Schlicting waves (see Chap. 9). To predict transition by this method for two-dimensional flows, it is necessary to calculate R_{θ_i}. That can be done by means of the curve of Fig. 6.4. Once R_{θ_i} is known, the values of $(R_\theta - R_{\theta_i})$ and $\bar{\lambda}$ are calculated from the point of instability, Re_{x_i}, along the body, until they intersect the universal transition curve of Fig. 6.3.

Recently, Granville (1974) proposed another correlation for predicting transition on bodies of revolution without a parallel middle body. According to this correlation,

$$R_{\theta_{tr}} - R_{\theta_i} = f_1\left[\frac{2(r_0)_{max}}{r_0}\frac{dr_0}{dx}\right]_{tr} \tag{6.1.3}$$

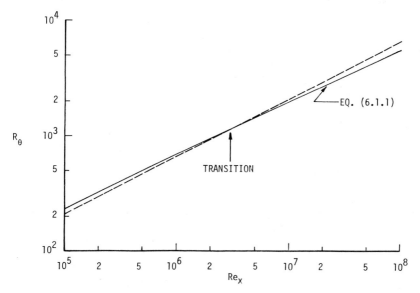

Fig. 6.2 Prediction of transition for Eq. (6.1.1) graphically. The dashed line represents Eq. (E6.1.1).

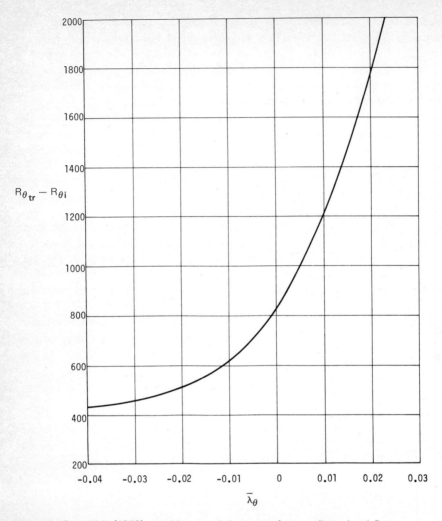

Fig. 6.3 Granville's (1953) transition correlation curve for two-dimensional flows.

Here R_{θ_i} is a function of H and is given by the relation in Fig. 6.4. The function f_1 is shown in Fig. 6.5. According to this method, we first compute the laminar boundary layer development on the axisymmetric flow, say, by Thwaites' method, and find the values of R_θ and H (computed by the method) that satisfy the relation shown in Fig. 6.4. Once R_{θ_i} is found, we continue calculating the development of R_θ and $[2(r_0)_{max}/r_0]\,dr_0/dx$ and look for the values of R_θ and $[2(r_0)_{max}/r_0]\,dr_0/dx$ that satisfy the relation shown in Fig. 6.5.

In a turbulent flow the numerical problems are virtually the same as in laminar flow, but we face the additional problem that there is no reliable general formula for the extra turbulent (Reynolds) stresses. There is indeed a set of exact "transport" equations (see Sec. 8.5) with left-hand sides each representing the rate of change of a Reynolds-stress component along a mean streamline, but these equations contain further unknown turbulence quantities on the right-hand sides. The terms on the right represent the *generation* of Reynolds stress by interaction between the mean rate of strain and the existing Reynolds stresses, the *destruction* of Reynolds stress by fluctuating pressures or viscous stresses, and the *spatial transport* of Reynolds stress by the diffusive action of the turbulence itself. Therefore the Reynolds-stress transport equations cannot be used to predict Reynolds stresses unless they are "closed" by replacing the unknown turbulence quantities on the right-hand sides of the equations by empirical combinations of the existing variables, that is, the Reynolds stresses and, where they are appropriate, the mean-rate-of-strain components.

Several methods in which this is done explicitly are now in regular use, and it is often helpful to regard any Reynolds-stress model as an implicit closure of the Reynolds-stress transport equations so as to assess its plausibility. For instance, any relation between Reynolds stresses and the mean rate of strain at the same point in space can be interpreted qualitatively as a closure of the Reynolds-stress transport equations, which neglects the transport terms (the mean-transport term, i.e., the left-hand side, and the diffusive turbulent-transport term on the right). The transport

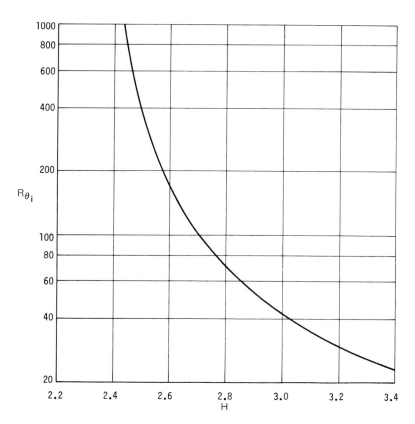

Fig. 6.4 Variation of critical Reynolds number R_{θ_i} with shape factor, according to stability theory.

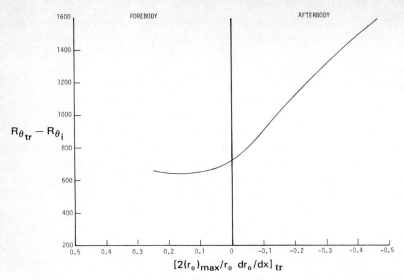

Fig. 6.5 Granville's (1974) transition correlation curve for prediction of transition on bodies of revolution.

equations then reduce to one or more of the so-called local-equilibrium relations: "generation equals destruction." These simplified equations contain the mean rate of strain and turbulence properties and can therefore be formally reduced to relations between Reynolds stresses or local rate of strain, but not relations involving spatial derivatives of the latter, by the general closure procedure just outlined.

We saw in Chap. 5 that there are many approximate methods of solution of the laminar TSL equations that incorporate the exact stress/rate-of-strain relation for viscous fluids. In turbulent flow, integral methods and other methods containing *numerical* approximations are again used, and it is sometimes difficult to distinguish these numerical simplifications from the approximations inherent in the turbulence model. The first question to be asked about a calculation method for turbulent flows in general or TSLs in particular is the following: "What closure assumptions are made about the Reynolds stresses (particularly the shear stress)?" There are two main types of Reynolds-stress hypotheses or "turbulence models," those that relate Reynolds stress to *mean-flow* properties at the same point or the same x-position, by a local-equilibrium relation or otherwise, and those that use transport equations for Reynolds stresses or other turbulence properties. In either type the numerical procedure used to solve the equations may be a solution of the PDEs or, in an integral method, a solution of ODEs derived either from the PDEs (by MIR) or from empirical data fits. We therefore have four combinations for complete calculation methods (turbulence model plus numerical procedure), for which mnemonics are mean-flow/PDE, mean-flow/ODE, transport-equation/PDE, and transport-equation/ODE. Methods of all four types exist (e.g., Bradshaw, 1972).

In laminar flows the numerical methods can again be classified into field and integral methods (see Sec. 5.6), but of course, the classification of stress hypotheses is not relevant.

The most common mean-flow models are variants of the "eddy-viscosity" formula, which, for a thin shear layer, defines a quantity ϵ_m, having the dimensions of (viscosity)/(density), by

$$-\rho\overline{u'v'} = \rho\epsilon_m \frac{\partial u}{\partial y} \tag{6.1.4}$$

where ϵ_m is related empirically to the mean-flow velocity and length scales. This can be done by a local-equilibrium relation in appropriate circumstances. More generally, Eq. (6.1.4) defines a quantity ϵ_m that is expected to vary less rapidly, or more comprehensibly, than $\overline{u'v'}$ and is therefore a better subject than $\overline{u'v'}$ itself for empirical formulas. In Sec. 6.3 we shall discuss the basis of eddy-viscosity formulas and the related "mixing-length" formula in more detail. For now, we note that *any* turbulence model for local Reynolds shear stress can be recast to give ϵ_m by using Eq. (6.1.4) as an equation of definition, and the momentum equation (3.2.4) then becomes

$$u\frac{\partial u}{\partial x} + v\frac{\partial u}{\partial y} = -\frac{1}{\rho}\frac{dp}{dx} + \frac{1}{r^k}\frac{\partial}{\partial y}\left[r^k(\nu + \epsilon_m)\frac{\partial u}{\partial y}\right] \tag{6.1.5}$$

which can be solved by any method suitable for a laminar flow with known but variable viscosity.

A detailed discussion of transport-equation methods would be outside the scope of this book. For such a discussion, see Bradshaw (1972), Cebeci and Smith (1974), or Bradshaw (1976), and for a self-contained research paper describing one of the most refined of these methods, see Hanjalic and Launder (1972), modified by Launder et al. (1975).

In all cases the numerical problem is the solution of coupled PDEs, since the velocity and Reynolds stress each appear in the equation for the other. In some cases, transport-equation methods are arranged to yield ϵ_m for substitution into the mean-motion equation, as just outlined. Alternatively, $\overline{u'v'}$, or the whole set of Reynolds stresses, can be retained in explicit form. The difference is more notional than real, and a good idea of the numerical procedure needed to handle coupled PDEs in general can be gained from the discussion of compressible heat-transfer problems covered elsewhere (Cebeci and Bradshaw, in preparation).

Because the closure assumptions are suggested by experiment, we shall first discuss the general behavior of turbulent shear layers and study various empirical laws established from experiments. We shall restrict our discussion of methods for predicting turbulent shear layers to simple formulas or to integral methods for two-dimensional boundary layers. Later, in Chaps. 7 and 8, we shall discuss differential methods for predicting laminar and turbulent shear layers.

6.2 COMPOSITE NATURE OF A TURBULENT BOUNDARY LAYER

Consider a constant-property steady two-dimensional flow past a flat plate (u_e = const). For a laminar boundary-layer flow, the velocity profiles are geometrically similar and reduce to a single curve if u/u_e is plotted against a dimensionless y-coordinate, $\eta = \sqrt{u_e/\nu x}\, y$. This is the well-known Blasius profile discussed in Sec. 5.5. The geometrical similarity is maintained, regardless of the Reynolds number of the flow or of the local skin friction. In a turbulent boundary layer there is no choice of dimensionless y-coordinate that leads to the collapse of the complete set of velocity profiles into a single curve because the viscous-dependent part of the profile, very close to the surface, and the remaining Reynolds-stress-dependent part of the profile require different length scaling parameters. For that reason it is helpful to treat a turbulent boundary layer as a composite layer consisting of inner and outer regions (Fig. 6.6).

The thickness of the inner region of a turbulent boundary layer (which contains, but extends far beyond, the region dominated by viscous stresses) is about 10–20% of the shear-layer thickness. It is generally assumed that, if the total (viscous plus turbulent) shear stress varies only slowly with distance from the surface, the mean-velocity distribution in this region is completely determined by the wall shear stress, τ_w; density, ρ; viscosity, μ; and the distance, y, from the wall. Thus the flow is independent of conditions in the outer part of the boundary layer and indeed of whether the flow is a boundary layer or some other type of wall shear layer. Therefore all the inner-layer formulas derived below are *valid for*

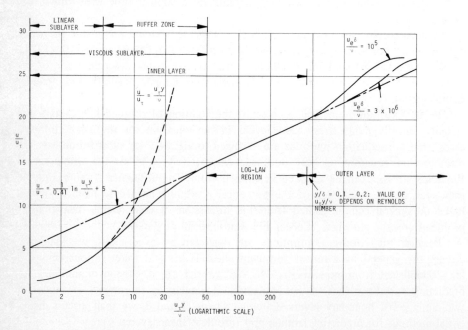

Fig. 6.6 Regions of a turbulent boundary layer. Outer-layer profile shown is for u_e = const.

internal flows as well as boundary layers if the explicit restrictions are met by the former. They will not be valid in either case if the flow changes rapidly in the x- or z-directions (e.g., near sudden changes of surface roughness or close to the corners of noncircular ducts).

Elementary dimensional analysis (Sec. 4.1) shows that the mean velocity is given by the "law of the wall,"

$$u^+ \equiv \frac{u}{u_\tau} = \phi_1(y^+) \tag{6.2.1}$$

where $u_\tau \equiv (\tau_w/\rho)^{1/2}$ is called the "friction velocity" and $y^+ \equiv u_\tau y/\nu$ is a Reynolds number based on a typical velocity scale of the turbulence, u_τ, and a typical length scale of the larger eddies at a distance y from the wall, namely y itself. When this Reynolds number is large (in practice, more than 30–50; below, we shall assume 50), we expect the relation between the Reynolds stress, now equal to τ_w, and the mean rate of strain, $\partial u/\partial y$, to be independent of viscosity. The only dimensionally correct relation is

$$\frac{\partial u}{\partial y} = \frac{u_\tau}{\kappa y} \tag{6.2.2}$$

where κ is found experimentally to be about 0.41. Equation (6.2.2) is a special case of Eq. (6.2.1) with $d\phi_1/dy^+ = 1/(\kappa y^+)$, and it follows that for $50\nu/u_\tau < y < 0.1 - 0.2\delta$,

$$u^+ = \frac{1}{\kappa} \ln y^+ + c \tag{6.2.3}$$

Here c is constant (about 5.0–5.2) for a smooth surface, according to the above simple analysis, and a function of roughness height and geometry in the case of a rough surface. This is the famous logarithmic law. Other methods of deriving it add nothing to the dimensional arguments given here, although they may—or may not—add to an understanding of the physics: the basic assumption is that conditions outside the inner region influence it only via the wall shear stress τ_w, and although this assumption can hardly be exact, it seems to be a remarkably good approximation.

If the total shear stress τ varies with y, we expect u to depend on the whole profile of τ between the surface and the position y. Equation (6.2.2) could be rewritten formally as

$$\frac{\partial u}{\partial y} = \frac{(\tau/\rho)^{1/2}}{\kappa y} f\left(\frac{y}{\tau}\frac{\partial \tau}{\partial y}, \frac{y^2}{\tau}\frac{\partial^2 \tau}{\partial y^2}, \cdots\right) \tag{6.2.4a}$$

but even in the case of uniform transpiration (suction or injection) through the surface where τ varies rapidly with y, $f = 1$ seems to be a good approximation, so

$$\frac{\partial u}{\partial y} = \frac{(\tau/\rho)^{1/2}}{\kappa y} \tag{6.2.4b}$$

Equation (6.2.4b) fails in flows that are changing rapidly in the x-direction because the local-equilibrium assumptions underlying the above analysis are not valid.

In three-dimensional flows that are changing slowly in the z-direction, there is moderate support for the extension of Eq. (6.2.4) to imply that the direction of the velocity gradient (components $\partial u/\partial y$ and $\partial w/\partial y$) is the same as that of the shear stress (components $\mu \partial u/\partial y - \rho \overline{u'v'}$ and $\mu \partial w/\partial y - \rho \overline{v'w'}$) and that its magnitude is related to the magnitude of the shear stress as in Eq. (6.2.4b). It also seems to be a good approximation where ρ varies with y, but we postpone discussion of the results to another work (Cebeci and Bradshaw, in preparation). Integrals of Eq. (6.2.4) for various profiles $\tau(y)$ will be discussed below.

For $u_\tau y/\nu < 50$ (the viscous sublayer), viscous stresses are significant, and the velocity profile departs from Eq. (6.2.3). For $u_\tau y/\nu < 5$ approximately, the turbulent shear stress $-\rho \overline{u'v'}$ is negligible because u' and v' are constrained to be zero at the wall, and the velocity profile follows the purely viscous law

$$\tau_{\mathrm{w}} = \mu \frac{\partial u}{\partial y} \tag{6.2.5a}$$

or

$$u^+ = y^+ \tag{6.2.5b}$$

in the present notation. This region is called the linear sublayer, and its outer edge will be denoted by $y = y_{\mathrm{s}}$. The remainder of the viscous sublayer, in which the velocity profile changes smoothly from Eq. (6.2.5) to Eq. (6.2.3), is the "buffer zone," sometimes misleadingly called the "transition region." These regions are shown in Fig. 6.6. It is important to note that the total thickness of the inner layer depends on δ and not directly on u_τ/ν. In most cases where τ varies fairly slowly with y, Eq. (6.2.3) is a good approximation at least in a small region for $u_\tau y/\nu > 50$. The inner-layer profile, either in the simple form for a constant-stress layer [Eq. (6.2.3)] or the more general form obtained by integrating Eq. (6.2.4) for given $\tau(y)$, is used explicitly in many calculation methods for turbulent wall layers, and nearly all other methods reproduce it implicitly. Virtually all the velocity-profile families that have been suggested for use in MIR employ Eq. (6.2.3).

EXAMPLE 6.2

Show that the viscous shear stress at $y^+ = 50$ is about 5% of the wall shear stress.

Solution We know that at this value of y^+, Eq. (6.2.2) applies to a good approximation. Therefore the viscous stress $\mu \, \partial u/\partial y$ is just $\mu u_\tau/(\kappa y)$ or, rewriting, $\tau_{\mathrm{w}}/(\kappa y^+)$. Since $\kappa \simeq 0.41$, the viscous shear stress at $y^+ = 50$ is about $0.05\tau_{\mathrm{w}}$.

The outer region of a turbulent boundary layer contains 80–90% of the boundary-layer thickness (see Fig. 6.6). In this region we expect the velocity profile to be free from the direct effects of viscosity, and in zero pressure gradient flows (u_e = const) the only relevant scales are u_τ and δ. Measuring velocity with respect to the external stream, because the difference between the outer-region velocity and the velocity at the surface (zero) depends on the viscous sublayer, we get the only dimensionally correct form of the velocity-defect law

$$\frac{u_e - u}{u_\tau} = f_1\left(\frac{y}{\delta}\right) \tag{6.2.6}$$

which is a special case of the self-preserving formula of Sec. 5.2. A possible extra parameter is $d\delta/dx$, but its effects are small. The effect of moderate wall roughness is negligible (Fig. 6.7), and $u_\tau\delta/\nu$ has a negligible effect if it exceeds about 2000 ($u_e\theta/\nu > 5000$). A formula analogous to Eq. (6.2.6) holds for circular pipes if we replace δ by the pipe radius, r_0; f_1 is rather different from a boundary layer, and by coincidence and *not* because the law-of-the-wall arguments are valid, the velocity profile follows the logarithmic law [Eq. (6.2.3)] fairly accurately almost to the center of the pipe. In a boundary layer, f_1 is markedly affected by the history of the pressure gradient and except for specially tailored pressure gradients of which zero is one, $f_1(y/\delta)$ depends on x. It should be noted that the arguments leading to Eq. (6.2.6) for constant-pressure flow apply anywhere outside the viscous sublayer: in the logarithmic region where Eq. (6.2.3) holds, Eq. (6.2.6) becomes

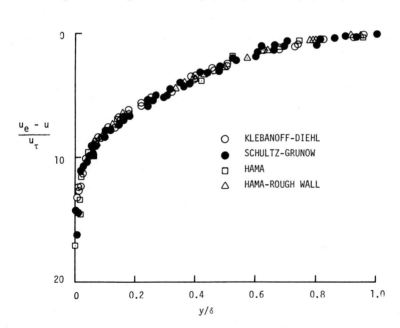

Fig. 6.7 Universal plot of turbulent boundary-layer profiles in zero pressure gradient, after Clauser (1956). The symbols denote the experimental data.

$$\frac{u_e - u}{u_\tau} = -\frac{1}{\kappa} \ln \frac{y}{\delta} + c' \tag{6.2.7}$$

where c' is an absolute constant for $u_e \theta / \nu > 5000$, independent even of wall roughness. In contrast to the law of the wall [Eq. (6.2.3)] the velocity-defect law [Eq. (6.2.6)] is of rather restricted application. Of course, Eq. (6.2.7) is always valid for some value of c', as long as a logarithmic region exists, but in a general boundary layer, c', like f_1, will be a function of x.

A class of boundary layers, in which the mainstream velocity distribution is characterized by the parameter

$$\beta = \frac{\delta^*}{\tau_w} \frac{dp}{dx} = \text{const} \tag{6.2.8}$$

consists of closely self-preserving flows. Here β represents the ratio of pressure forces to shear forces in a section of the boundary layer. These flows are analogous to similar Falkner-Skan flows in laminar layers, in that for a given flow the parameter β is constant, but are self-preserving flows rather than "similar" flows because Eq. (6.2.6) is of the form of Eq. (5.2.3), not Eq. (5.2.2), the form of the Falkner-Skan flows. A flat-plate (zero pressure gradient) flow is a special case of a self-preserving flow. The function f_1 in Eq. (6.2.6) is different for each value of β. Figure 6.8 shows the velocity-defect profiles for two different experimental pressure distributions corresponding to two self-preserving boundary layers together with the velocity-defect profile for zero pressure gradient. As can be seen, there is a marked difference between the velocity profiles with pressure gradient and those with no pressure gradient. Furthermore, the difference increases with increasing pressure-gradient parameter β. [Compare Fig. 5.4, noting that in the laminar (similar) boundary layer, β is uniquely related to m.]

According to experimental observations, as the free stream is approached, the flow at a given point becomes intermittently turbulent. This on-off character of turbulence is also observed in wake and jet flows. Figure 6.9 shows the boundary between turbulent and nonturbulent regions, traced from a photograph of a smoke-filled boundary layer. Drawings in textbooks usually underestimate the irregularity of the interface, which at low Reynolds numbers becomes even more pronounced than that shown here. The boundary is drawn as sharp, but in fact there is a viscous superlayer, probably slightly thicker than the viscous sublayer, in which viscous stresses transfer fluctuating vorticity to the outer "irrotational" flow.

6.3 EDDY-VISCOSITY AND MIXING-LENGTH CONCEPTS

Eddy-viscosity and mixing-length concepts are among the more popular and extensively used mean-flow models of turbulence (Sec. 6.1) relating Reynolds stress to local mean-velocity gradient. The main objection to the eddy-viscosity and mixing-length concepts is that they lack generality—they are based on local-equilibrium ideas that assume the transport terms in the governing equations are small. However,

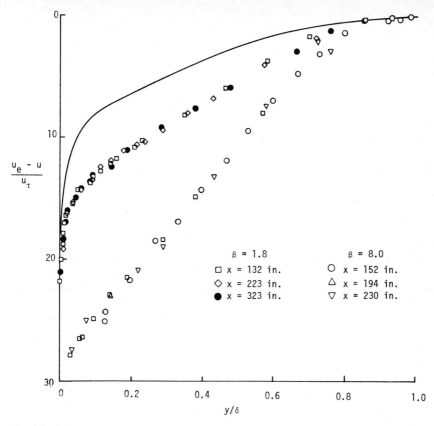

Fig. 6.8 Effect of adverse pressure gradient on the velocity-defect profiles. The data for $\beta = 1.8$ and 8.0 are from Clauser (1954).

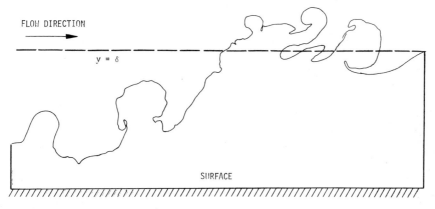

Fig. 6.9 Interface between turbulent and nonturbulent fluid in a boundary layer (side view, traced from a smoke photograph).

regarded as vehicles for empirical constants obtained from experimental data, they have proved very useful for calculations in restricted ranges of flows.

Boussinesq (1877; see Schlichting, 1968) was the first to attack the problem of finding a model for the Reynolds shear stress by introducing the concept of eddy viscosity. He assumed that the turbulent stresses act like the viscous stresses, an assumption that implies the turbulent stresses are proportional to the velocity gradient. The coefficient of proportionality was called the eddy viscosity and was defined, as in Eq. (6.1.4), by

$$-\rho \overline{u'v'} = \rho \epsilon_m \frac{\partial u}{\partial y} \tag{6.3.1}$$

Here, ϵ_m, like the kinematic viscosity, ν, can be thought of as the product of a velocity and a length, that is,

$$\epsilon_m \sim \text{length} \times \text{velocity} \tag{6.3.2}$$

The mixing-length concept was first proposed by Prandtl (1925; see Schlichting, 1968). According to this concept the Reynolds shear stress is to be calculated from

$$-\rho \overline{u'v'} = \rho l^2 \left| \frac{\partial u}{\partial y} \right| \frac{\partial u}{\partial y} \tag{6.3.3}$$

The basis of Prandtl's mixing-length hypothesis is an analogy with the kinetic theory of gases, based on the assumption that turbulent eddies, like gas molecules, are discrete entities that collide and exchange momentum at discrete intervals. This is not a realistic model, and it has been suggested (Bradshaw, 1974) that it arose out of a misinterpretation of the Ahlborn method of visualizing the motion at the free surface in a turbulent water flow.

By Eq. (6.3.1) we can write a relation between eddy viscosity and mixing length:

$$\epsilon_m = l^2 \left| \frac{\partial u}{\partial y} \right| \tag{6.3.4}$$

The length l defined by Eq. (6.3.3) is, of course, a quantity whose value is yet to be found. According to Von Karman's hypothesis (1931; see Schlichting, 1968), l is given by

$$l = \kappa \left| \frac{\partial u/\partial y}{\partial^2 u/\partial y^2} \right| \tag{6.3.5}$$

where κ is an empirical constant known as Von Karman's constant. This merely reproduces the expected result $l = \kappa y$ in the logarithmic part of the inner layer where Eq. (6.3.3) must be compatible with Eq. (6.2.2) and is not necessarily reliable elsewhere. It is unfortunate that the most accessible testing ground for mixing length formulas such as Eq. (6.3.5) was the logarithmic region where, *for purely dimensional reasons*, any viscosity-independent quantity having the dimensions of length must be proportional to y, whether it is a meaningful eddy size or not. Where Eq. (6.2.2) holds, it can be seen from Eq. (6.3.4) that

$$\epsilon_m = \kappa u_\tau y \tag{6.3.6}$$

because

$$l = \kappa y \tag{6.3.7}$$

EXAMPLE 6.3

For the inner region of a turbulent boundary layer, Reichardt (1951; see Schlichting, 1968) used the eddy-viscosity formula given by Eq. (6.3.6) and modified it to account for the viscous sublayer [see another modification proposed by Van Driest (1956), Eq. (6.4.11)]. This expression is

$$\epsilon_m = \kappa u_\tau y \left[1 - \left(\frac{y_l}{y} \right) \tanh\left(\frac{y}{y_l} \right) \right] \tag{E6.3.1}$$

Here y_l denotes the viscous sublayer thickness. Show that ϵ_m is proportional to y^3 for $(y/y_l) \ll 1$.

Solution Expanding $\tanh (y/y_l)$ close to the wall, we get

$$\tanh\left(\frac{y}{y_l} \right) = \frac{y}{y_l} - \frac{1}{3}\left(\frac{y}{y_l} \right)^3 + \cdots \tag{E6.3.2}$$

Inserting Eq. (E6.3.2) into Eq. (E6.3.1) we get

$$\epsilon_m = \frac{\kappa u_\tau}{3 y_l{}^2} y^3 + \cdots \tag{E6.3.3}$$

Figures 6.10 and 6.11 show the distributions of ϵ_m and l in a constant-pressure boundary layer according to Klebanoff's measurements (1955); they have been confirmed by later experiments. The results shown in those figures indicate that in the region $50\nu/u_\tau < y < 0.15$ to 0.2δ the eddy viscosity and mixing length vary linearly with distance y from the wall, in agreement with Eqs. (6.3.6) and (6.3.7). Both quantities appear to have a maximum value anywhere from $y/\delta = 0.20$–0.30. For y/δ greater than approximately 0.20 the eddy viscosity begins to decrease slowly, but the mixing length remains approximately constant. The latter can be approximated by

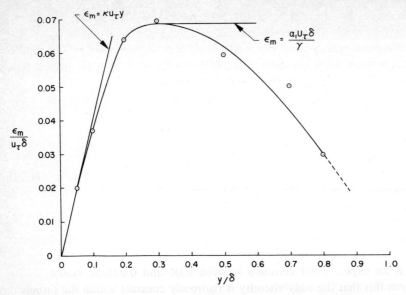

Fig. 6.10 Dimensionless eddy-viscosity distribution across a turbulent boundary layer in zero pressure gradient, $u_e\theta/\nu \simeq 8000$. The symbols denote experimental data from Klebanoff (1955).

$$\frac{l}{\delta} = \text{const} \tag{6.3.8}$$

The constant used by different workers varies from 0.075 to 0.09, necessarily depending on the definition of boundary-layer thickness, δ. Using our definition, $\delta_{.995}$, $l/\delta \simeq 0.08$.

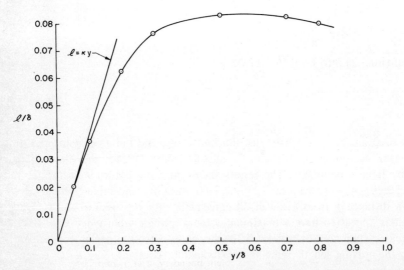

Fig. 6.11 Dimensionless mixing-length distribution across a turbulent boundary layer in zero pressure gradient, $u_e\theta/\nu \simeq 8000$. The symbols denote experimental data from Klebanoff (1955).

Let us now define the "intermittency factor," $\gamma(y)$, as the fraction of time for which the flow at height y is turbulent (see Fig. 6.9). Figure 6.12 shows γ in a constant-pressure boundary layer according to Klebanoff's measurements (1955). It can be fitted approximately by the expression

$$\gamma = \frac{1}{2}\left[1 - \operatorname{erf} 5\left(\frac{y}{\delta} - 0.78\right)\right] \tag{6.3.9}$$

which implies that the probability distribution of the instantaneous edge of the boundary layer is nearly Gaussian with mean 0.78δ and standard deviation 0.14δ.

If the distribution of eddy viscosity is corrected for the effect of intermittency, the dimensionless eddy viscosity $\epsilon/u_\tau\delta$ becomes nearly constant across the main outer part, as is shown in Fig. 6.10. It can be approximated by

$$\epsilon_m = \alpha_1 u_\tau \delta \tag{6.3.10a}$$

where α_1 is an experimental constant between 0.06 and 0.075. It should *not* be inferred from this that the eddy viscosity is rigorously constant within the turbulent fluid, but clearly the falloff is in some way due to intermittency.

It should be pointed out that the length and the velocity scales used to normalize the eddy viscosity in Fig. 6.10 are not the only possible characteristic scales. Other length and velocity scales such as δ^* and u_e, respectively, can also be used. Equation (6.3.7) can also be written in the form

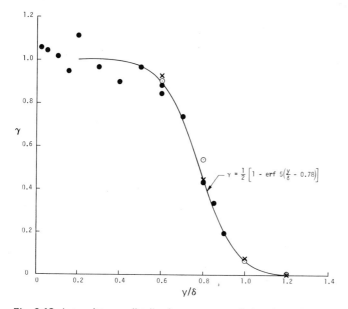

Fig. 6.12 Intermittency distribution across a turbulent boundary layer in zero pressure gradient, according to the data of Klebanoff (1955). Data represent three different techniques of measurement.

$$\epsilon_m = \alpha u_e \delta^* \tag{6.3.10b}$$

where α is a constant approximately equal to 0.0168 in quite a wide range of boundary layers (see Cebeci and Smith, 1974).

Analogous empirical formulas can be written for other types of shear layers. According to the measurements of Nikuradse (1935; see Schlichting, 1968), the mixing-length distribution across a pipe can be expressed with good approximation by the following equation:

$$\frac{l}{r_0} = 0.14 - 0.08\left[1 - \left(\frac{y}{r_0}\right)\right]^2 - 0.06\left[1 - \left(\frac{y}{r_0}\right)\right]^4 \tag{6.3.11}$$

Developing l as a series gives

$$l = 0.4y - 0.44\left(\frac{y^2}{r_0}\right) + \cdots \tag{6.3.12}$$

A helpful way of deriving the eddy-viscosity and mixing-length formulas for self-preserving flows is to note that the arguments used to derive Eq. (6.2.6) can be equally well used to give

$$\frac{-\rho\overline{u'v'}}{\tau_w} = \frac{\tau}{\tau_w} = f\left(\frac{y}{\delta}\right) \tag{6.3.13}$$

outside the viscous sublayer. [Actually the equations of motion permit Eqs. (6.2.6) and (6.3.13) to hold simultaneously only at infinite Reynolds number—in other words, $d\delta/dx$ and $d\tau_w/dx$ should appear in general—but discrepancies are small in practice.] Using Eqs. (6.2.6) and (6.3.13) and the definition of eddy viscosity given by Eqs. (6.1.4) or (6.3.1), we get

$$\frac{\epsilon_m}{u_\tau\delta} = f_1\left(\frac{y}{\delta}\right) \tag{6.3.14}$$

Now from the definition of δ^*, Eq. (1.2.19), and from Eq. (6.2.6) we see that

$$\frac{u_e\delta^*}{u_\tau\delta} = \int_0^\infty f\left(\frac{y}{\delta}\right) d\frac{y}{\delta} \tag{6.3.15}$$

Therefore Eq. (6.3.14) can be rewritten as

$$\frac{\epsilon_m}{u_e\delta^*} = f_2\left(\frac{y}{\delta}\right) \tag{6.3.16}$$

Because δ^*/θ varies with Reynolds number, using θ instead of δ^* in Eq. (6.3.16) would introduce a false Reynolds-number dependence of ϵ_m. By what can only be described as good luck, the function f_2 turns out to be nearly the same in all self-preserving boundary layers over a wide range of $(\delta^*/\tau_w)\,dp/dx$. Also, simply because the different velocity-defect profiles in Fig. 6.8, and the corresponding shear-stress profiles, are nearly geometrically similar, l/δ is closely the same function of y/δ for a range of values of $(\delta^*/\tau_w)\,dp/dx$, so Eq. (6.3.8) is a good approximation for a range of self-preserving boundary layers. It will be clear that this "success" of the mixing-length and eddy-viscosity formulas, even outside the inner layer for which they can be rigorously derived, owes *nothing* to the erroneous physical arguments originally used to derive these formulas but is a combination of dimensional analysis [giving Eq. (6.3.16)] and good fortune. It is further found that Eqs. (6.3.8) and (6.3.16) are fair approximations for general boundary layers not too far from self-preservation [i.e., with $(\delta^*/\tau_w)\,dp/dx$ varying only slowly with x]; the quantitative meaning of "not too far" depends on the accuracy demanded of Eqs. (6.3.8) and (6.3.16).

6.4 MEAN–VELOCITY DISTRIBUTION ON SMOOTH SURFACES

In two-dimensional flows the mean-velocity distribution across the boundary layer can be represented approximately by several empirical formulas. Here we shall discuss a formula proposed by Coles (1956). According to this formula the dimensionless velocity distribution is given by

$$u^+ = \phi_1(y^+) + \frac{\Pi(x)}{\kappa}\,w\!\left(\frac{y}{\delta}\right) \tag{6.4.1}$$

where ϕ_1 is the function appearing in Eq. (6.2.1) and w is effectively zero in the inner layer.

Equation (6.4.1) is applicable to flows with and without pressure gradient. If we exclude the viscous sublayer, the law of the wall function, $\phi_1(y^+)$, is given by Eq. (6.2.3).

The quantity Π in Eq. (6.4.1) is a profile parameter that is, in general, a function of x. The function $w(y/\delta)$, representing the deviation of the outer-layer profile from the law of the wall, is called the *law-of-the-wake* function and is of nearly universal character, according to experiments. However, it must be clearly understood that the function $w(y/\delta)$ is just an empirical fit to measured velocity profiles, yielding the useful algebraic results below. It does not imply any universal similarity of the sort implied by the velocity-defect function [Eq. (6.2.3)] for zero-pressure gradient or equilibrium boundary layers. It is given to a good approximation by the *empirical fit*

$$w\!\left(\frac{y}{\delta}\right) = 2\sin^2\frac{\pi}{2}\!\left(\frac{y}{\delta}\right) = 1 - \cos\!\left(\pi\frac{y}{\delta}\right) \tag{6.4.2a}$$

although other simple algebraic forms have been used successfully. In all cases, $w(1)$

is taken as 2, so that w is the normalized shape function and the x-wise variation is represented by Π. Note that this δ is a freely chosen parameter and is not the same as the choice $\delta = \delta_{995}$ used elsewhere in this book.

Equation (6.4.2a) for $w(y/\delta)$ has the disadvantage that $\partial u/\partial y$, evaluated from Eq. (6.4.1), is nonzero at $y = \delta$: it equals $u_\tau/(\kappa\delta)$. A better choice for the wake, due originally to Finley et al. (1966), is to replace the term Πw in Eq. (6.4.1) by

$$\Pi w = \eta^2(1 - \eta) + 2\Pi\eta^2(3 - 2\eta) \tag{6.4.2b}$$

where $\eta = y/\delta$ and δ is closely equal to the value implied by Eq. (6.4.2a). In effect, the first group is an addition to the logarithmic law-of-the-wall term in Eq. (6.4.1), which forces $\partial u/\partial y = 0$ at $y = \delta$. The second group is the wake component proper, which varies with Π.

For flows with zero pressure gradient, the profile parameter Π is a constant equal to about 0.55, provided that the momentum-thickness Reynolds number R_θ is greater than 5000. For $R_\theta < 5000$, the variation of Π with R_θ is as shown in Fig. 6.13. In self-preserving boundary layers, by definition, Π is constant, its value depending on the strength of the pressure gradient. In this case, Eq. (6.4.1) can be rewritten as a quantitative form of Eq. (6.2.6):

$$\frac{u_e - u}{u_\tau} = \phi_1(\delta^+) - \phi_1(y^+) + \frac{\Pi}{\kappa}\left[w(1) - w\left(\frac{y}{\delta}\right)\right] \tag{6.4.3}$$

from which the viscosity dependence implicit in y^+ and $\delta^+ \equiv \delta u_\tau/\nu$ vanishes when we use the logarithmic form [Eq. (6.2.3)], yielding

$$\frac{u_e - u}{u_\tau} = -\frac{1}{\kappa}\ln\frac{y}{\delta} + \frac{\Pi}{\kappa}\left[w(1) - w\left(\frac{y}{\delta}\right)\right] \tag{6.4.4}$$

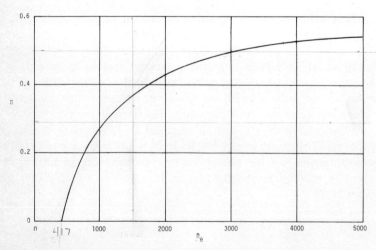

Fig. 6.13 Variation of Coles' (1962) profile parameter, Π, with momentum-thickness Reynolds number, R_θ, for zero pressure gradient flow.

for $50\nu/u_\tau < y < \delta$. Of course, Eq. (6.4.4) is valid in *any* flow [assuming that Eq. (6.4.1) is valid] but conforms to Eq. (6.2.6) only if Π is independent of x.

The profile parameter, Π, can be related to the local skin-friction coefficient, $c_f = 2\tau_w/\rho u_e^2 = 2(u_\tau/u_e)^2$, by evaluating Eq. (6.4.1) at the edge of the boundary layer, $u = u_e$, $y = \delta$,

$$\sqrt{\frac{2}{c_f}} \equiv \frac{u_e}{u_\tau} = \frac{1}{\kappa} \ln \frac{\delta u_\tau}{\nu} + c + \frac{2\Pi}{\kappa} \tag{6.4.5}$$

It can be related to the displacement thickness, δ^*, and to the momentum thickness, θ, by

$$\kappa \frac{\delta^* u_e}{\delta u_\tau} = 1 + \Pi \tag{6.4.6}$$

$$\kappa^2 \frac{(\delta^* - \theta) u_e^2}{\delta u_\tau^2} = 2 + 2\left[1 + \frac{1}{\pi} Si(\pi)\right]\Pi + \frac{3}{2}\Pi^2 \tag{6.4.7}$$

Then

$$\frac{H}{H-1} \frac{u_\tau}{\kappa u_e} \equiv \frac{1}{\kappa G} = F(\Pi) \tag{6.4.8a}$$

$$F(\Pi) = \frac{1 + \Pi}{2 + 2[1 + 1/\pi \, Si(\pi)]\Pi + 3/2\,\Pi^2} \tag{6.4.8b}$$

where $Si(\pi) = \displaystyle\int_0^\pi [(\sin u)/u] \; du = 1.8519$ and G is the Clauser (1956) shape parameter.

EXAMPLE 6.4

Use Coles' relation [Eq. (6.4.5)] to find the skin-friction coefficient in a constant-pressure boundary layer at $u_e\delta^*/\nu = 15{,}000$, and then use Eq. (6.4.8) to calculate $u_e\theta/\nu$. Take $\kappa = 0.41$ and $c = 5.0$.

Solution Equation (6.4.6) gives $u_\tau\delta/\nu = 3968$, taking $\Pi = 0.55$ since $u_e\theta/\nu$ is roughly $\frac{2}{3}$ of $u_e\delta^*/\nu$ ($H \approx 1.5$), and therefore exceeds 5000. Then Eq. (6.4.5) gives $c_f = 0.00257$, with $u_\tau/\kappa u_e = 0.0874$. Next, Eq. (6.4.8b) gives $F(0.55) = 0.369$, and finally Eq. (6.4.8a) gives $H = 1.31$, so $u_e\theta/\nu = 11{,}450$.

The law-of-the-wall equation (6.2.1) can be extended to include the viscous sublayer by using one of several empirical equations. Here we shall present an equation proposed by Van Driest (1956) for all wall flows (ducts or boundary layers) with a negligible shear-stress gradient in the viscous sublayer. If we assume

$\tau = \tau_w$, we get

$$\nu\frac{du}{dy} - \overline{u'v'} = \frac{\tau_w}{\rho} = u_\tau{}^2 \tag{6.4.9}$$

Prandtl's mixing-length formula [Eq. (6.3.3)] *defines* l anywhere, but the particular choice is $l = \kappa y$. Equation (6.3.7) applies only to the fully turbulent part of the inner layer and does not include the viscous sublayer. By an analogy with the laminar flow on an oscillating plate (which is probably not a good quantitative model of the sublayer but yields a solution with the right qualitative behavior), Van Driest suggested modifying Eq. (6.3.7) to

$$l = \kappa y\left[1 - \exp\left(-\frac{y}{A}\right)\right] \tag{6.4.10}$$

where A is a damping-length constant, for which the best dimensionally correct empirical choice is about $26\nu(u_\tau)^{-1/2}$. It can be seen that $l \rightarrow \kappa y$ for $y \gg A$. If we now use Eq. (6.3.3) together with Eq. (6.4.10), we can write Eq. (6.4.9) as

$$\nu\frac{du}{dy} + (\kappa y)^2\left[1 - \exp\left(-\frac{y}{A}\right)\right]^2\left(\frac{du}{dy}\right)^2 = u_\tau{}^2 \tag{6.4.11}$$

In terms of dimensionless quantities this equation can be written as

$$a(y^+)\left(\frac{du^+}{dy^+}\right)^2 + b\left(\frac{du^+}{dy^+}\right) - 1 = 0 \tag{6.4.12}$$

or

$$\frac{du^+}{dy^+} = \frac{-b + \sqrt{b^2 + 4a}}{2a} \tag{6.4.13}$$

where $a = (\kappa y^+)^2\,[1 - \exp(-y^+/A^+)]^2$, $A^+ = 26$, and $b = 1$. Multiplying both numerator and denominator of du^+/dy^+ by $(b + \sqrt{b^2 + 4a})$ and formally integrating the resulting expression, we obtain

$$u^+ = \int_0^{y^+} \frac{2}{1 + \sqrt{1 + 4a}}\,dy^+ \tag{6.4.14}$$

since $u^+ = 0$ at $y^+ = 0$.

Equation (6.4.14) defines a continuous velocity distribution in the inner region of the turbulent boundary layer and applies to the viscous sublayer, to the buffer layer, and to the region of fully turbulent flow. For example, in the viscous sublayer, $a = 0$. Then from Eq. (6.4.14) the mean-velocity distribution in the viscous sublayer is, as required,

$$u^+ = y^+ \qquad\qquad (6.4.15)$$

As y^+ increases, the exponential term in $a(y^+)$ becomes smaller. For example, at $y^+ = 60$, exp $(-y^+/A^+)$ is 0.099, and at $y^+ = 100$, exp $(-y^+/A^+)$ is 0.026. Thus in the fully turbulent region ($y^+ > 50$, approximately) and with $\kappa = 0.41$ and $A^+ = 24.4$ we obtain Eq. (6.2.3) with $c = 5.0$ (as was suggested by Coles and Hirst, 1969), while to get $c = 5.2$, we require $A^+ = 25.6$. Van Driest's original suggestion of $A^+ = 26$, still widely used, yields the rather high value $c = 5.3$. If either τ or the fluid temperature varies significantly across the sublayer, A can be redefined as— say—$26\nu(\tau/\rho)^{-1/2}$, where *local* values of the variables are understood, but in general the optimum numerical value of the factor here written as 26 will depend on the dimensionless gradients of τ, ρ, and ν in the sublayer.

Figure 6.14 shows that the mean-velocity distribution calculated by Eq. (6.4.14) agrees quite well with the experimental data of Laufer (1954) in a pipe and with the flat-plate data of Klebanoff (1955) and of Wieghardt (1944).

Of the empirical curve-fitting results in this section, Eq. (6.4.1) applies to almost any boundary layer and could be applied to ducts with a different choice of $w(y/\delta)$; Eq. (6.4.4) applies to almost any boundary layer but is especially useful in the self-preserving flow, where Π and G are constant; and Eq. (6.4.14) applies to any wall flow over a smooth surface as long as τ is effectively constant across the viscous sublayer (small-pressure gradient and/or small $d\tau_w/dx$).

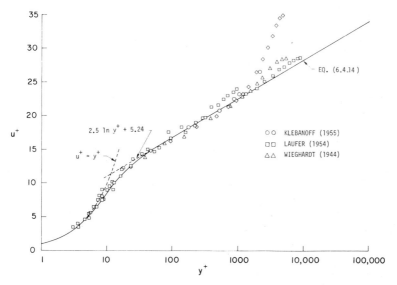

Fig. 6.14 Mean-velocity distribution in the inner region as calculated by Eq. (6.4.14) with $\kappa = 0.4$, $A^+ = 26$.

EXAMPLE 6.5

Find the velocity profile, in the inner layer but outside the viscous sublayer, if $\tau = \tau_w + \alpha y$, where α is a constant. On what dimensionless parameter does the final constant of integration depend?

Solution Since

$$\frac{\partial u}{\partial y} = \frac{(\tau/\rho)^{1/2}}{\kappa y} \qquad \tau = \tau_w + \alpha y$$

then

$$\frac{\partial u/u_\tau}{\partial y} = \frac{(1 + \alpha y/\tau_w)^{1/2}}{\kappa y}$$

$$u^+ \equiv \frac{u}{u_\tau} = \int \frac{(1 + \alpha y/\tau_w)^{1/2}}{\kappa y} \, dy$$

Substituting $1 + \alpha y/\tau_w = z^2$, we finally obtain

$$u^+ = \frac{1}{\kappa} \left(2z + \ln \frac{z-1}{z+1} \right) + \text{const}$$

The logarithmic term reduces, on multiplying top and bottom by $z + 1$ and noting that $\ln (1 + z)^2 = 2 \ln (1 + z)$, to

$$\ln \frac{\alpha y}{\tau_w} - 2 \ln \left(1 + \sqrt{1 + \frac{\alpha y}{\tau_w}} \right)$$

Now for $\alpha \to 0$ we require compatibility with the logarithmic law [Eq. (6.2.3)], and the constant must therefore be chosen by inspecting each term in turn. We get

$$u^+ = \frac{1}{\kappa} \ln \frac{u_\tau y}{\nu} + c - \frac{2}{\kappa} \ln \left(\frac{1 + \sqrt{1 + \alpha y/\tau_w}}{2} \right) + \frac{2}{\kappa} \left(\sqrt{1 + \frac{\alpha y}{\tau_w}} - 1 \right)$$

Here c must depend on α made dimensionless with the inner-layer scales u_τ, ρ, and ν (y should obviously not appear, since c is constant with respect to y): thus $c = f[\alpha\nu/(\rho u_\tau^3)]$. Note that α, taken as an average value of $\partial\tau/\partial y$ across the inner layer, is not necessarily equal to dp/dx.

6.5 MEAN-VELOCITY DISTRIBUTION ON ROUGH SURFACES

As was stated in Sec. 6.2, the velocity-defect law [Eq. (6.2.6)] is valid for both smooth and rough surfaces. Thus roughness affects only the inner region. It is, of course, impossible to make a surface absolutely smooth, but the wall is aerodynamically smooth for a turbulent boundary layer if the height of the roughness elements, k, is much less than the thickness of the viscous sublayer. Since in most cases the viscous sublayer is extremely thin, less than 1% of the shear-layer thickness in a boundary layer with $u_e\delta/\nu > 10^5$, the roughness elements must be very small if the surface is to be aerodynamically smooth. On a given surface, as the boundary-layer thickness and Reynolds number change, the surface may change from effectively rough to aerodynamically smooth.

According to dimensional analysis, the law-of-the-wall for a surface with uniform roughness of given geometrical shape is

$$u^+ = \phi_2(y^+, k^+) \tag{6.5.1}$$

Here k^+ is a roughness Reynolds number defined by

$$k^+ = \frac{ku_\tau}{\nu} \tag{6.5.2}$$

In the fully turbulent part of the inner region, the law of the wall for a uniform rough surface is similar to that for a smooth surface except that the additive constant, c in Eq. (6.2.3), is a function of the roughness Reynolds number, k^+, and of the roughness geometry. In that region the law of the wall can be written as

$$\phi_2 = \frac{1}{\kappa} \ln y^+ + B_1(k^+) \tag{6.5.3}$$

for a given shape of roughness elements. It is reasonable to assume κ to be the same as on smooth surfaces, on the argument that κ is unaffected by sublayer conditions in the latter case and should be similarly unaffected here. If we let

$$B_2 = \frac{1}{\kappa} \ln k^+ + B_1(k^+) \tag{6.5.4}$$

we can write Eq. (6.5.3) as

$$\phi_2 = \frac{1}{\kappa} \ln y^+ - \frac{1}{\kappa} \ln k^+ + B_2 = \frac{1}{\kappa} \ln \frac{y}{k} + B_2 \tag{6.5.5}$$

Here B_1 and B_2 are functions of roughness geometry and density, and in general of k^+, and must be determined from experiments.

If we let

$$B_3 = c - B_2 \tag{6.5.6}$$

and

$$\Delta u^+ = \frac{\Delta u}{u_\tau} = \frac{1}{\kappa} \ln k^+ + B_3 \qquad\qquad (6.5.7)$$

then we can write Eq. (6.5.5) as

$$u^+ = \frac{1}{\kappa} \ln y^+ + c - \Delta u^+ \qquad\qquad (6.5.8)$$

since $\phi_2 = u^+$.

The relation between Δu^+ and k^+ has been determined empirically for various types of roughness geometry. Some results are shown in Fig. 6.15.

We see from Eq. (6.5.8) that, since for a given roughness Δu^+ is a known function of k^+, the sole effect of the roughness is to shift the intercept, $c - \Delta u^+$, as a function of k^+. For values of k^+ below approximately 5, the vertical shift Δu^+ approaches zero, except for those roughnesses having such a wide distribution of particle sizes that there are some particles large enough to protrude from the sublayer even though the average size is considerably less than the thickness of the sublayer. For large values of k^+ the vertical shift is proportional to $\ln k^+$, with the constant of proportionality equal to $1/\kappa$, and by comparing Eqs. (6.5.5) and (6.5.8) we can see that this implies that B_2 is independent of k^+. This means that the drag of the roughness elements is independent of viscosity, a reasonable result for nonstreamlined obstacles at large Reynolds number.

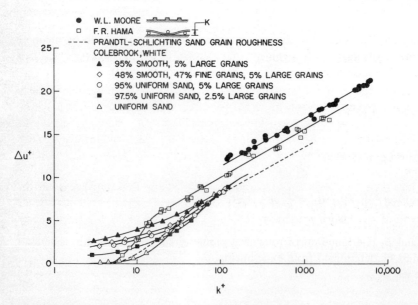

Fig. 6.15 Effect of wall roughness on universal velocity profiles, after Clauser (1956), who gives the references in full.

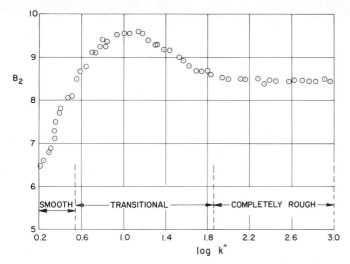

Fig. 6.16 Variation of B_2 with k^+.

Figure 6.16 shows the variation of B_2 with k^+ in sand-roughened pipes (Nikuradse, 1933; see Schlichting, 1968). Ioselevich and Pilipenko (1974) give an analytic fit to the sand-roughness data of Nikuradse: for $k^+ < 2.25$, $B_1 = c \approx 5.2$ (negligible roughness effect); for $2.25 < k^+ < 90$, $B_1 = c + [8.5 - c - (1/\kappa) \ln k^+]$ $\sin 0.4258 (\ln k^+ - 0.811)$; for $k^+ > 90$, $B_1 = 8.5 - (1/\kappa) \ln k^+$.

The hydraulically smooth condition exists when roughness heights are so small that the roughness is buried in the viscous sublayer. The fully rough flow condition exists when the roughness elements are so large that the sublayer is completely eliminated and the flow can be considered to be independent of molecular viscosity, that is, the velocity shift is proportional to $\ln k^+$ and Eq. (6.5.5) holds with B_2 constant. Because molecular viscosity still has some role in the transitional regime, the geometry of roughness elements has a relatively large effect on the velocity shift, as can be seen in Fig. 6.15. The effect of certain types of roughness, notably square-section spanwise grooves with a groove width about half the pitch, scales on the shear-layer thickness rather than the roughness height. The reason appears to be that the flow would pass smoothly over the grooves if it were not for disturbances caused by outer-layer eddies. For details see Perry et al. (1969), who call this "d-type" roughness.

The fact that the shifts in velocity for fully rough flow are linear on the semilogarithmic plot, Fig. 6.15, can be used to express different roughness geometries in terms of a reference roughness. It follows from Eq. (6.5.7) that for the same velocity shift,

$$\frac{k_s}{k} = \exp [\kappa(B_3 - B_{3_s})] \tag{6.5.9}$$

where the subscript s refers to a reference roughness, commonly taken as uniform sand-grain roughness.

EXAMPLE 6.6

Determine the equivalent sand-grain height of the square-bar roughness distribution tested by Moore and shown in Fig. 6.15. Assume fully rough conditions.

Solution By making use of Eq. (6.5.7) we can express Eq. (6.5.9) for a fixed value of k^+ as

$$\frac{k_s}{k} = \exp\left[\kappa(\Delta u^+ - \Delta u_s^+)\right] \tag{E6.6.1}$$

According to Fig. 6.15, for fully rough conditions, $\Delta u^+ - \Delta u_s^+ = 3.25$. Thus with $\kappa = 0.41$ it follows from Eq. (E6.6.1) that

$$k_s = 3.79k \tag{E6.6.2}$$

Equation (6.5.8) applies only in the inner region of the boundary layer. For application to the entire boundary layer it must be corrected for the wakelike behavior of the outer region. That can be done by using Coles's wake expression. With the correction we can write

$$u^+ = \phi_2(y^+) + \frac{\Pi}{\kappa}\, w\!\left(\frac{y}{\delta}\right) \tag{6.5.10}$$

where ϕ_2 is given by Eq. (6.5.8) outside the immediate neighborhood of the roughness.

6.6 MEAN-VELOCITY DISTRIBUTION WITH SURFACE MASS TRANSFER

The law of the wall for two-dimensional incompressible turbulent boundary layers on porous surfaces is

$$\frac{2}{v_w^+}\left[(1 + v_w^+ u^+)^{1/2} - 1\right] = \frac{1}{\kappa}\ln y^+ + c \tag{6.6.1}$$

with the usual convention that a plus superscript denotes a quantity made dimensionless by inner-layer variables, so that $v_w^+ = v_w/u_\tau$, and v_w is the transpiration velocity at the surface. We shall use u_p^+ to denote the left-hand side of Eq. (6.6.1).

It can be derived by noting that if x-derivatives (including the pressure gradient) are negligible, the x-component momentum equation reduces, in two-dimensional incompressible flow, to

$$\rho v_w \frac{\partial u}{\partial y} = \frac{\partial \tau}{\partial y} \tag{6.6.2}$$

Here v is everywhere equal to the wall transpiration velocity, v_w, because the continuity equation is $\partial v/\partial y = -\partial u/\partial x$ and we are neglecting x-derivatives. Thus

$$\tau = \tau_w + \rho u v_w \tag{6.6.3}$$

By substituting in Eq. (6.2.4) with $f = 1$, integrating, and requiring compatibility with Eq. (6.2.3) in the limit of zero v_w, the representation given in Eq. (6.6.1) follows. A more easily comprehensible form due to B. E. Launder (personal communication) obtained by multiplying numerator and denominator by $[(1 + v_w^+ u^+)^{1/2} + 1]$, is

$$\frac{u^+}{(1/2)[(1 + v_w^+ u^+)^{1/2} + 1]} = \frac{1}{\kappa} \ln y^+ + c \tag{6.6.4}$$

which clearly reduces to the logarithmic law [Eq. (6.2.3)] for $v_w = 0$. In general, c is now a function of v_w^+, to which the dimensionless shear-stress gradient in the sublayer can be related. In the discussion below, it will be convenient to regard the left-hand side of Eq. (6.6.1) as a "pseudovelocity," u_p^+. In the region of validity of Eq. (6.6.1), roughly $50\nu/u_\tau < y < 0.2\delta$, we can rearrange it as a velocity-defect law:

$$u_{pe}^+ - u_p^+ = u_{pe}^+ - \left(\frac{1}{\kappa} \ln y^+ + c\right)$$

and a *sufficient*, though not necessary, condition for this to be compatible with the velocity-defect law [Eq. (6.2.6)], in the limit of zero v_w^+ and for zero pressure gradient, is that the right-hand side shall be the same function, $f(y/\delta)$, that appears in Eq. (6.2.6). The results in Fig. 6.17 show that the function $f(y/\delta)$ in the equation of the modified velocity-defect law [Eq. (6.2.6)] is indeed closely independent of v_w and u_τ everywhere outside the viscous sublayer.

According to extensive measurements on flat plates with mass transfer made by Simpson et al. (1967), good correlation with experiment, at least in the range $-0.004 < v_w/u_e < 0.01$, is obtained if c in Eq. (6.6.1) is represented by

$$c = c_0 + \frac{2}{v_w^+}[(1 + K v_w^+)^{1/2} - 1] - K \tag{6.6.5}$$

where $K \simeq 11$. [This implies that Eq. (6.6.1) intersects the sublayer velocity profile for pure viscous stress at $y^+ \simeq 11$.] It is purely by chance that the intersection point is independent of v_w^+. Using Eq. (6.6.5) for c, Coles (1971) has shown that Eq. (6.4.1), when it is generalized to the form

$$u_p^+ = \frac{1}{\kappa} \ln y^+ + c + \frac{\Pi}{\kappa} w\left(\frac{y}{\delta}\right) \tag{6.6.6}$$

with $c_0 = 5$ and $K = 10.8$, describes the experimental data for flows with mass

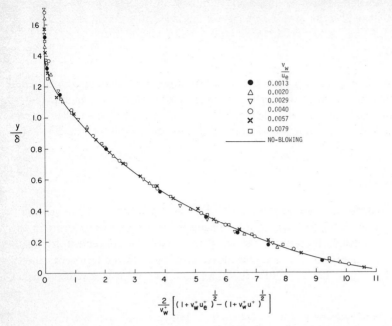

Fig. 6.17 The velocity-defect law for zero pressure gradient flow with mass transfer. After Stevenson (1963).

transfer very well, at least in zero pressure gradient. As in flows with zero mass transfer, Π varies with Reynolds number. Figure 6.18 shows the prediction of pseudovelocity profiles by Eq. (6.6.6) for the experimental data of Simpson et al. (1967).

Equation (6.6.1) excludes the viscous sublayer and the buffer layer. It can be generalized to include these two regions by using Eq. (6.4.12), with $-(1 + v_w^+ u^+)$ replacing -1, and with A^+ expressed as a function of mass transfer, v_w^+, as shown in Fig. 6.19. Once u^+ is calculated, we can determine u_p^+ from its definition. For further details, see Cebeci (1973a).

It must be remembered that the results of Secs. 6.4–6.6 are all based on the dimensional arguments that led to the logarithmic law in Sec. 6.2. The local-equilibrium assumptions underlying those arguments cannot be exactly correct, even when the necessary condition of slow streamwise change is met. Therefore the above formulas should be used with caution. Also, of course, the analytic functions (mainly logarithms) that result from integration of Eq. (6.2.4) with $f = 1$ and with various profiles of τ are to be distinguished from analytic curve-fits like Eqs. (6.4.2), (6.4.11), and (6.6.5). The main uses of the above formulas are in generating velocity-profile families and skin-friction laws for use in MIR (Sec. 4.3.4), in deducing skin friction from inner-layer measurements once the various constants have been established by prior experiment, and in providing inner-boundary conditions for finite-difference field calculations (Sec. 4.3.3) that would otherwise waste computer time in reproducing the known universal profile all the way to the surface.

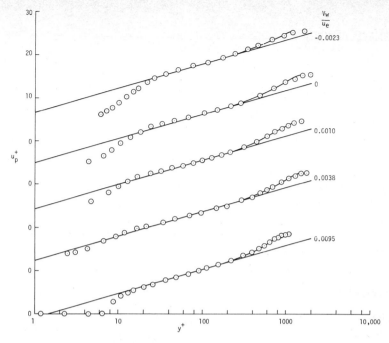

Fig. 6.18 Prediction of velocity profiles by Coles' (1971) expression, Eq. (6.6.6), for mass transfer, from the data of Simpson et al. (1967).

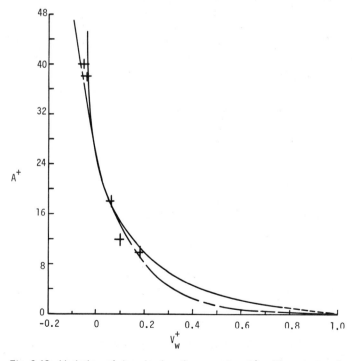

Fig. 6.19 Variation of damping-length parameter, A^+, with mass transfer parameter, v_w^+, for an incompressible zero pressure gradient flow (solid line), Cebeci (1973a); $A^+ = 26 \exp(-5.9 v_w^+)$, (dashed line), Cebeci (1970a); $+$, Bushnell and Beckwith (1970).

6.7 TURBULENT BOUNDARY LAYERS ON SMOOTH AND ROUGH SURFACES WITH ZERO PRESSURE GRADIENT

Once a satisfactory model for Reynolds shear stress is devised, differential methods similar to those of Chap. 5, with ν replaced by $(\nu + \epsilon_m)$, can be used to compute the development of external boundary layers on two-dimensional and axisymmetric bodies with different boundary conditions. One such method is discussed in Chap. 8. However, all these methods require the use of large computers. Sometimes it is useful to have simple methods or formulas that do not require the use of computers and can be used with reasonable accuracy for computing boundary layers. In this section we shall discuss such practical formulas, which can be used for turbulent flows over smooth and rough surfaces with zero pressure gradient. In Sec. 6.8 we shall discuss an integral method for computing two-dimensional turbulent flows over smooth surfaces with pressure gradient.

6.7.1 Smooth Flat Plate

Consider constant-property flow over a smooth flat plate (zero pressure gradient). If the Reynolds number is sufficiently large, we can identify three different flow regimes on such a surface (see Fig. 6.20 and recall Sec. 1.3). Starting from the leading edge, there is first a region $(0 < \mathrm{Re}_x < \mathrm{Re}_{x_{tr}})$ in which the flow is laminar, or laminar with small instability oscillations (Chap. 9). Further downstream, there is a region $(\mathrm{Re}_{x_{tr}} < \mathrm{Re}_x < \mathrm{Re}_{x_t})$ in which transition from laminar to turbulent flow takes place, following the first appearance of isolated spots of turbulence at x_{tr}. In the third region $(\mathrm{Re}_x \geqslant \mathrm{Re}_{x_t})$ the flow is fully turbulent. The transition Reynolds number $\mathrm{Re}_{x_{tr}}$ depends partly upon the turbulence in the free stream; $\mathrm{Re}_{x_{tr}}$ may be as low as 5×10^5 or as high as 5×10^6. In Sec. 6.1 we presented simple correlations for transition in boundary layers adjacent to streams with low turbulence intensity, with and without pressure gradient. Later, in Chap. 9, we shall

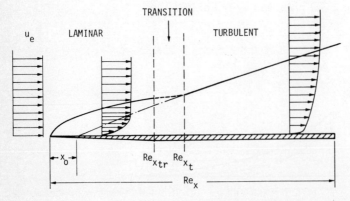

Fig. 6.20 Boundary layer on a smooth flat plate at sufficiently large Reynolds numbers. (The vertical scale is greatly enlarged.)

discuss the prediction of transition in flows with and without pressure gradient by using the stability theory. We should point out, however, that the lower value of $\mathrm{Re}_{x_{tr}}$ corresponds to $R_{\theta_{tr}} = 425$, which is the approximate minimum value of R_θ for turbulent flow. The value of $\mathrm{Re}_{x_{tr}} = 3 \times 10^6$ is a typical natural transition Reynolds number on a smooth flat plate in low-turbulence test rigs. The dashed line in Fig. 6.20 represents the development of an imaginary turbulent boundary layer, starting with zero thickness at $x = x_0$ and matching the thickness of the real boundary layer at $x = x_t$. The point $x = x_0$ is called the "effective" or "virtual" origin of the turbulent boundary layer. In practice, it is estimated by upstream extrapolations of the formulas for growth of the real boundary layer. Strictly speaking, of course, these formulas are not valid below about $R_\theta = 400$, but the effective-origin concept is useful though not completely rigorous.

For a two-dimensional zero pressure gradient flow, by using the definition of local skin-friction coefficient, c_f, the momentum integral equation (3.50) can be written as

$$\frac{dR_\theta}{d\mathrm{Re}_x} = \frac{c_f}{2} \tag{6.7.1}$$

where $R_\theta = u_e\theta/\nu$ and $\mathrm{Re}_x = u_e x/\nu$. Denote $(2/c_f)^{1/2}$ by z, assuming that the transition from laminar to turbulent flow takes place instantaneously, ($\mathrm{Re}_{x_{tr}} = \mathrm{Re}_{x_t}$), and using integration by parts, we can express Eq. (6.7.1) in the form

$$\mathrm{Re}_x = z^2 R_\theta - 2\int_{z_{tr}}^{z} R_\theta z \, dz + \text{const} \tag{6.7.2}$$

Equation (6.7.2) can be used to obtain a relationship between Re_x and, say, c_f, if we express R_θ as a function of z and integrate the integral in Eq. (6.7.2). That can be done in a number of ways. Cebeci and Smith (1974) describe a procedure that utilizes Coles' (1971) expression [Eq. (6.4.5)], together with the integral of the velocity-defect law for a flat plate, and write Eq. (6.7.2) as

$$(\mathrm{Re}_x - A_2)c_f = 0.324 \, \exp\left(\frac{0.58}{\sqrt{c_f}}\right)(1 - 8.125\sqrt{c_f} + 22.08c_f) \tag{6.7.3}$$

Here A_2 is an integration constant that depends on where transition occurs. It is given by

$$A_2 = \mathrm{Re}_{x_{tr}} - \frac{2R_{\theta_{tr}}}{c_{f_{tr}}} + \left(\frac{1.12}{\sqrt{c_{f_{tr}}}} - 7.16\right)\exp\left(\frac{0.58}{\sqrt{c_{f_{tr}}}}\right) \tag{6.7.4}$$

where $c_{f_{tr}}$ is the local skin-friction coefficient for turbulent flow calculated at the transition Reynolds number.

Putting $A_2 = 0$ in Eq. (6.7.3), taking logarithms, and making further approximations leads to formulas like

$$\frac{1}{\sqrt{c_f}} = a + b \log{(c_f \, \mathrm{Re}_x)} \tag{6.7.5a}$$

where a and b are constants chosen to get the best agreement with experiment. Such formulas have been derived by many previous workers. Von Karman (1932; see Schlichting, 1968) took $a = 1.7$ and $b = 4.15$ and approximated Eq. (6.7.3) as

$$\frac{1}{\sqrt{c_f}} = 1.7 + 4.15 \log{(c_f \, \mathrm{Re}_x)} \tag{6.7.5b}$$

A formula for the streamwise-average skin friction that makes use of Eq. (6.7.5b) was also obtained by Schoenherr (1932; also see Schlichting, 1968):

$$\frac{1}{\sqrt{\bar{c}_f}} = 4.13 \log{(\bar{c}_f \, \mathrm{Re}_x)} \tag{6.7.6}$$

Schoenherr also related $\bar{c}_f$ to R_θ and expressed the relationship as

$$\frac{1}{\sqrt{\bar{c}_f}} = 1.24 + 4.13 \log{R_\theta} \tag{6.7.7}$$

Equation (6.4.8), with $\Pi = 0.55$ for a constant-pressure boundary layer, gives

$$H = \frac{1}{1 - 4.67\sqrt{c_f}} \tag{6.7.8}$$

which can be used to replace R_θ by $R_{\delta *} \equiv u_e \delta^*/\nu = HR_\theta$ in the above formulas.

Much simpler, but less accurate, relations between δ, c_f, δ^*, θ, and H can also be obtained if we use the power-law assumption for the velocity distribution, namely, Eq. (1.3.16). The exponent n is about 7 in a constant-pressure boundary layer, increasing slowly with Reynolds number. Using Eq. (1.3.16) and the definitions of δ^*, θ, and H, we can show that

$$\begin{aligned} \frac{\delta^*}{\delta} &= \frac{1}{1 + n} \\[2mm] \frac{\theta}{\delta} &= \frac{n}{(1 + n)(2 + n)} \\[2mm] H &= \frac{2 + n}{n} \end{aligned} \tag{6.7.9}$$

Other formulas obtained from power-law assumptions, given by Schlichting (1968), are the following:

$$c_f = \frac{0.059}{\mathrm{Re}_x^{1/5}} \tag{6.7.10}$$

$$\bar{c}_f = \frac{0.074}{\mathrm{Re}_x^{1/5}} \tag{6.7.11}$$

$$\frac{\delta}{x} = \frac{0.37}{\mathrm{Re}_x^{1/5}} \tag{6.7.12}$$

$$\frac{\theta}{x} = \frac{0.036}{\mathrm{Re}_x^{1/5}} \tag{6.7.13}$$

Those equations are valid only for Reynolds numbers, Re_x, between 5×10^5 and 10^7, covering the laboratory range of most engineering applications. At high Reynolds numbers the boundary-layer thickness can be calculated more accurately by the following empirical formula given by Granville (1959):

$$\frac{\delta}{x} = \frac{0.0598}{\log \mathrm{Re}_x - 3.170} \tag{6.7.14}$$

Equation (6.7.6) was obtained on the assumption that the boundary layer is turbulent from the leading edge onward, that is, that the effective origin defined above is at $x = 0$. If the flow is turbulent but the Reynolds number is moderate, we should consider the portion of the laminar flow that precedes the turbulent flow. There are several empirical formulas for $\bar{c}_f$ that account for this effect. One is the formula quoted by Schlichting (1968). It is given by

$$\bar{c}_f = \frac{0.455}{(\log \mathrm{Re}_x)^{2.58}} - \frac{A}{\mathrm{Re}_x} \tag{6.7.15}$$

and another is

$$\bar{c}_f = \frac{0.074}{\mathrm{Re}_x^{1/5}} - \frac{A}{\mathrm{Re}_x} \qquad 5 \times 10^5 < \mathrm{Re}_x < 10^7 \tag{6.7.16}$$

Here A is a constant that depends on the transition Reynolds number Re_{x_t}. It is given by

$$A = \mathrm{Re}_{x_{tr}}(\bar{c}_{f_t} - \bar{c}_{f_l}) \tag{6.7.17}$$

where $\bar{c}_{f_t}$ and $\bar{c}_{f_l}$ correspond to the values of average skin-friction coefficient for turbulent and laminar flow at $\mathrm{Re}_{x_{tr}}$. We note that while Eq. (6.7.16) is restricted to the indicated Re_x range, Eq. (6.7.15) is valid for a wide range of Re_x and has given good results up to $\mathrm{Re}_x = 10^9$.

EXAMPLE 6.7

A thin flat plate is immersed in a stream of air at atmospheric pressure and at $25°C$ moving at a velocity of 50 m s^{-1}. Calculate the momentum thickness, boundary-layer thickness, local skin-friction coefficient, and average skin-friction coefficient at $x = 3$ m. Assume that $\nu = 1.5 \times 10^{-5} \text{ m}^2 \text{ s}^{-1}$ and $\text{Re}_{x_t} = 3 \times 10^6$.

Solution We assume that the momentum thickness is continuous at transition and that x_{tr} and x_t are close together. Then for laminar flow, from Eq. (5.5.8), we can write

$$R_{\theta_t} = 0.664 \sqrt{\text{Re}_{x_{tr}}} = 1150 \qquad\qquad Re_x = \frac{50 \times 3}{1.5 \times 10^{-5}} = 10^7 \qquad\qquad \text{(E6.7.1)}$$

To find the effective origin (see Fig. 6.20) of the turbulent boundary layer, we use Eq. (6.7.13) and write it, basing the Reynolds number on $x_t - x_o$, as

$$R_{\theta_t} = 0.036(\Delta\text{Re}_{x_t})^{4/5} \qquad\qquad \text{(E6.7.2)}$$

Substituting Eq. (E6.7.1) into Eq. (E6.7.2), we get

$$\Delta\text{Re}_{x_t} = 0.427 \times 10^6 \qquad\qquad \text{(E6.7.3)}$$

Thus the effective turbulent Reynolds number, based on $x - x_o$, is

$Re_x = Re_{x_t} + \Delta Re_{x_t}$

$$\text{Re}_x = 10^7 - 3 \times 10^6 + 0.427 \times 10^6 = 7.427 \times 10^6$$

$\dfrac{U_\infty(x-x_o)}{\nu} = \dfrac{U_\infty x}{\nu} - \dfrac{U_\infty x_T}{\nu} + \dfrac{U_\infty(x_T - x_o)}{\nu} = \dfrac{U_\infty}{\nu}\left[x - x_T + x_T - x_o\right] = \dfrac{U_\infty(x-x_o)}{\nu}$

To calculate the local boundary-layer parameters at $x = 3$ m, we use the power-law formulas. From Eq. (6.7.13),

$$\theta = \frac{0.036 \times 1.928}{(7.427 \times 10^6)^{0.20}} = 0.00293 \text{ m}$$

$\dfrac{U_\infty(x-x_o)}{\nu} \cdot \dfrac{\nu}{U_\infty x} = \dfrac{x-x_o}{x} = \dfrac{7.427 \times 10^6}{10^7} = .742$

$x - x_o = .7427\, x = .7427(3) = 2.2281$

From Eq. (6.7.12),

$$\delta = \frac{0.37 \times 1.928}{(7.427 \times 10^6)^{0.20}} = 0.030 \text{ m}$$

From Eq. (6.7.10),

$$c_f = \frac{0.0592}{(7.427 \times 10^6)^{0.20}} = 0.0025$$

To find $\bar{c}_f$, we use Eq. (6.7.15). With $\text{Re}_{x_t} = 3 \times 10^6$, from Eq. (6.7.11),

$$\bar{c}_{f_t} = \frac{0.074}{(3 \times 10^6)^{0.20}} = 0.00375$$

Since for a flat plate, $\bar{c}_f/2 = \theta/x$, from Eq. (5.5.8),

$$\bar{c}_{f_1} = 2\left(\frac{0.664}{\sqrt{Re_x}}\right) = \frac{1.328}{(3 \times 10^6)^{0.5}} = 0.000767$$

Thus from Eq. (6.7.16),

$$\bar{c}_f = \frac{0.074}{(10^7)^{0.20}} - \frac{3 \times 10^6}{10^7}(0.00375 - 0.000767) = 0.00205$$

6.7.2 Rough Flat Plate

A procedure similar to that used for a smooth flat plate can be used to obtain the boundary-layer parameters for an incompressible turbulent flow over a rough flat plate; see, for example, Cebeci and Smith (1974). The results can be represented in two graphs. Figures 6.21 and 6.22 show the variation of c_f and $\bar{c}_f$ with Re_x for sand-roughened flat plates. Also shown in these figures are the lines for constant-roughness Reynolds number, $R_k = u_e k/\nu$, and for constant relative roughness, x/k. As in the case of the smooth flat-plate problem, the origin of the turbulent boundary layer is assumed to be close to the leading edge of the plate.

In the completely rough regime it is possible to make use of the following interpolation formulas given by Schlichting (1968) for the coefficients of skin friction in terms of relative roughness:

$$c_f = \left(2.87 + 1.58 \log \frac{x}{k_s}\right)^{-2.5} \tag{6.7.18}$$

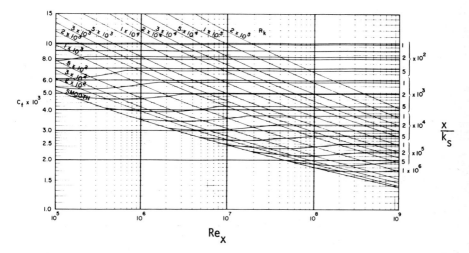

Fig. 6.21 Local skin-friction coefficient on sand-roughened flat plate.

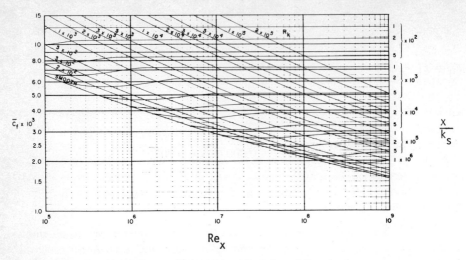

Fig. 6.22 Average skin-friction coefficient on sand-roughened flat plate.

$$\bar{c}_f = \left(1.89 + 1.62 \log \frac{x}{k_s}\right)^{-2.5} \tag{6.7.19}$$

which is valid for $10^2 < x/k_s < 10^6$.

EXAMPLE 6.8

Consider the flat-plate problem in Example 6.6, that is, a flat plate covered with square-bar roughness distribution. Compute the local skin-friction coefficient at $x = 1$ m for $u_e/\nu = 10^7$ m^{-1} and k (the roughness height) $= 0.1$ cm.

Solution From Example 6.6,

$$k_s = 3.97k = 3.97(0.1) = 0.397 \text{ cm}$$

Then

$$\frac{x}{k_s} = \frac{100}{0.397} = 264$$

From Fig. 6.21,

$$c_f = 8.6 \times 10^{-3}$$

Since fully rough conditions exist, we can also use Eq. (6.7.18), which gives

$$c_f = 8.6 \times 10^{-3}$$

The values of c_f and $\bar{c}_f$ presented in Figs. 6.21 and 6.22 are for sand type of roughness only. Furthermore, the values are for maximum sand density, that is, on the rough plates of Nikuradse the grains of sand are glued to the wall as close as possible to each other. In many practical applications the roughness density is considerably less. Also the roughness would be of different shapes. As was mentioned in Sec. 6.5, it is convenient to classify all types of roughness, except the anomalous d-type, by their "equivalent sand roughness," where equivalent sand roughness is defined as that value of k that gives the same skin-friction coefficient as the actual roughness (when it is substituted in the equations for c_f presented in Figs. 6.21 and 6.22, say). Schlichting (1968) and Hoerner (1958) report experimentally determined values of equivalent sand roughness for various types of rough surfaces. Some of the values determined for practical physical surfaces are given in Table 6.1. The d-type groove roughness, the effects of which do not scale on roughness height, is rare in practice.

EXAMPLE 6.9

Consider the flat-plate problem in Example 6.7, but assume that (a) the plate surface is covered with camouflage paint applied in mass production condition and (b) the plate surface is a dip-galvanized metal surface. Calculate the momentum thickness, boundary-layer thickness, local skin-friction coefficient, and average skin-friction coefficient at $x = 3$ m. As a simplification assume that roughness causes the transition to be at the leading edge so that we can neglect the contribution of laminar flow.

Solution (a) Table 6.1 gives the equivalent sand-grain roughness, k_s, as 3×10^{-3} cm. Then at $x = 3$ m, we have $x/k_s = 300/(3 \times 10^{-3}) = 10^5$. It follows from Fig. 6.22 that the painted surface is hydraulically smooth. Hence all the smooth-surface formulas apply. From Eqs. (6.7.13), (6.7.12), (6.7.10), and (6.7.11), we find, respectively, that

$$\theta = \frac{0.036(3)}{(10^7)^{0.20}} = 0.0043 \text{ m}$$

Table 6.1   Equivalent Sand Roughness for Several Types of Surfaces

Type of surface	k_s, cm
Aerodynamically smooth surface	0
Polished metal or wood	$0.05\text{--}0.2 \times 10^{-3}$
Natural sheet metal	0.4×10^{-3}
Smooth matte paint, carefully applied	0.6×10^{-3}
Standard camouflage paint, average application	1×10^{-3}
Camouflage paint, mass-production spray	3×10^{-3}
Dip-galvanized metal surface	15×10^{-3}
Natural surface of cast iron	25×10^{-3}

$$\delta = \frac{0.37(3)}{(10^7)^{0.20}} = 0.044 \text{ m}$$

$$c_f = \frac{0.0592}{(10^7)^{0.20}} = 0.00236$$

$$\bar{c}_f = \frac{0.074}{(10^7)^{0.20}} = 0.00295$$

(b) Table 6.1 gives the equivalent sand-grain roughness as 15×10^{-3} cm. Then at $x = 3$ m, $x/k_s = 300/(15 \times 10^{-3}) = 2 \times 10^4$. Figure 6.21 gives $c_f = 0.00305$, and Fig. 6.22 gives $\bar{c}_f = 0.00360$. Note that roughness effect is first felt at $\text{Re}_x = 2 \times 10^6$, or 0.6 m from the leading edge. To calculate the momentum thickness, we observe that for a flat plate

$$\frac{\theta}{x} = \frac{\bar{c}_f}{2}$$

or

$$\theta = \tfrac{3}{2}(0.0036) = 0.0054 \text{ m}$$

If we make the power law assumption and take $n = 7$, then from Eq. (6.7.9),

$$\delta = \tfrac{72}{7}\theta = 0.055 \text{ m}$$

6.8 TURBULENT BOUNDARY LAYERS WITH PRESSURE GRADIENT: HEAD'S METHOD

Most of the more refined methods for calculating turbulent boundary layers are based on the solution of PDEs, with the shear stress represented either by an eddy-viscosity formula like those discussed above or by an empirical transport equation. The proceedings of the 1968 Stanford conference on computation of turbulent boundary layers (Kline et al., 1969) gives a comprehensive review of the main types of method. More elaborate ones have appeared since, but they obey the same general principles. We postpone discussion of PDE-based methods until Chap. 8; here we present a simple integral method.

There are a number of integral methods for calculating two-dimensional turbulent boundary layers in pressure gradient, all of which use the momentum-integral equation, a formula relating c_f to the Reynolds number and a profile shape parameter, and an ODE for the rate of change of profile shape parameter with x. The skin-friction formula can be Eq. (6.4.5) or a wholly empirical data correlation. The shape-parameter equation can be seen to contain all the information about the Reynolds stress within the layer. It is usually a linear relation between $\theta \, dH/dx$ and

$(\theta/u_e) \, du_e/dx$, the coefficients being in general functions of $u_e\theta/\nu$ and H. A shape parameter equation of this kind (probably nonlinear) could be obtained from Eq. (6.4.1) with an eddy-viscosity or mixing-length assumption like those shown in Figs. 6.10 and 6.11 by MIR (method 2 of Sec. 5.6). The existing shape-parameter equations are nearly all wholly empirical (method 2a of Sec. 5.6). Note the conceptual difference between these methods and the Pohlhausen method (Sec. 5.6.1) in which the shape parameter, H, was related to the pressure gradient by an algebraic formula instead of an ODE.

A notably successful integral method of this kind is that of Head (1958); see also the improved version of Head and Patel (1969) and of Granville (1976a) for two-dimensional flows and (1976b) for thick axisymmetric flows. Head assumes that the dimensionless entrainment velocity, v_E/u_e [see Eq. (3.4.20)], is a function of a shape factor, H_1:

$$\frac{v_E}{u_e} = \frac{1}{u_e}\frac{d}{dx}[u_e(\delta - \delta^*)] = F(H_1) \tag{6.8.1}$$

Here H_1 is defined by

$$H_1 = \frac{\delta - \delta^*}{\theta} \tag{6.8.2}$$

Using Eq. (6.8.2) we can write Eq. (6.8.1) as

$$\frac{d}{dx}(u_e\theta H_1) = u_e F \tag{6.8.3}$$

Head also assumes that H_1 is related to the shape factor H by

$$H_1 = G(H) \tag{6.8.4}$$

The functions F and G are determined from experiment. A best fit to several sets of experimental data (see Cebeci et al., 1970) is

$$F = 0.0306(H_1 - 3.0)^{-0.6169} \tag{6.8.5}$$

$$G = \begin{cases} 0.8234(H - 1.1)^{-1.287} + 3.3 & H \leqslant 1.6 \\ 1.5501(H - 0.6778)^{-3.064} + 3.3 & H \geqslant 1.6 \end{cases} \tag{6.8.6}$$

If we write the momentum integral equation (3.4.8) as

$$\frac{d\theta}{dx} + (H + 2)\frac{\theta}{u_e}\frac{du_e}{dx} = \frac{c_f}{2} \tag{6.8.7}$$

we see that it has three unknowns: θ, H, and c_f for a given external-velocity distribution. Equation (6.8.3), with F, H, and G defined by Eqs. (6.8.4)-(6.8.6), provides a relationship between θ and H. Another equation relating c_f to θ and/or H is needed. Head used the c_f law given by Ludwieg and Tillmann (1949),

$$c_f = 0.246 \times 10^{-0.678H} R_\theta^{-0.268} \tag{6.8.8}$$

where $R_\theta = u_e \theta / \nu$. The system [Eqs. (6.8.3)-(6.8.8)], which includes two ODEs, can be solved numerically for a specified external-velocity distribution to obtain the boundary-layer development. We note that to start the calculations, say, at $x = x_0$, initial values of two of the three quantities θ, H, and c_f must be specified, the third following from Eq. (6.8.8). In practice, it may be easier to work H_1, rather than H, as a variable, as is done in the program below.

This method, like most integral methods, uses the shape factor, H, as the criterion for separation. Equation (6.8.8) predicts $c_f = 0$ only if H tends to infinity. It is not possible to give an exact value of H corresponding to separation, and a range between 1.8 and 2.4 is commonly quoted. The difference between the lower and upper limits of H makes only little difference in locating the separation point, since close to separation dH/dx is large.

6.8.1 Fortran Program for Head's Method

In this section we present a Fortran program for predicting the turbulent boundary-layer development on two-dimensional bodies by Head's method discussed in Sec. 6.8. The system of equations given in that section is integrated by the Runge-Kutta subroutine (RKM) (see Sec. 4.2.3) for a given set of initial conditions. It is programmed in Fortran IV for the IBM 370/165; it can be used on other computers with minor changes.

The input to Head's method is similar to Thwaites' method described in Sec. 5.6.3. The derivative du_e/dx is computed by using the three-point Lagrange interpolation formulas discussed in Sec. 8A.1.2. The input to the method consists of the following.

Card 1 The first two variables are integers with a field length of 3; remaining variables are F10.0 format.

NXT   Total number of x-stations, not to exceed 40.
KDIST   Index for surface distance; 2 when surface distance is input, $\neq 2$ when surface distance is calculated.
RTH(1) R_θ for first station.

NXT	KDIST	RTH(1)	T(1)	H(1)

Load sheet for card 1.

T(1) θ for first station, feet or meters.
H(1) H for first station.

Cards 2 to NXT+1   Variables are punched in F10.0 format, one set of three per card. Number of cards equal NXT of card 1.

X   Dimensional chordwise or surface distance, feet or meters. If KDIST=2, then X is the surface distance.
UE   Dimensional velocity, u_e, feet per second or meters per second.

```
      COMMON/SHARE/ NXT,VISC,X(40),UE(40),DUEDX(40),T(40),S(40),
     1             H(40),RTH(40),CF(40)
      DIMENSION Y(40)
C - - - - - - - - - - - - - - - - - - - - - - - - - - - - - - - -
      READ(5,2) NXT,KDIST,RTH(1),T(1),H(1)
      READ(5,3) (X(I),UE(I),Y(I),I=1,NXT)
      IF(KDIST .EQ. 2) GO TO 8
C  CALCULATE CHORDWISE DISTANCE
      XTEMP1= X(1)
      DO 5 I=2,NXT
      XTEMP2= X(I)
      X(I)  = X(I-1)+SQRT((XTEMP2-XTEMP1)**2+(Y(I)-Y(I-1))**2)
      XTEMP1= XTEMP2
    5 CONTINUE
      X(NXT+1) =2.*X(NXT)-X(NXT-1)
C  CALCULATE THE VELOCITY GRADIENT BY THREE POINT DIFFERENCE FORMULA
    8 NXT1  = NXT-1
      DX2   = X(2)-X(1)
      DO 10 I=2,NXT1
      DX1   = DX2
      DX2   = X(I+1)-X(I)
      DX3   = DX2+DX1
   10 DUEDX(I) = -DX2*UE(I-1)/(DX1*DX3)+(DX2-DX1)*UE(I)/(DX1*DX2)+
     1           DX1*UE(I+1)/(DX2*DX3)
      T1    = (2.0*X(1)-X(2)-X(3))/((X(1)-X(2))*(X(1)-X(3)))
      T2    = (X(1)-X(3))/((X(2)-X(1))*(X(2)-X(3)))
      T3    = (X(1)-X(2))/((X(3)-X(1))*(X(3)-X(2)))
      DUEDX(1) = T1*UE(1)+T2*UE(2)+T3*UE(3)
      T1    = (X(NXT)-X(NXT-1))/((X(NXT-2)-X(NXT-1))*(X(NXT-2)-X(NXT)))
      T2    = (X(NXT)-X(NXT-2))/((X(NXT-1)-X(NXT-2))*(X(NXT-1)-X(NXT)))
      T3    = (2.*X(NXT)-X(NXT-1)-X(NXT-2))/((X(NXT)-X(NXT-1))*
     1        (X(NXT)-X(NXT-2)))
      DUEDX(NXT)=T1*UE(NXT-2)+T2*UE(NXT-1)+T3*UE(NXT)
      VISC  = UE(1)*T(1)/RTH(1)
      S(1)  = UE(1)*T(1)*HOFH1(-H(1))
      CALL STNDRD
C
      WRITE(6,20)
      DO 14 I=1,NXT
      H1    = S(I)/UE(I)/T(I)
      DELST = T(I)*H(I)
      DELTA = T(I)*H1+DELST
      WRITE(6,15) I,X(I),T(I),H(I),DELST,DELTA,CF(I),RTH(I)
   14 CONTINUE
      STOP
C - - - - - - - - - - - - - - - - - - - - - - - - - - - - - - - -
    2 FORMAT(2I3,3F10.0)
    3 FORMAT(3F10.0)
   15 FORMAT(1H ,I3,8E14.6)
   20 FORMAT(1H0,1X,2HNX,6X,1HX,11X,5HTHETA,11X,1HH,11X,5HDELST,9X,
     1        5HDELTA,11X,2HCF,10X,6HRTHETA/)
      END
```

1	2	3	4	5	6	7	8	9	10	11	12	13	14	15	16	17	18	19	20	21	22	23	24	25	26	27	28	29	30
		X											UE										Y						

Load sheet for cards 2 to NXT + 1.

Y   Dimensional body ordinate, feet or meters. If KDIST=2, then it can be put as zero.

EXAMPLE 6.10

Consider the airfoil shown in Fig. 6.23a. Here the measurements, made by Newman (1951), include experimental velocity distribution, $u_e(x)$, various boundary-layer parameters such as c_f, θ, and H, and the separation point for a turbulent flow. Using Head's method, compute c_f, θ, H, and separation point and compare the

```
      SUBROUTINE STNDRD
      COMMON/SHARE/ NXT,VISC,X(40),UE(40),DUEDX(40),T(40),S(40),
     1             H(40),RTH(40),CF(40)
      DIMENSION C(2),B(2),Z(2),G(8)
C - - - - - - - - - - - - - - - - - - - - - - - - - - - - - - - - -
      N     = 2
      XX    = X(1)
      IS    = 0
      B(1)  = T(1)
      B(2)  = S(1)
      DX    = X(2)-X(1)
      UI    = UE(1)
      UP    = DUEDX(1)
      NXTP1 = NXT+1
      DO 100 I=2,NXTP1
      DO 110 LL=1,4
      GO TO (18,16,18,17), LL
   16 UI    = (UI+UE(I))/2.0+DX*(UP-DUEDX(I))/8.0
      UP    = (UP+DUEDX(I))/2.0
      GO TO 18
   17 UI    = UE(I)
      UP    = DUEDX(I)
   18 CONTINUE
      H1    = B(2)/B(1)/UI
      HB    = HOFH1(H1)
      RTHE  = UI*B(1)/VISC
      CFO2  = 0.123/10.0**(0.678*HB)/RTHE**0.268
      C(1)  = -(HB+2.0)*B(1)/UI*UP+CFO2
      C(2)  = UI*0.0306/(H1-3.0)**0.6169
      IF(LL .GT. 1) GO TO 19
      H(I-1)= HB
      RTH(I-1) = RTHE
      CF(I-1)  = 2.0*CFO2
      IF(I .GT. NXT) GO TO 100
   19 CONTINUE
  110 CALL RKM(XX,B,C,DX,N,Z,G,IS)
      T(I)  = B(1)
      S(I)  = B(2)
      DX    = X(I+1)-X(I)
  100 CONTINUE
      RETURN
      END
```

```
        FUNCTION HOFH1(A)
C    H = FUNCT(H1=A), INVERSE IF A NEGATIVE,
C    H1= FUNCT(H=-A)
        REAL C1/1.5501/,C2/0.6778/,C3/-3.064/,C4/3.3/,C5/0.8234/,
     1      C6/1.1/,C7/-1.287/
C - - - - - - - - - - - - - - - - - - - - - - - - - - - - - - -
        HOFH1 = 0.0
        IF(A .LT. -C6) GO TO 2
        IF(A .LE. C4) RETURN
        IF(A .LT. 5.3) GO TO 3
        HOFH1 = ((A-C4)/C5)**(1.0/C7)+C6
        RETURN
      3 HOFH1 = ((A-C4)/C1)**(1.0/C3)+C2
        RETURN
      2 IF(A .LT. -1.6) GO TO 4
        HOFH1 = C5*(-A-C6)**C7+C4
        RETURN
      4 HOFH1 = C1*(-A-C2)**C3+C4
        RETURN
        END
```

results with the experimental data. Assume initial conditions to be $R_\theta = 5509$, $\theta = 0.0676$ in., and $H = 1.5953$ at $x = 2.009$ ft. Data are tabulated by Coles and Hirst (1969) as run 3500.

Solution We use the experimental velocity distribution shown in Fig. 6.23a given as a function of surface distance, x, and choose approximately 22 x-stations to perform the boundary-layer calculations. We input these data under X, UE, and Y ($\equiv 0$) as required in the load sheets. Thus, with NXT = 22, KDIST = 2, RTH(1) = 5509, T(1) = 5.63 × 10^{-3} ft, and $H = 1.5953$, we perform the boundary-layer calculations with our computer program. Figure 6.23b shows a comparison of calculated and experimental boundary-layer parameters. It is seen from the computed results that as separation is approached, H increases sharply. The computed separation point ($x = 5.1$ ft) agrees well with the measured separation point ($x = 4.95$ ft). In fact, Newman's data were used by Head in choosing his correlations for F and G.

6.9 TURBULENT INTERNAL FLOWS

At sufficiently high Reynolds numbers, the flow in a duct becomes turbulent. According to experiments in pipes with disturbed entry conditions, the transition Reynolds number based on pipe diameter, d, that is,

$$R_d = \frac{u_{av}d}{\nu} \qquad (6.9.1)$$

is roughly 2000. However, as in external flows, the transition Reynolds number depends on many factors, the most important being the surface roughness and the entry conditions. According to the experiments of Pfenninger (1951) the upper limit of Reynolds number to maintain a laminar flow is as high as 500,000. Fully developed pipe flow is stable to infinitesimal disturbances, but the boundary layer in the entry region is not.

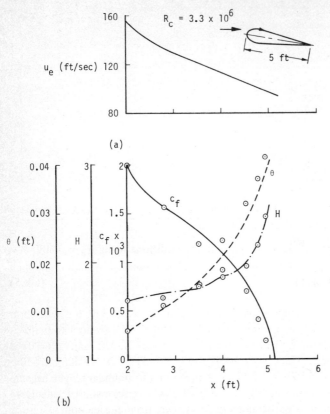

Fig. 6.23 (a) Experimental velocity distribution. (b) Comparison of calculated and experimental results. Circles denote the experimental data of Newman (1951).

Prediction of turbulent flows in the entrance region of internal flows requires the solution of the continuity and momentum equations together with a model for the Reynolds stresses. The numerical procedure can be similar to that outlined in Sec. 5.7; we shall discuss a complete calculation method in Chap. 8. In the fully developed part of the internal flow (no change with x) the problem is much simpler, and some important parameters such as friction factor and mean-velocity distribution can be obtained accurately and conveniently by using empirical laws and formulas.

6.9.1 Fully Developed Pipe Flows on Smooth Surfaces

Consider a fully developed flow in a circular pipe. In order to get a relationship between the friction factor, f, and Reynolds number, R_d, similar to the one obtained for laminar flow in Sec. 5.7.2, we first consider Eq. (6.2.3) and write it as

$$u = u_\tau \left[\frac{1}{\kappa} \ln \left(\frac{y u_\tau}{\nu} \right) + c \right] \qquad (6.9.2)$$

By arguments analogous to those used in self-preserving boundary layers in Sec. 6.2 we can write a velocity-defect law for a pipe as

$$\frac{u_{max} - u}{u_\tau} = f\left(\frac{y}{r_0}\right) \tag{6.9.3}$$

corresponding to Eq. (6.2.6) but with a different function f. Further, we can write the whole profile as

$$\frac{u}{u_\tau} = \frac{1}{\kappa} \ln \frac{u_\tau y}{\nu} + c + W\left(\frac{y}{r_0}\right) \tag{6.9.4}$$

where $W(y/r_0)$ corresponds to $\Pi(x)w(y/\delta)$ in Eq. (6.4.1). W is almost negligibly small in pipe flow, by coincidence and *not* because the inner-layer arguments are valid throughout the pipe. Then

$$\frac{u_{max}}{u_\tau} = \frac{1}{\kappa} \ln \frac{u_\tau r_0}{\nu} + c + W(1) \tag{6.9.5}$$

is a skin-friction law for the pipe. This form is not very convenient, but we can use Eq. (6.9.3) to relate the average velocity u_{av}, as defined in Eq. (5.7.16), to u_{max}; this move is the same in spirit as the replacement of $u_\tau \delta$ by $u_e \delta^*$ in the boundary-layer skin-friction formula of Sec. 6.4. Experimentally,

$$u_{av} = u_{max} - 4.07 u_\tau \tag{6.9.6}$$

(if the logarithmic law really held to the center of the pipe, the constant would be about 3.7). In terms of the pipe "friction factor," f, defined by Eq. (5.7.26) we can write

$$u_{max} = u_{av}(1 + 1.44\sqrt{f}) \tag{6.9.7}$$

Substituting into Eq. (6.9.5) gives

$$\sqrt{\frac{8}{f}}(1 + 1.33\sqrt{f}) = \frac{1}{0.41} \ln \left(\frac{u_{av} r_0}{\nu} \sqrt{\frac{f}{8}}\right) + 5 + W(1) \tag{6.9.8}$$

and taking $W(1) = -0.55$ gives

$$\frac{1}{\sqrt{f}} = 0.87 \ln R_d \sqrt{f} - 0.8 \tag{6.9.9}$$

which is Prandtl's friction law for smooth pipes. The corresponding law for parallel-plate ducts, where $u_{max} = u_{av} + 2.64 u_\tau$ and $W(1)$ is slightly positive, has an additive constant of -0.41 instead of -0.8. Equation (6.9.9) has been verified by

Nikuradse's experiments up to a Reynolds number of 3.4×10^6, and the agreement is seen to be excellent (see Schlichting, 1968). For ducts, see Dean (1977).

Another empirical formula that can be used to calculate the friction factor in smooth pipes has been obtained by Blasius (Schlichting, 1968); it is given by

$$f = \frac{0.3164}{R_d^{0.25}} \quad R_d \leqslant 10^5 \tag{6.9.10}$$

A lower limit of R_d is set by the requirement that the flow be turbulent. At $R_d = 10^5$, Eq. (6.9.10) gives $f = 0.0178$, while the more exact relation gives 0.0179; at $R_d = 2 \times 10^3$, the values are 0.0473 and 0.0497, respectively.

EXAMPLE 6.11

For air at atmospheric pressure and at $25°C$ flowing with $u_{av} = 15$ m s^{-1} in a long smooth pipe of 5 cm internal diameter, find the total dimensionless pressure drop Δp^* ($\equiv \Delta p / \rho u_{av}^2$) over a length of 20 m. Assume that $\nu = 1.5 \times 10^{-5}$ m^2 s^{-1}.

Solution

$$R_d = \frac{u_{av} d}{\nu} = \frac{(15)(0.05)}{1.5 \times 10^{-5}} = 5 \times 10^4 \tag{E6.11.1}$$

Since $R_d > 2000$, the flow is turbulent, and since the pipe is long, we can use the formula for fully developed flow. From Eq. (5.7.26) we can write

$$f = 2 \frac{d}{L} \frac{\Delta p}{\rho u_{av}^2} = 2 \frac{d}{L} \Delta p^*$$

or

$$\Delta p^* = \frac{f}{2} \frac{L}{d} \tag{E6.11.2}$$

From Eq. (6.9.10), $f = 0.021$. Then from Eq. (E6.11.2),

$$\Delta p^* = \frac{0.021}{2} \frac{20}{0.05} = 4.20$$

6.9.2 Fully Developed Pipe Flow on Rough Surfaces

According to the experiments of Nikuradse the variation of friction factor for fully developed pipe flow on rough surfaces depends on three distinct flow regimes discussed in Sec. 6.5. In the hydraulically smooth regime the variation of f is the same as the one for a smooth surface. In the transitional regime, f depends on the relative roughness, k/d, as well as on the Reynolds number, R_d. In the fully rough regime, f varies only with relative roughness and is independent of the Reynolds number.

Accurate friction factor formulas have also been obtained for flows in rough pipes. These formulas again make use of the law of the wall and were obtained in a manner similar to the one discussed in Sec. 6.9.1. We shall discuss only the one for the fully rough regime and present the others.

The law of the wall for a surface with uniform roughness is given by Eqs. (6.5.1) and (6.5.5). For the fully rough regime, with sand-grain roughness, Eq. (6.5.5) can be written as

$$u = u_\tau \left[\frac{1}{\kappa} \ln\left(\frac{y}{k}\right) + 8.5 \right] = u_\tau \left[\frac{2.303}{\kappa} \log\left(\frac{y}{k}\right) + 8.5 \right] \qquad (6.9.11)$$

Evaluating this at the pipe centerline, $y = r_0$, and making use of the relations given by Eq. (6.9.6), we obtain

$$f = \frac{1}{[2 \log (r_0/k) + 1.68]^2} \qquad (6.9.12a)$$

a relationship of similar form to Eq. (6.9.9), first obtained by Von Karman. According to the experimental data of Nikuradse a closer agreement with experiment is obtained if the constant 1.68 is changed to 1.74 in effect to allow for $W(1)$ in Eq. (6.9.4). With this change we get the following friction factor formula for the fully rough region (for sand-grain roughness):

$$f = \frac{1}{[2 \log (r_0/k) + 1.74]^2} \qquad (6.9.12b)$$

In practice, pipes have some degree of roughness different from that with sand roughness as prepared and measured by Nikuradse. The roughness in commercial pipes is not geometrically similar to the sand grains, and such roughness cannot be described by the roughness height alone. In addition, it is not practical to measure roughness of each pipe. Moody (1944; see Schlichting, 1968) has established an equivalent relative sand roughness, k_s/d, for each of the most frequently used types of pipe, corresponding to the Nikuradse sand roughness. Figure 6.24 shows the variation of friction factor, f, with Reynolds number, R_d, for different values of relative sand roughness, k_s/d, and Fig. 6.25 shows k_s/d as a function of pipe diameter, d, for different types of roughness in commercial pipes.

EXAMPLE 6.12

Determine the dimensionless pressure drop, Δp^* $(\equiv \Delta p/\rho u_{av}^2)$, due to the flow of 0.4 m³ s⁻¹ of oil through 400 m of 20 cm diameter cast-iron pipe. Assume that $\nu = 1 \times 10^{-5}$ m² s⁻¹.

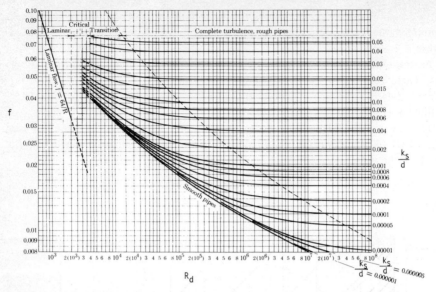

Fig. 6.24 Moody (1944) diagram.

Solution To find Δp^*, we need to know f according to Eq. (E6.11.2) in Example 6.11, that is,

$$\Delta p^* = \frac{f}{2}\frac{L}{d} \tag{E6.12.1}$$

The average velocity, u_{av}, and Reynolds number, R_d are

$$u_{av} = \frac{Q}{\pi r_0^2} = \frac{0.4}{\pi(0.10)^2} = 12.73 \text{ m s}^{-1}$$

$$R_d = \frac{u_{av}d}{\nu} = \frac{(12.73)(0.20)}{1 \times 10^5} = 2.546 \times 10^5$$

From Fig. 6.25 the relative roughness, k_s/d, is approximately 1.7×10^{-3}. From Fig. 6.24 the friction factor $f \approx 0.023$. Therefore from Eq. (E6.12.1),

$$\Delta p^* = \frac{0.023}{2}\frac{400}{0.20} = 23$$

Although it is not exact, the concept of equivalent diameter (see Sec. 5.7) can also be used to compute turbulent flows through ducts of cross section other than circular with reasonable accuracy. The Moody (1944) diagram applies as before.

EXAMPLE 6.13

Determine the dimensionless pressure drop, Δp^*, for flow of 3 m³ s⁻¹ of air at 25°C and 1 atm through a rectangular galvanized duct 50 cm wide, 25 cm high, and 500 m long. Assume that $\nu = 1 \times 10^{-5}$ m² s⁻¹.

Solution We must first determine whether the flow is laminar or turbulent. For a rectangular duct of sides a and b, the equivalent diameter, d_e (see Sec. 5.7), is

$$d_e = \frac{2\,ab}{a + b} = \frac{2(1/2)(1/4)}{1/2 + 1/4} = \frac{1}{3} \text{ m}$$

The cross-sectional area of flow is $ab = (\tfrac{1}{2})(\tfrac{1}{4}) = \tfrac{1}{8}$ m², therefore

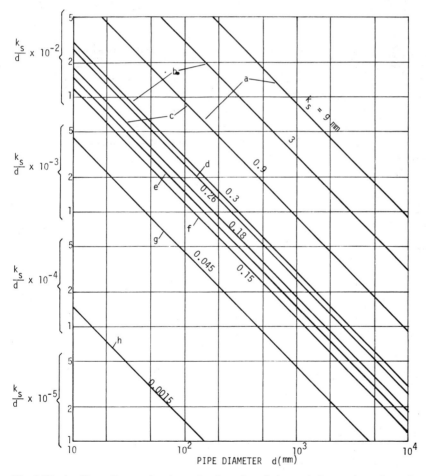

Fig. 6.25 Auxiliary diagram for the evaluation of equivalent relative sand roughness for commercial pipes, after Moody (1944): (a) riveted steel, (b) reinforced concrete, (c) wood, (d) cast iron, (e) galvanized steel, (f) bitumen-coated steel, (g) structural and forged steel, (h) drawn pipes.

$$u_{av} = \frac{3}{1/8} = 24 \text{ m s}^{-1}$$

The Reynolds number, R_{d_e}, is

$$R_{d_e} = \frac{24}{1 \times 10^{-5}} \frac{1}{3} = 8 \times 10^5$$

Since $R_{d_e} > 2000$, the flow may be taken to be turbulent. From Fig. 6.25 the relative roughness, k_s/d, is approximately 5×10^{-4}. From Fig. 6.24, the friction factor $f \simeq 0.0180$. Therefore from Eq. (E6.12.1) of Example 6.12,

$$\Delta p^* = \frac{0.0180}{2} \frac{500}{1/3} = 13.5$$

6.10 FREE SHEAR LAYERS

Turbulent free shear layers resemble the corresponding laminar flow more closely than is the case with boundary layers. It happens that the eddy viscosity in a fully developed jet or wake is roughly constant across the width of the layer except for the extreme edges. It follows that the velocity profile is quite close to the laminar one when it is normalized by the width and maximum-velocity difference. As usual, the eddy viscosity is large in comparison to the molecular viscosity and, purely for dimensional reasons, is proportional to the product of the width and the maximum-velocity difference. Therefore the growth rates of turbulent free shear layers are large in comparison to the laminar case and vary according to different power laws.

In this section we consider similar or self-preserving free shear layers, parallel to the discussion in Chap. 5. In practice, the main interest in free shear layers concerns either their early stage of development, for example, the wake near the trailing edge of an airfoil, or their distortion by a nonuniform flow or a solid surface, for example, the impinging jets in Fig. 1.2. We shall return to these more realistic, but more complicated, problems in Chap. 11.

Because free-shear-layer problems are either comparatively simple or very complicated, there are few advanced calculation methods analogous to those for boundary layers. Even to calculate self-preserving free shear layers, it is at present necessary to change the empirical constants in the calculation method from one flow to another, or to insert purely arbitrary interpolations. For example, the dimensionless eddy viscosity is twice as large in a wake as in a jet.

6.10.1 The Axisymmetric Jet

We recall that Eq. (5.8.11), namely,

$$\rho u_c^2 \frac{d\delta}{dx} \left[\frac{(f')^2}{\eta} + f \left(\frac{f'}{\eta} \right)' \right] + (\tau \eta)' = 0 \tag{5.8.11}$$

applies to both laminar and turbulent jets. For turbulent jets the contribution of the laminar shear stress to τ defined by Eq. (5.8.3) is small and can be neglected. Assuming that

$$\frac{\tau}{\rho} = -\overline{u'v'} = u_c^2 g(\eta) \tag{6.10.1}$$

we can write Eq. (5.8.11) as

$$\frac{d\delta}{dx}\left[\frac{(f')^2}{\eta} + f\left(\frac{f'}{\eta}\right)\right] + (g\eta)' = 0 \tag{6.10.2}$$

For similarity we must have

$$\frac{d\delta}{dx} = \text{const} = B$$

or

$$\delta = Bx \tag{6.10.3}$$

Therefore we can write Eq. (5.8.10) as

$$u_c = \frac{M}{Bx} \tag{6.10.4}$$

We have obtained the power laws for growth rate and centerline velocity decay rate without introducing a turbulence model, but to integrate Eq. (6.10.2), a relation between f' and g' is needed. If we use the eddy viscosity concept [see Eq. (6.1.1)] and let

$$\frac{\tau}{\rho} = \epsilon \frac{\partial u}{\partial r} = \frac{\epsilon u_c}{\delta} F' = \frac{\epsilon u_c}{\delta}\left(\frac{f'}{\eta}\right) \tag{6.10.5}$$

so that $g = \epsilon/u_c\delta(f'/\eta)'$ with ϵ independent of y—a good approximation in practice—we can write Eq. (6.10.2) as

$$\frac{u_c\delta}{\epsilon}\frac{d\delta}{dx}\left[\frac{(f')^2}{\eta} + f\left(\frac{f'}{\eta}\right)\right] + \left[\eta\left(\frac{f'}{\eta}\right)'\right]' = 0 \tag{6.10.6}$$

For similarity we must have

$$\frac{MB}{\epsilon} = \text{const} = A \tag{6.10.7}$$

If we define

$$\zeta = a\eta \qquad f(\eta) = \beta\phi(\zeta) \tag{6.10.8}$$

then we can express Eq. (6.10.6) in the same form as Eq. (5.8.15) by taking

$$\beta = \frac{1}{A} \tag{6.10.9}$$

and write its solution as

$$\phi(\zeta) = \frac{1/2\zeta^2}{1 + 1/8\zeta^2} \tag{6.10.10}$$

$$\frac{\phi'(\zeta)}{\zeta} = \frac{1}{(1 + 1/8\zeta^2)^2} \tag{6.10.11}$$

In terms of f and η, Eq. (6.10.11) can be written as

$$\frac{u}{u_c} = \frac{f'}{\eta} = \frac{1}{(1 + 1/8A\eta^2)^2} \tag{6.10.12}$$

Here we have chosen $a^2\beta = 1$ so that $u = u_c$ at $\eta = 0$, and from Eq. (6.10.9) determined that $a = \sqrt{A}$.

To determine the two empirical constants A and B, we use experimental data. If we write Eq. (6.10.7) as

$$\epsilon = \frac{B}{A}u_c\delta$$

and define δ as the y-distance where $u/u_c = \frac{1}{2}$, then according to the experimental data of Reichardt (1951; see Schlichting, 1968), $B = 0.0848$ and $A = 3.31$. Thus

$$\delta = 0.0848x \tag{6.10.13}$$

$$\epsilon = 0.0256\delta u_c \tag{6.10.14}$$

Inserting Eq. (6.10.12) into Eq. (5.8.9), we find

$$M = \frac{\sqrt{3}}{2}\sqrt{\frac{J/\rho}{2\pi}}\sqrt{A} \tag{6.10.15}$$

Therefore, from Eq. (6.10.4), with $A = 3.31$,

$$u_c = 7.41\frac{\sqrt{J/\rho}}{x} \tag{6.10.16}$$

The mass flow rate

$$\dot{m} = 2\pi \int_0^\infty \rho u r \, dr = 2\pi \rho \delta^2 u_c \int_0^\infty f' \, d\eta$$

can be obtained using the expressions for δ, u_c, and f'. It is

$$\dot{m} = 0.404x \sqrt{J\rho} \tag{6.10.17}$$

We note that, in contrast to laminar flow, $\dot{m}$ depends on J.

EXAMPLE 6.14

The jet in the "turbulence amplifier" shown in the sketch accompanying Example 5.10 is now forced to go turbulent by injecting a small mass flow through the control jet. Using the same figures as those in Example 5.10, calculate the pressure at the pressure tapping.

Solution From Eq. (6.10.16) we have

$$u = 7.41 \frac{\sqrt{J/\rho}}{x} = 1.49 \text{ m s}^{-1} \quad \text{at } x = 1 \text{ cm}$$

using $J = 4.81 \times 10^{-6}$ N from Example 5.10. Therefore the dynamic pressure (and the static-pressure rise at the pressure tapping) is now 1.32 N m^{-2} compared with 6.13 N m^{-2} in the laminar case. The object of the turbulence amplifier is to use a small mass flow (through the control jet) to alter the characteristics of a larger mass flow. With suitable proportions the change in pressure at the tapping can be made large enough to drive mechanical switches or other devices.

6.10.2 Two-Dimensional Wake

Equation (5.8.33) applies to both laminar and turbulent wakes. For turbulent wakes the contribution of the laminar shear stress to τ defined by Eq. (5.8.26) is small and can be neglected. Using Eq. (6.10.1), we can write Eq. (5.8.33) as

$$\frac{u_\infty \delta}{u_c^2} \left(f \frac{du_c}{dx} - u_c f' \frac{\eta}{\delta} \frac{d\delta}{dx} \right) - g' = 0 \tag{6.10.18}$$

As in the laminar case this applies only for $u_c \ll u_\infty$. For similarity we must have

$$\frac{u_\infty \delta}{u_c^2} \frac{du_c}{dx} = \text{const} \qquad \frac{u_\infty}{u_c} \frac{d\delta}{dx} = \text{const} \tag{6.10.19}$$

Since $u_c\delta = $ const [see Eq. (5.8.38)], we can integrate the second relation in Eq. (6.10.19) to get

$$\delta \sim x^{1/2} \tag{6.10.20}$$

Therefore

$$u_c \sim x^{-1/2} \tag{6.10.21}$$

Equations (6.10.20) and (6.10.21) are the power laws for growth rate and wake centerline velocity defect rate, obtained from Eq. (6.10.18) without introducing a turbulence model, as in the previous example concerning the axisymmetric turbulent jet. To integrate (6.10.18), we again use the eddy-viscosity concept and let $\tau/\rho = \epsilon\, \partial u/\partial y = \epsilon u_c/\delta f'$ so that $g = \epsilon/\delta u_c f'$. With ϵ independent of y, we get

$$-\frac{u_\infty \delta^2}{\epsilon u_c}\left(f\frac{du_c}{dx} - \frac{u_c}{\delta}f'\eta\frac{d\delta}{dx} \right) + f'' = 0 \tag{6.10.22}$$

For similarity we must have

$$\frac{u_\infty \delta^2}{\epsilon u_c}\frac{du_c}{dx} = \text{const} \qquad \frac{u_\infty \delta}{\epsilon}\frac{d\delta}{dx} = \text{const} = C_2 \tag{6.10.23}$$

Integration of the second relation in Eq. (6.10.23) gives

$$\delta = \left(\frac{2C_2\epsilon x}{u_\infty} \right)^{1/2} \tag{6.10.24}$$

Therefore

$$u_c = Ax^{-1/2} \tag{6.10.25}$$

Using Eqs. (6.10.24) and (6.10.25), we find that the constant in the first expression of (6.10.23) is $-C_2$. This enables us to write Eq. (6.10.22) as

$$f'' + C_2(\eta f)' = 0 \tag{6.10.26}$$

If we define

$$\varsigma = a\eta \qquad f(\eta) = \beta\phi(\varsigma) \tag{6.10.27}$$

then we can express Eq. (6.10.26) in the same form as Eq. (5.8.40) by taking $a = \sqrt{C_2}$ and writing its solution as

$$-\frac{u_1}{u_c} = f(\eta) = Be^{-C_2\eta^2/2} \tag{6.10.28}$$

or

$$u_1 = -ABe^{-C_2\eta^2/2}x^{-1/2} \tag{6.10.29}$$

To evaluate AB, we use Eqs. (5.8.38). Inserting f into Eq. (5.8.38) and using the definitions of δ and u_c as given by Eqs. (6.10.24) and (6.10.25), respectively, we get

$$AB = -\frac{F}{\rho b}(2\epsilon u_\infty)^{-1/2}(2\pi)^{-1/2} \tag{6.10.30}$$

We note that Eq. (6.10.30) is the same as Eq. (5.8.42) except that ϵ has replaced ν.

The expressions Eq. (6.10.29) and Eq. (6.10.30) give the mean-velocity distribution in the wake of a two-dimensional body provided that ϵ and C_2 are determined. According to the measurements conducted behind circular cylinders of diameter d (see Schlichting, 1968),

$$\delta = \tfrac{1}{4}(xC_Dd)^{1/2} \tag{6.10.31}$$

and

$$\frac{\epsilon}{u_\infty C_Dd} = 0.0222 \tag{6.10.32}$$

Here C_D is the drag coefficient for the cylinder defined by

$$C_D = \frac{F}{1/2\,\rho u_\infty^2\,bd} \tag{6.10.33}$$

Using Eqs. (6.10.31) and (6.10.32), we find from (6.10.24) that $C_2 = 1.41$, so that δ, defined as the value of y at $u_1 = 0.5u_c$, can be written as

$$\delta = 1.675\left(\frac{\epsilon x}{u_\infty}\right)^{1/2} \tag{6.10.34}$$

Solving Eq. (6.10.33) for F and inserting the resulting expression into Eqs. (6.10.30) and (6.10.29), we get the mean-velocity distribution behind a circular cylinder as

$$\frac{u_1}{u_\infty} = 0.141\left(\frac{u_\infty C_Dd}{\epsilon}\right)^{1/2}\left(\frac{x}{C_Dd}\right)^{-1/2}\exp\left(-0.70\eta^2\right) \tag{6.10.35}$$

where $\eta = y/\delta$ and δ is given by Eq. (6.10.34).

EXAMPLE 6.15

Two tubes in a cross-flow heat exchanger can be idealized as parallel circular cylinders of 1 cm diameter, 10 cm apart. Find the distance downstream at which the two wakes meet, taking the sectional drag coefficient of each cylinder as 1.0.

Solution From Eq. (6.10.34) we have

$$\delta = 1.675 \times (0.0222 u_\infty C_D \ dx/u_\infty)^{1/2}$$

which is independent of u_∞, although unlike the laminar case it does depend on C_D. Taking the edge of the wake as the point where $\exp(-1.41\eta^2/2) = 0.01$ for consistency with Example 5.11, we have $y/\delta = 2.56$. Requiring $2.56\delta = 5$ cm we get

$$x = \left(\frac{5 \times 10^{-2}}{2.56 \times 1.675}\right)^2 \frac{1}{0.0222 \times 10^{-2}} = 0.61 \text{ m}$$

This is somewhat too small a distance for accurate use of the formulas.

Table 6.2 Power Laws for Width and Centerline Velocity of Turbulent Similar Free Shear Layers

Flow	Sketch	Width, δ	Centerline velocity, $u_c(x)$ or u_1
Two-dimensional jet		x	$x^{-1/2}$
Axisymmetric jet		x	x^{-1}
Two-dimensional wake		$x^{1/2}$	$x^{-1/2}$
Axisymmetric wake		$x^{1/3}$	$x^{-2/3}$
Two uniform streams		x	x^0

6.10.3 Power Laws for the Width and the Centerline Velocity of Similar Free Shear Layers

Similar to laminar flows (see Sec. 5.8.7) the variations of the width, δ, and the centerline velocity, u_c or u_1, of several turbulent shear layers are summarized in Table 6.2.

PROBLEMS

6.1. Water at $20°C$ flows at a velocity of 3 m s^{-1} past a flat plate. Use Eq. (6.1.1) to find the transition position, and then use Eq. (6.7.15) and the results of Prob. 5.9 to determine the average skin-friction drag of the first 10 m of the plate. Check the contribution of the turbulent portion by Head's method, assuming that $H = 1.5$ at the end of transition.

6.2. Using the mixing-length formula [Eq. (6.2.4b)], find u/u_τ outside the viscous sublayer but inside the limit of validity of Eq. (6.2.4), say $y = 0.2\delta$, in the case where $\tau = \tau_w + \alpha y$. Hint: Require compatibility with the logarithmic law in the limit $\alpha \to 0$, analogous to the derivation of Eq. (6.6.4) for a flow with transpiration.

6.3. By using Eq. (6.4.10), show that ϵ_m is proportional to y^4 for $(y/y_l) \ll 1$.

6.4. Show that for equilibrium boundary layers at high Reynolds numbers, if α in Eq. (6.3.10b) is 0.0168, then α_1 in Eq. (6.3.10a) must be 0.0635 for $\kappa = 0.41$.

6.5. Derive Eq. (6.4.6). Hint: Write the definition of δ^* as

$$\delta^* = \int_0^\infty \left(\frac{u_e - u}{u_e}\right) dy = \frac{\delta u_\tau}{u_e} \int_0^1 \left(\frac{u_e}{u_\tau} - \frac{u}{u_\tau}\right) d\left(\frac{y}{\delta}\right)$$

Substitute Eqs. (6.4.1) and (6.4.5) into the above expression and integrate the resulting expression.

6.6. Derive Eqs. (6.4.7), (6.4.8a), and (6.4.8b).

6.7. Express Eq. (6.4.5) in terms of (a) $u_e\delta^*/\nu$ and (b) $u_e\theta/\nu$ for a boundary layer in zero pressure gradient, inserting numerical values where possible.

6.8. Compare the prediction of Eq. (6.7.13) for c_f at $u_e\theta/\nu = 10^4$ with that obtained in Prob. 6.7.

6.9. Show that Head's assumptions predict that in order to maintain constant H in a strongly retarded boundary, u_e must vary as a negative power of x. Neglect the contribution of the skin-friction coefficient to the momentum integral.

6.10. The external velocity distribution for the symmetrical NACA 0012 airfoil is given in Prob. 5.20. Compute the complete boundary-layer development (laminar and turbulent) on this airfoil for two chord Reynolds number, $R_c = 6 \times 10^6$ and 9×10^6 by using Thwaites' and Head's methods. Compute the transition position by Eq. (6.1.1). To start the turbulent flow calculation in Head's method, assume continuity in momentum thickness and take $H = 1.5$. Discuss the effect of R_c on the location of transition.

6.11. Determine the equivalent sand-grain height for the wire netting tested by Hama (1954) and shown in Fig. 6.15. Assume fully rough conditions.

6.12. Show that the velocity defect in the wake of a tall building, approximating a two-dimensional cylinder with a diameter of 100 ft (30 m), exceeds 10% of the wind velocity for a distance of 1 mile downstream of the building.

6.13. Trailing vortices from an airliner can endanger following aircraft. Do the jet exhausts significantly affect the decay of the trailing vortices by enhancing turbulent mixing? A simplified version of this question is to ask whether the jet velocity at the vortex position (say 20 nozzle diameters outboard of the jet axis) ever exceeds, say, 5% of the exhaust velocity. Answer the question, assuming that the velocity profile in a circular jet in still air can be approximated by $u/u_c = \frac{1}{2}(1 + \cos \pi r/2R)$, where R is the radius at which $u/u_c = 0.5$ and the approximation applies for $r < 2R$ only. State the main assumptions made in simplifying the question and any further assumptions you make.

6.14. Derive the power laws for the thickness and centerline velocity, or velocity defect, for the two-dimensional and axisymmetric jet and wake, using mass and momentum conservation requirements without direct reference to the equations of motion. Hint: Assume that the entrainment velocity, like all other velocity scales in a similar or self-preserving flow, is proportional to the centerline velocity or velocity defect.

Chapter 7

Numerical Methods for Thin Shear Layers

7.1 INTRODUCTION

There are several well-developed computer programs available for solving the shear-layer equations (with given Reynolds stresses in the turbulent case). Some at least are capable of producing solutions for a wide range of shear layers (i.e., a wide range of boundary conditions). Since the TSL equations and their boundary conditions constitute a classical problem in parabolic PDEs, a full discussion belongs in a treatise on PDEs rather than in the present work. Packages available for public use include various versions of MIR and finite-difference methods, a typical example being the well-documented and widely used finite-difference method of Patankar and Spalding (1970). It is our opinion that the finite-difference methods are the most flexible, practical, and efficient tools for the solution of shear-layer equations. Since one of the objects of this book is to introduce readers to the numerical solution of shear-layer problems, we have decided to present one finite-difference method in detail and then use it in demonstration solutions of real problems. The student may care to reread Chap. 4 quickly before proceeding.

We have chosen to present the Keller and Cebeci (1971) "box" method. We chose this method partly because it is familiar to us and partly because it seems to be the most flexible of the common methods, being easily adaptable to solving equations of any order, whereas many other methods are designed for the solution of the second-order equations (like the laminar TSL equations) that appear when heat and momentum transfer rates are assumed to depend on diffusivities. We shall use this method in Chap. 8 to obtain the solutions of TSL equations for a variety of laminar and turbulent flows and in Chap. 9 to solve the stability equation.

One of the basic ideas of the box method, originating with Keller (1970), is to write the governing system of equations in the form of a first-order system (Sec. 4.2.1). In the TSL equations some second derivatives with respect to the cross-stream variable, such as $\partial^2 u/\partial \eta^2$, appear. First derivatives of u and other quantities with respect to the cross-stream variable must therefore be introduced as new unknown functions. Derivatives with respect to all other, streamwise or spanwise, variables occur only to first order as a consequence of the TSL approximations. With the resulting first-order equations and on an arbitrary rectangular net, Fig. 7.1, we use simple "centered-difference" derivatives and averages at the midpoints of net rectangles and net segments, as they are required, to get finite-difference equations with a truncation error of order $(\Delta \eta)^2$.

This scheme is unconditionally stable, but of course the equations are implicit and nonlinear. Newton's method is employed to solve them. In order to do this with an efficient and stable computational scheme, a block-tridiagonal factorization scheme is employed on the coefficient matrix of the finite-difference equations for all η at a given ξ.

For internal flows, as discussed in Sec. 5.1, the pressure gradient is not known prior to the calculations and must be calculated during the calculations. In Sec. 7.3 we consider the numerical solution of such flows.

7.2 NUMERICAL FORMULATION OF MOMENTUM EQUATION FOR EXTERNAL FLOWS

It is helpful to transform the equations so that the boundary-layer thickness in transformed coordinates (ξ, η) is nearly independent of streamwise distance and can therefore be represented by a fixed number of profile points at fixed spacing. In laminar flows this is achieved by the Falkner-Skan transformation (Sec. 5.5), which reduces the apparent growth rate in turbulent flow also. It is therefore convenient, though not necessary, to apply the box method to Eq. (5.3.5) rather than to Eq.

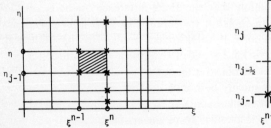 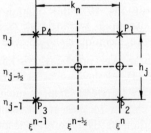

Fig. 7.1 Net rectangle for difference approximations.

(5.3.9), and the resulting program can be used for axisymmetric as well as two-dimensional flows.

We write Eq. (5.3.5) in terms of a *first-order system* of PDEs, closely following the procedure used for the Falkner-Skan ODE for similar flows in Sec. 4.2.2. For this purpose we introduce new dependent variables, $u(\xi, \eta)$ and $v(\xi, \eta)$, so that Eq. (5.3.5) can be written as

$$f' = u \tag{7.2.1a}$$

$$u' = v \tag{7.2.1b}$$

$$(bv)' + \left(\frac{m+1}{2}\right)fv + m(1 - u^2) = \xi\left(u\frac{\partial u}{\partial \xi} - v\frac{\partial f}{\partial \xi}\right) \tag{7.2.1c}$$

with $\xi = \bar{x}$ and $b = (1 + t)^{2k}$. Recall from Sec. 3.2 that $k = 0$ for plane flows and $k = 1$ for axisymmetric flows, and recall from Sec. 4.2.2 that while u is related to the usual x-component velocity, v is not the y-component velocity. The boundary conditions [Eq. (5.3.8)], three in number for this third-order system, are

$$f(\xi, 0) = f_w(\xi) \qquad u(\xi, 0) = 0 \qquad u(\xi, \eta_\infty) = 1 \tag{7.2.2}$$

We now consider the net rectangle shown in Fig. 7.1. We denote the net points by

$$\begin{aligned}\xi_0 &= 0 \qquad \xi^n = \xi^{n-1} + k_n \qquad n = 1, 2, \ldots, N \\ \eta_0 &= 0 \qquad \eta_j = \eta_{j-1} + h_j \qquad j = 1, 2, \ldots, J \qquad \eta_J \equiv \eta_\infty\end{aligned} \tag{7.2.3}$$

Here n and j are just sequence numbers, *not* tensor indices or exponents. The use of a superscript for the timelike direction is conventional.

We approximate the quantities (f, u, v) at points (ξ^n, η_j) of the net by (f_j^n, u_j^n, v_j^n), which we shall call net functions. We also employ the notation g_j^n for points and quantities midway between net points and for *any* net function:

$$\xi^{n-1/2} \equiv \tfrac{1}{2}(\xi^n + \xi^{n-1}) \qquad \eta_{j-1/2} \equiv \tfrac{1}{2}(\eta_j + \eta_{j-1}) \tag{7.2.4a}$$

$$g_j^{n-1/2} \equiv \tfrac{1}{2}(g_j^n + g_j^{n-1}) \qquad g_{j-1/2}^n \equiv \tfrac{1}{2}(g_j^n + g_{j-1}^n) \tag{7.2.4b}$$

We now write the difference equations that are to approximate Eq. (7.2.1) by considering one mesh rectangle as in Fig. 7.1. The reader who is unfamiliar with finite-difference techniques should read the following passage slowly, referring constantly to Fig. 7.1. We start by writing the finite-difference approximations of Eqs. (7.2.1a) and (7.2.1b) for the midpoint $(\xi^n, \eta_{j-1/2})$ of the segment $P_1 P_2$, using centered-difference derivatives: this process is called "centering about $(\xi^n, \eta_{j-1/2})$." We get

$$\frac{f_j^n - f_{j-1}^n}{h_j} = u_{j-1/2}^n \tag{7.2.5a}$$

$$\frac{u_j^n - u_{j-1}^n}{h_j} = v_{j-1/2}^n \tag{7.2.5b}$$

Similarly, Eq. (7.2.1c) is approximated by centering about the midpoint ($\xi^{n-1/2}$, $\eta_{j-1/2}$) of the rectangle $P_1P_2P_3P_4$. This can be done in two steps. In the first step we center Eq. (7.2.1c) about the point ($\xi^{n-1/2}, \eta$) without specifying η. If we denote the left-hand side of Eq. (7.2.1c) by L, then the difference approximation to Eq. (7.2.1c) is

$$\tfrac{1}{2}(L^n + L^{n-1}) = \xi^{n-1/2}\left[u^{n-1/2}\left(\frac{u^n - u^{n-1}}{k_n}\right) - v^{n-1/2}\left(\frac{f^n - f^{n-1}}{k_n}\right)\right]$$

Rearranging this equation and using Eq. (7.2.4b), we can write

$$[(bv)']^n + \alpha_1(fv)^n - \alpha_2(u^2)^n + \alpha(v^{n-1}f^n - f^{n-1}v^n) = R^{n-1} \tag{7.2.6}$$

where we have used the abbreviations

$$\alpha = \frac{\xi^{n-1/2}}{k_n} \qquad \alpha_1 = \frac{m^n + 1}{2} + \alpha \qquad \alpha_2 = m^n + \alpha \tag{7.2.7a}$$

$$R^{n-1} = -L^{n-1} + \alpha[(fv)^{n-1} - (u^2)^{n-1}] - m^n \tag{7.2.7b}$$

$$L^{n-1} \equiv \left[(bv)' + \frac{m+1}{2}fv + m(1 - u^2)\right]^{n-1} \tag{7.2.7c}$$

and where the identity sign introduces a useful shorthand: $[\quad]^{n-1}$ means that the quantity in square brackets is evaluated at $\xi = \xi^{n-1}$.

We next center Eq. (7.2.6) about the point ($\xi^{n-1/2}, \eta_{j-1/2}$), that is, we choose $\eta = \eta_{j-1/2}$ and get

$$h_j^{-1}(b_j^n v_j^n - b_{j-1}^n v_{j-1}^n) + \alpha_1(fv)_{j-1/2}^n - \alpha_2(u^2)_{j-1/2}^n$$
$$+ \alpha(v_{j-1/2}^{n-1}f_{j-1/2}^n - f_{j-1/2}^{n-1}v_{j-1/2}^n) = R_{j-1/2}^{n-1} \tag{7.2.8}$$

where

$$R_{j-1/2}^{n-1} = -L_{j-1/2}^{n-1} + \alpha[(fv)_{j-1/2}^{n-1} - (u^2)_{j-1/2}^{n-1}] - m$$
$$\equiv \{-L_{j-1/2} + \alpha[(fv)_{j-1/2} - u_{j-1/2}^2] - m\}^{n-1}$$

$$L_{j-1/2}^{n-1} = \left\{h_j^{-1}(b_j v_j - b_{j-1}v_{j-1}) + \frac{m+1}{2}(fv)_{j-1/2} + m[1 - (u^2)_{j-1/2}]\right\}^{n-1}$$

Equations (7.2.5) and (7.2.8) are imposed for $j = 1, 2, \ldots, J - 1$ at given n, and the transformed boundary-layer thickness, η_J, is to be sufficiently large so that

$u \to 1$ asymptotically. The latter is usually satisfied when $v(\eta_\infty)$ is less than approximately 0.001.

The boundary conditions [Eq. (7.2.2)] yield, at $\xi = \xi^n$,

$$f_0^n = f_w(\xi) \qquad u_0^n = 0 \qquad u_J^n = 1 \tag{7.2.9}$$

7.2.1 Newton's Method

If we assume f_j^{n-1}, u_j^{n-1}, and v_j^{n-1} to be known for $0 \leqslant j \leqslant J$, then Eqs. (7.2.5), (7.2.8), and (7.2.9) are a system of $3J + 3$ equations for the solution of $3J + 3$ unknowns (f_j^n, u_j^n, v_j^n), $j = 0, 1, \ldots, J$. We shall solve this nonlinear system by means of Newton's method, already introduced for a simpler problem in Sec. 4.2.2. For simplicity of notation we shall write the unknowns at $\xi = \xi^n$ as $(f_j^n, u_j^n, v_j^n) \equiv (f_j, u_j, v_j)$. Then the system [Eq. (7.2.5)] and [Eq. (7.2.8)] can be written as

$$f_j - f_{j-1} - \frac{h_j}{2}(u_j + u_{j-1}) = 0 \tag{7.2.10a}$$

$$u_j - u_{j-1} - \frac{h_j}{2}(v_j + v_{j-1}) = 0 \tag{7.2.10b}$$

$$h_j^{-1}(b_j v_j - b_{j-1} v_{j-1}) + \alpha_1(fv)_{j-1/2} - \alpha_2(u^2)_{j-1/2}$$
$$+ \alpha(v_{j-1/2}^{n-1} f_{j-1/2} - f_{j-1/2}^{n-1} v_{j-1/2}) = R_{j-1/2}^{n-1} \tag{7.2.10c}$$

We note that $R_{j-1/2}^{n-1}$ involves only known quantities if we assume that the solution is known on $\xi = \xi^{n-1}$. To solve Eq. (7.2.10), with Eq. (7.2.9), by Newton's method, we introduce the iterates $[f_j^{(i)}, u_j^{(i)}, v_j^{(i)}]$, $i = 0, 1, 2, \ldots$, with initial values, say,

$$
\begin{aligned}
f_0^{(0)} &= f_w(\xi_n) & u_0^{(0)} &= 0 & v_0^{(0)} &= v_0^{n-1} \\
f_j^{(0)} &= f_j^{n-1} & u_j^{(0)} &= u_j^{n-1} & v_j^{(0)} &= v_j^{n-1} & 1 \leqslant j \leqslant J-1 \\
f_J^{(0)} &= f_J^{n-1} & u_J^{(0)} &= 1 & v_J^{(0)} &= v_J^{n-1}
\end{aligned}
\tag{7.2.11}
$$

The superscripts in parentheses have of course no relation to the superscript n. For the higher-order iterates we set

$$f_j^{(i+1)} = f_j^{(i)} + \delta f_j^{(i)} \qquad u_j^{(i+1)} = u_j^{(i)} + \delta u_j^{(i)} \qquad v_j^{(i+1)} = v_j^{(i)} + \delta v_j^{(i)} \tag{7.2.12}$$

Then we insert the right-hand sides of these expressions in place of f_j, u_j, and v_j in Eq. (7.2.10) and drop the terms that are *quadratic* in $(\delta f_j^{(i)}, \delta u_j^{(i)}, \delta v_j^{(i)})$. This procedure yields the following *linear* system (in which we drop the superscript i in δ quantities for simplicity):

$$\delta f_j - \delta f_{j-1} - \frac{h_j}{2}(\delta u_j + \delta u_{j-1}) = (r_1)_j \tag{7.2.13a}$$

$$\delta u_j - \delta u_{j-1} - \frac{h_j}{2}(\delta v_j + \delta v_{j-1}) = (r_3)_{j-1} \tag{7.2.13b}$$

$$(s_1)_j \delta v_j + (s_2)_j \delta v_{j-1} + (s_3)_j \delta f_j + (s_4)_j \delta f_{j-1} + (s_5)_j \delta u_j + (s_6)_j \delta u_{j-1} = (r_2)_j \quad (7.2.13c)$$

where

$$(r_1)_j = f_{j-1}^{(i)} - f_j^{(i)} + h_j u_{j-1/2}^{(i)} \tag{7.2.14a}$$

$$(r_3)_{j-1} = u_{j-1}^{(i)} - u_j^{(i)} + h_j v_{j-1/2}^{(i)} \tag{7.2.14b}$$

$$(r_2)_j = R_{j-1/2}^{n-1} - [h_j^{-1}(b_j^{(i)} v_j^{(i)} - b_{j-1}^{(i)} v_{j-1}^{(i)}) + \alpha_1 (fv)_{j-1/2}^{(i)} - \alpha_2 (u^2)_{j-1/2}^{(i)}$$
$$+ \alpha(v_{j-1/2}^{n-1} f_{j-1/2}^{(i)} - f_{j-1/2}^{n-1} v_{j-1/2}^{(i)})] \tag{7.2.14c}$$

$$(s_1)_j = h_j^{-1} b_j^{(i)} + \frac{\alpha_1}{2} f_j^{(i)} - \frac{\alpha}{2} f_{j-1/2}^{n-1} \tag{7.2.14d}$$

$$(s_2)_j = -h_j^{-1} b_{j-1}^{(i)} + \frac{\alpha_1}{2} f_{j-1}^{(i)} - \frac{\alpha}{2} f_{j-1/2}^{n-1} \tag{7.2.14e}$$

$$(s_3)_j = \frac{\alpha_1}{2} v_j^{(i)} + \frac{\alpha}{2} v_{j-1/2}^{n-1} \tag{7.2.14f}$$

$$(s_4)_j = \frac{\alpha_1}{2} v_{j-1}^{(i)} + \frac{\alpha}{2} v_{j-1/2}^{n-1} \tag{7.2.14g}$$

$$(s_5)_j = -\alpha_2 u_j^{(i)} \tag{7.2.14h}$$

$$(s_6)_j = -\alpha_2 u_{j-1}^{(i)} \tag{7.2.14i}$$

To complete the system [Eq. (7.2.13)] we recall the boundary condition [Eq. (7.2.9)], which can be satisfied exactly with no iteration. In Eq. (7.2.11) we have altered the initial values of f_0, u_0, and u_J accordingly (i.e., we have set $f_0 = f_w(\xi_n)$, $u_0 = 0$, and $u_J = 1$), so to maintain these correct values in all the iterates we take

$$\delta f_0 = 0 \qquad \delta u_0 = 0 \qquad \delta u_J = 0 \tag{7.2.15}$$

The linear system [Eqs. (7.2.13) and (7.2.15)] can be solved in a very efficient manner by using the block-elimination method discussed in Appendix 7A. Beginners, or readers concerned only with external flows, can skip Sec. 7.3; readers uninterested in numerical details can skip the Appendices.

7.3 NUMERICAL FORMULATION OF MOMENTUM EQUATION FOR INTERNAL FLOWS

The numerical formulation of the momentum equations given by Eqs. (5.7.3) and (5.7.5) or by Eqs. (5.7.9) and (5.7.11) for internal flows resembles that of external flows except for the dp^*/dx term. As was discussed in Sec. 5.7, the presence of that term introduces an additional unknown to the system given by Eqs. (5.7.3) and (5.7.5) or by Eqs. (5.7.9) and (5.7.11). Thus another equation is needed, and we use conservation of mass. For *flow between parallel plates*, mass balance gives

$$u_0 h = \int_0^h u \, dy \tag{7.3.1}$$

In terms of dimensionless variables, Y, and stream function, $F(x, Y)$, defined by Eqs. (5.7.7) and (5.7.8), and with $L \equiv h$ we can write Eq. (7.3.1) as

$$1 = \frac{F(x, \sqrt{R_L})}{\sqrt{R_L}} \tag{7.3.2}$$

When the stream function is expressed in transformed variables, we can write Eq. (7.3.1) in a form similar to that of Eq. (7.3.2), that is,

$$1 = \frac{f(x, \eta_{sp})}{\eta_{sp}} \tag{7.3.3}$$

where $\eta_{sp} = \sqrt{R_L}(x/L)^{-1}$. Similarly, *for the case of flow in a pipe* we can write equations similar to Eqs. (7.3.2) and (7.3.3), respectively. They are

$$1 = \frac{2F(\bar{x}, \sqrt{R_L})}{\sqrt{R_L}} \tag{7.3.4}$$

$$1 = \frac{2f(\bar{x}, \eta_{sp})}{\eta_{sp}} \tag{7.3.5}$$

In Eqs. (7.3.4) and (7.3.5) we have taken $L = r_0$. The extension to ducts of varying height is straightforward.

Since $dp^*/d\xi$ is not known, it is obvious that an iteration procedure is necessary, in addition to the iterative solution of the equations for given pressure gradient discussed above. In our study this is done by Newton's method. We first solve Eqs. (5.7.3) and (5.7.5) or Eqs. (5.7.9) and (5.7.11) for an assumed dp^*. This is a typical boundary-layer problem; for brevity we shall call it the "*standard problem.*" The next value of dp^* is obtained from Newton's method. For example, if we write Eq. (7.3.3) as

$$\phi(\beta^\nu) \equiv 1 - \frac{f(x, \eta_{sp})}{\eta_{sp}} = 0 \tag{7.3.6}$$

where we have denoted dp^* by $-\beta$ for convenience (it is not the same as the pressure gradient parameter β of Chap. 5) then the next value of $\beta^{\nu+1}$ is obtained from

$$\beta^{\nu+1} = \beta^\nu - \frac{\phi(\beta^\nu)}{\partial\phi(\beta^\nu)/\partial\beta} \qquad \nu = 0, 1, 2, \ldots \tag{7.3.7}$$

The derivative of ϕ with respect to β is obtained from Eq. (7.3.6):

$$\frac{\partial \phi}{\partial \beta}(\beta^\nu) = -\frac{1}{\eta_{sp}}\frac{\partial f}{\partial \beta}(x, \eta_{sp}) \tag{7.3.8}$$

The derivative of f with respect to β is obtained by solving a system of variational equations described in Appendix 7B. The iteration procedure is repeated until

$$|\beta^{\nu+1}(x_n) - \beta^\nu(x_n)| < \gamma_1 \tag{7.3.9}$$

where γ_1 is a small error tolerance.

7.3.1 Difference Equations for the Standard Problem When the Momentum Equation Is Expressed in Transformed Variables

Let us consider Eq. (5.7.3) and use the same dependent variables, u and v, defined by Eqs. (7.2.1a) and (7.2.1b) to write Eq. (5.7.3) in a form similar to that of Eq. (7.2.1c), that is,

$$(bv)' + \tfrac{1}{2}fv = \xi\frac{dp^*}{d\xi} + \xi\left(u\frac{\partial u}{\partial \xi} - v\frac{\partial f}{\partial \xi}\right) \tag{7.3.10}$$

with $\xi = \bar{x}$ and $b = (1-t)^{2k}$.

The boundary conditions are

$$f(\xi, 0) = f_w(\xi) \qquad u(\xi, 0) = 0 \tag{7.3.11a}$$

$$\eta = \eta_\infty \qquad \frac{d}{d\xi}\left(\frac{u^2}{2}\right) = -\frac{dp^*}{d\xi} \tag{7.3.11b}$$

The difference equation for Eq. (7.3.10) is

$$h_j^{-1}(b_j^n v_j^n - b_{j-1}^n v_{j-1}^n) + \alpha_1(fv)_{j-1/2}^n + \alpha[v_{j-1/2}^{n-1}f_{j-1/2}^n - (u^2)_{j-1/2}^n - f_{j-1/2}^{n-1}v_{j-1/2}^n]$$

$$= R_{j-1/2}^{n-1} \tag{7.3.12}$$

where

$$\alpha_1 = \tfrac{1}{2} + \alpha \qquad \beta = (p^*)^{n-1} - (p^*)^n \tag{7.3.13a}$$

$$R_{j-1/2}^{n-1} = -L_{j-1/2}^{n-1} + \alpha[(fv)_{j-1/2}^{n-1} - (u^2)_{j-1/2}^{n-1} - 2\beta] \tag{7.3.13b}$$

$$L_{j-1/2}^{n-1} = [h_j^{-1}(b_j v_j - b_{j-1}v_{j-1}) + \tfrac{1}{2}(fv)_{j-1/2}]^{n-1} \tag{7.3.13c}$$

The boundary conditions [Eq. (7.3.11)] at $\xi = \xi^n$ are

$$f_0^n = f_w(\xi) \qquad u_0^n = 0 \tag{7.3.14a}$$

$$u_J^n = \sqrt{2\beta + (u_J^{n-1})^2} \tag{7.3.14b}$$

Using Newton's method, we can now linearize the differenced equations. Obviously, Eqs. (7.2.13a) and (7.2.13b) remain the same. Equation (7.3.12) can be linearized and expressed in the same form as that of Eq. (7.2.13c). The coefficients $(r_2)_j$ and $(s_k)_j$ ($k = 1$-6), in Eq. (7.2.14), remain the same provided that we set

$$\alpha_1 = \tfrac{1}{2} + \alpha \qquad \alpha_2 = \alpha \tag{7.3.15}$$

The boundary conditions [Eq. (7.3.14)] become the same as those in Eq. (7.2.15), that is,

$$\delta f_0^n = 0 \qquad \delta u_0^n = 0 \qquad \delta u_J^n = 0 \tag{7.3.16}$$

7.3.2 Difference Equations for the Standard Problem When the Momentum Equation Is Expressed in Physical Variables

We now consider Eq. (5.7.9) and introduce the new variables $U(\xi, Y)$, $V(\xi, Y)$, with $\xi = \bar{x}$ and Y defined by Eq. (5.7.7), to write it as

$$F' = U \tag{7.3.17a}$$

$$U' = V \tag{7.3.17b}$$

$$(bV)' = \frac{dp^*}{d\xi} + U\frac{\partial U}{\partial \xi} - V\frac{\partial F}{\partial \xi} \tag{7.3.17c}$$

The boundary conditions [Eq. (5.7.8)] are

$$F(\xi,0) = F_w(\xi) \qquad U(\xi,0) = 0 \qquad V(\xi,\bar{y}_c) = -\frac{\sqrt{R_L}}{2}\left[\frac{dp^*}{d\xi} + \frac{d}{d\xi}\left(\frac{U^2}{2}\right)\right] \tag{7.3.18}$$

The difference approximations of Eq. (7.3.17) are obtained as they were before and can be written as

$$\frac{F_j^n - F_{j-1}^n}{h_j} = U_{j-1/2}^n \tag{7.3.19a}$$

$$\frac{U_j^n - U_{j-1}^n}{h_j} = V_{j-1/2}^n \tag{7.3.19b}$$

$$h_j^{-1}(b_j^n V_j^n - b_{j-1}^n V_{j-1}^n) + \alpha[(FV)_{j-1/2}^n + V_{j-1/2}^{n-1}F_{j-1/2}^n - F_{j-1/2}^{n-1}V_{j-1/2}^n$$
$$- (U^2)_{j-1/2}^n] = R_{j-1/2}^{n-1} \tag{7.3.19c}$$

where

$$R_{j-1/2}^{n-1} = -h_j^{-1}(b_j^{n-1}V_j^{n-1} - b_{j-1}^{n-1}V_{j-1}^{n-1}) + \alpha[(FV)_{j-1/2}^{n-1} - (U^2)_{j-1/2}^{n-1} - 2\beta)$$

$$\text{(7.3.20a)}$$

$$\alpha = \frac{1}{k_n} \tag{7.3.20b}$$

with β defined as in Eq. (7.3.13a). The boundary conditions [Eq. (7.3.18)] at $\xi = \xi^n$ are

$$F_0^n = F_w(\xi) \qquad U_0^n = 0$$

$$V_J^n = \frac{\alpha}{2}\sqrt{R_L}[2\beta - (U^2)_J^n + (U^2)_J^{n-1}] - V_J^{n-1} \tag{7.3.21}$$

Using Newton's method, we now linearize Eq. (7.3.19) to get

$$\delta F_j - \delta F_{j-1} - \frac{h_j}{2}(\delta U_j + \delta U_{j-1}) = (r_1)_j \tag{7.3.22a}$$

$$\delta U_j - \delta U_{j-1} - \frac{h_j}{2}(\delta V_j + \delta V_{j-1}) = (r_3)_{j-1} \tag{7.3.22b}$$

$$(s_1)_j\delta V_j + (s_2)_j\delta V_{j-1} + (s_3)_j\delta F_j + (s_4)_j\delta F_{j-1} + (s_5)_j\delta U_j + (s_6)_j\delta U_{j-1}$$
$$= (r_2)_j \tag{7.3.22c}$$

where

$$(r_1)_j = F_{j-1}^{(i)} - F_j^{(i)} + h_jU_{j-1/2}^{(i)} \tag{7.3.23a}$$

$$(r_3)_{j-1} = U_{j-1}^{(i)} - U_j^{(i)} + h_jV_{j-1/2}^{(i)} \tag{7.3.23b}$$

$$(r_2)_j = R_{j-1/2}^{n-1} - \{h_j^{-1}(b_j^{(i)}V_j^{(i)} - b_{j-1}^{(i)}V_{j-1}^{(i)}) + \alpha[(FV)_{j-1/2}^{(i)} + V_{j-1/2}^{n-1}F_{j-1/2}^n$$
$$- F_{j-1/2}^{n-1}V_{j-1/2}^n - (U^2)_{j-1/2}^n]\} \tag{7.3.23c}$$

$$(s_1)_j = h_j^{-1}b_j^{(i)} + \frac{\alpha}{2}(F_j^{(i)} - F_{j-1/2}^{n-1}) \tag{7.3.23d}$$

$$(s_2)_j = -h_j^{-1}b_{j-1}^{(i)} + \frac{\alpha}{2}(F_{j-1}^{(i)} - F_{j-1/2}^{n-1}) \tag{7.3.23e}$$

$$(s_3)_j = \frac{\alpha}{2}(V_j^{(i)} + V_{j-1/2}^{n-1}) \tag{7.3.23f}$$

$$(s_4)_j = \frac{\alpha}{2}(V_{j-1}^{(i)} + V_{j-1/2}^{n-1}) \tag{7.3.23g}$$

$$(s_5)_j = -\alpha U_j^{(i)} \tag{7.3.23h}$$

$$(s_6)_j = -\alpha U_{j-1}^{(i)} \tag{7.3.23i}$$

The boundary conditions [Eq. (7.3.21)] become

$$\delta F_0{}^n = 0 \qquad \delta u_0{}^n = 0 \qquad\qquad\qquad\qquad\qquad\qquad (7.3.24a)$$

$$\delta V_J{}^n + \alpha \sqrt{R_L}\, U_J{}^n \delta U_J{}^n = - V_J{}^n + \tfrac{1}{2}\alpha\sqrt{R_L}\, [2\beta - (U^2)_J{}^n + (U^2)_J{}^{n-1}] - V_J{}^{n-1}$$

$$(7.3.24b)$$

7.3.3 Variational Equations

In order to calculate $\partial\phi/\partial\beta$ in Eq. (7.3.8), it is necessary to know $\partial f/\partial\beta$. For this reason when the transformed coordinates are being used, we take the derivatives of Eqs. (7.2.10a), (7.2.10b), and (7.3.12) with respect to β. This leads to a system of three linear difference equations, known as the variational equations for Eqs. (7.2.10a), (7.2.10b), and (7.3.12). These, together with the variational equations for Eqs. (7.3.19) and (7.3.21), are presented in Appendix 7B. The resulting equations can again be solved by the block-elimination method of Appendix 7A, as discussed in Appendix 7B.

To summarize, one step of iteration of Newton's method for $\xi > 0$ is carried out as follows: We first solve the standard problem in transformed coordinates for an assumed value of β. This consists of solving the system given by Eqs. (7.2.13) and (7.2.15) with the new definitions and conditions given in Eq. (7.3.13)-(7.3.15). For example, $u_J{}^n$ is given by Eq. (7.3.14), α_1 has a new definition, etc. From the solution we can compute ϕ in Eq. (7.3.6). To find $\partial\phi/\partial\beta$, which is equal to $-1/\eta_{sp}(\partial f/\partial\beta)$ at $\eta = \eta_{sp}$, we solve the variational equations given by Eqs. (7B.2) and (7B.3) (see Appendix 7B). Then we use Eq. (7.3.7) to find a new value of β and check to see whether Eq. (7.3.9) is satisfied; if it is not, we repeat the above procedure until Eq. (7.3.9) is satisfied. When the governing equations are solved in physical coordinates, it is necessary this time to find $\partial F/\partial\beta$. So we solve the variational equations given by Eqs. (7B.5) and (7B.6) and repeat the above procedure.

APPENDIX 7A: Block Elimination Method

The solution of the linearized difference equations of the momentum equation for external and internal flows can be obtained in a very efficient manner by using the block elimination method as discussed by Keller (1974). In Sec. 7A.1 we discuss the method in detail for external flows; in Sec. 7A.2 we present a Fortran program for it; and in Sec. 7B.1 we apply the method to internal flows. The algebra is quite complex, but the program is very efficient.

7A.1 SOLUTION OF THE DIFFERENCE EQUATIONS OF THE MOMENTUM EQUATION FOR EXTERNAL FLOWS

The linearized difference equations of the momentum equation for external flows are given by the system Eqs. (7.2.13) and (7.2.15), which has a block tridiagonal structure. This is not obvious, and to clarify the solution, we write the system in matrix-vector form. We first define the three-dimensional vectors, δ_j and $\mathbf{r}_j$, for each value of j by

$$\delta_j \equiv \begin{pmatrix} \delta f_j \\ \delta u_j \\ \delta v_j \end{pmatrix} \quad 0 \leqslant j \leqslant J \quad \mathbf{r}_0 = \begin{pmatrix} 0 \\ 0 \\ (r_3)_0 \end{pmatrix}$$

$$\mathbf{r}_j = \begin{pmatrix} (r_1)_j \\ (r_2)_j \\ (r_3)_j \end{pmatrix} \quad 1 \leqslant j \leqslant J-1 \quad \mathbf{r}_J = \begin{pmatrix} (r_1)_J \\ (r_2)_J \\ 0 \end{pmatrix} \tag{7A.1.1}$$

and the 3×3 matrices, A_j, B_j, C_j, by

$$A_0 = \begin{bmatrix} 1 & 0 & 0 \\ 0 & 1 & 0 \\ 0 & -1 & -h_1/2 \end{bmatrix} \quad A_j \equiv \begin{bmatrix} 1 & -h_j/2 & 0 \\ (s_3)_j & (s_5)_j & (s_1)_j \\ 0 & -1 & -h_{j+1}/2 \end{bmatrix} \quad 1 \leqslant j \leqslant J-1$$

$$A_J \equiv \begin{bmatrix} 1 & -h_J/2 & 0 \\ (s_3)_J & (s_5)_J & (s_1)_J \\ 0 & 1 & 0 \end{bmatrix} \quad B_j = \begin{bmatrix} -1 & -h_j/2 & 0 \\ (s_4)_j & (s_6)_j & (s_2)_j \\ 0 & 0 & 0 \end{bmatrix} \quad 1 \leqslant j \leqslant J \tag{7A.1.2}$$

$$C_j \equiv \begin{bmatrix} 0 & 0 & 0 \\ 0 & 0 & 0 \\ 0 & 1 & -h_{j+1}/2 \end{bmatrix} \quad 0 \leqslant j \leqslant J-1$$

Note that the first two rows of A_0 and C_0 and the last row of A_J and B_J correspond to the boundary conditions [Eq. (7.2.15)]. To solve the same set of finite-difference equations with different boundary conditions, only the matrix rows mentioned above would need altering. In terms of the above definitions it can be shown (Keller, 1974) that the system Eqs. (7.2.13) and (7.2.15) can be written as

$$\mathcal{C} \delta = \mathbf{r} \tag{7A.1.3}$$

where

$$\mathcal{C} \equiv \begin{bmatrix} A_0 & C_0 & & & & & \\ B_1 & A_1 & C_1 & & & & \\ & \cdot & \cdot & \cdot & & & \\ & & \cdot & \cdot & \cdot & & \\ & & & B_j & A_j & C_j & \\ & & & & \cdot & \cdot & \cdot \\ & & & & & \cdot & \cdot & \cdot \\ & & & & & B_{J-1} & A_{J-1} & C_{J-1} \\ & & & & & & B_J & A_J \end{bmatrix} \tag{7A.1.4}$$

$$\delta \equiv \begin{bmatrix} \delta_0 \\ \delta_1 \\ \delta_2 \\ \cdot \\ \cdot \\ \cdot \\ \cdot \\ \delta_{J-1} \\ \delta_J \end{bmatrix} \qquad r = \begin{bmatrix} r_0 \\ r_1 \\ r_2 \\ \cdot \\ \cdot \\ \cdot \\ \cdot \\ r_{J-1} \\ r_J \end{bmatrix} \qquad (7A.1.5)$$

To solve Eq. (7A.1.3), we use the block elimination method as described by Keller (1974). According to this method we first seek a factorization of the form

$$\mathfrak{a} = \mathcal{L}\mathfrak{u} \qquad (7A.1.6a)$$

which can also be written in the form

$$\begin{bmatrix} A_0 & C_0 & & & & \\ B_1 & A_1 & C_1 & & & \\ & \cdot & \cdot & \cdot & & \\ & & \cdot & \cdot & \cdot & \\ & & & B_j & A_j & C_j \\ & & & & \cdot & \cdot & \cdot \\ & & & & & \cdot & \cdot & \cdot \\ & & & & & & B_{J-1} & A_{J-1} & C_{J-1} \\ & & & & & & & B_J & A_J \end{bmatrix} = \begin{bmatrix} I & & & & \\ \Gamma_1 & I & & & \\ & \cdot & \cdot & & \\ & & \cdot & \cdot & \\ & & & \Gamma_j & I & \\ & & & & \cdot & \cdot \\ & & & & & \cdot & \cdot \\ & & & & & & \Gamma_{J-1} & I \\ & & & & & & & \Gamma_J & I \end{bmatrix}$$

$$\times \begin{bmatrix} \Delta_0 & C_0 & & & & \\ & \Delta_1 & C_1 & & & \\ & & \cdot & \cdot & & \\ & & & \cdot & \cdot & \\ & & & & \Delta_j & C_j \\ & & & & & \cdot & \cdot \\ & & & & & & \cdot & \cdot \\ & & & & & & & \Delta_{J-1} & C_{J-1} \\ & & & & & & & & \Delta_J \end{bmatrix} \qquad (7A.1.6b)$$

Here I is the identity matrix of order 3:

$$I \equiv \begin{bmatrix} 1 & 0 & 0 \\ 0 & 1 & 0 \\ 0 & 0 & 1 \end{bmatrix}$$

and the Δ_j and Γ_j are 3×3 matrices. Proceeding formally, we find that

$$\Delta_0 = A_0 \tag{7A.1.7a}$$

$$\Gamma_j \Delta_{j-1} = B_j \qquad j = 1, 2, \ldots, J \tag{7A.1.7b}$$

$$\Delta_j = A_j - \Gamma_j C_{j-1} \qquad j = 1, 2, \ldots, J \tag{7A.1.7c}$$

Keller shows that the Γ_j matrix has the same structure as that of the B_j matrix. Therefore if we denote the elements of Γ_j by γ_{ik} $(i, k = 1, 2, 3)$, we can write Γ_j as

$$\Gamma_j \equiv \begin{bmatrix} (\gamma_{11})_j & (\gamma_{12})_j & (\gamma_{13})_j \\ (\gamma_{21})_j & (\gamma_{22})_j & (\gamma_{23})_j \\ 0 & 0 & 0 \end{bmatrix} \tag{7A.1.8a}$$

Similarly, if the elements of Δ_j are denoted by α_{ik}, we can write Δ_j as [note that the third row of Δ_j follows from the third row of A_j according to Eq. (7A.1.7c)]

$$\Delta_j \equiv \begin{bmatrix} (\alpha_{11})_j & (\alpha_{12})_j & (\alpha_{13})_j \\ (\alpha_{21})_j & (\alpha_{22})_j & (\alpha_{23})_j \\ 0 & -1 & -h_{j+1}/2 \end{bmatrix} \qquad 0 \leqslant j \leqslant J-1 \tag{7A.1.8b}$$

and for $j = J$, the first two rows are the same as the first two rows in Eq. (7A.1.8b), but the elements of the third row, which correspond to the boundary conditions at $j = J$, are $(0, 1, 0)$.

For $j = 0$, $\Delta_0 = A_0$; therefore the values of $(\alpha_{ik})_0$ are

$$(\alpha_{11})_0 = 1 \qquad (\alpha_{12})_0 = 0 \qquad (\alpha_{13})_0 = 0$$
$$(\alpha_{21})_0 = 0 \qquad (\alpha_{22})_0 = 1 \qquad (\alpha_{23})_0 = 0 \tag{7A.1.9a}$$

and the values of $(\gamma_{ik})_1$ are

$$(\gamma_{11})_1 = -1 \qquad (\gamma_{12})_1 = -\tfrac{1}{2} h_1 \qquad (\gamma_{13})_1 = 0$$
$$(\gamma_{21})_1 = (s_4)_1 \qquad (\gamma_{23})_1 = -2 \left[\frac{(s_2)_1}{h_1} \right] \qquad (\gamma_{22})_1 = (s_6)_1 + (\gamma_{23})_1 \tag{7A.1.9b}$$

The elements of the Δ_j matrices are calculated from Eq. (7A.1.7c). Using the definitions of A_j, Γ_j, and C_{j-1}, we find from Eq. (7A.1.7c) that for $j = 1, 2, \ldots, J$,

$$(\alpha_{11})_j = 1 \qquad (\alpha_{12})_j = -\frac{h_j}{2} - (\gamma_{13})_j \qquad (\alpha_{13})_j = \frac{h_{j+1}}{2} (\gamma_{13})_j$$

$$(\alpha_{21})_j = (s_3)_j \qquad (\alpha_{22})_j = (s_5)_j - (\gamma_{23})_j \qquad (\alpha_{23})_j = (s_1)_j + \frac{h_{j+1}}{2} (\gamma_{23})_j \tag{7A.1.10a}$$

To find the elements of the Γ_j matrices, we use Eq. (7A.1.7b). With Δ_j defined by

Eq. (7A.1.8b) and B_j by Eq. (7A.1.2) it follows that for $1 \leqslant j \leqslant J$,

$$(\gamma_{11})_j = \left\{ (\alpha_{23})_{j-1} + \frac{h_j}{2} \left[\left(\frac{h_j}{2}\right)(\alpha_{21})_{j-1} - (\alpha_{22})_{j-1} \right] \right\} / \Delta_0$$

$$(\gamma_{12})_j = -\left\{ \frac{h_j}{2}\frac{h_j}{2} + (\gamma_{11})_j \left[(\alpha_{12})_{j-1}\frac{h_j}{2} - (\alpha_{13})_{j-1} \right] \right\} / \Delta_1$$

$$(\gamma_{13})_j = [(\gamma_{11})_j(\alpha_{13})_{j-1} + (\gamma_{12})_j(\alpha_{23})_{j-1}]/\frac{h_j}{2}$$

$$(\gamma_{21})_j = \left\{ (s_2)_j(\alpha_{21})_{j-1} - (s_4)_j(\alpha_{23})_{j-1} + \frac{h_j}{2}[(s_4)_j(\alpha_{22})_{j-1} - (s_6)_j(\alpha_{21})_{j-1}] \right\} / \Delta_0$$

$$(\gamma_{22})_j = \left\{ (s_6)_j\frac{h_j}{2} - (s_2)_j + (\gamma_{21})_j \left[(\alpha_{13})_{j-1} - (\alpha_{12})_{j-1}\frac{h_j}{2} \right] \right\} / \Delta_1$$

$$(\gamma_{23})_j = (\gamma_{21})_j(\alpha_{12})_{j-1} + (\gamma_{22})_j(\alpha_{22})_{j-1} - (s_6)_j$$

$$\Delta_0 = (\alpha_{13})_{j-1}(\alpha_{21})_{j-1} - (\alpha_{23})_{j-1}(\alpha_{11})_{j-1} - \frac{h_j}{2}[(\alpha_{12})_{j-1}(\alpha_{21})_{j-1} - (\alpha_{22})_{j-1}(\alpha_{11})_{j-1}]$$

$$\Delta_1 = (\alpha_{22})_{j-1}\frac{h_j}{2} - (\alpha_{23})_{j-1}$$

(7A.1.10b)

To summarize the calculation of Γ_j and Δ_j matrices, we first calculate α_{ik} from Eq. (7A.1.9a) for $j = 0$, γ_{ik} from Eq. (7A.1.10) for $j = 1$, α_{ik} from Eq. (7A.1.9b) for $j = 1$, then γ_{ik} from Eq. (7A.1.10) for $j = 2$, α_{ik} from Eq. (7A.1.9b) for $j = 2$, then α_{ik} from Eq. (7A.1.10) for $j = 3$, etc.

Let us rewrite Eq. (7A.1.3), using Eq. (7A.1.6a):

$$\mathcal{L}\mathcal{U}\delta = r \tag{7A.1.11}$$

If we let

$$\mathcal{U}\delta = w \tag{7A.1.12}$$

then Eq. (7A.1.11) becomes

$$\mathcal{L}w = r \tag{7A.1.13a}$$

From $\mathcal{L}$ defined in Eq. (7A.1.6b) we can write Eq. (7A.1.13a) as

$$
\begin{bmatrix}
I \\
\Gamma_1 & I \\
& \cdot & \cdot \\
& & \cdot & \cdot \\
& & & \cdot & \cdot \\
& & & & \Gamma_j & I \\
& & & & & \cdot & \cdot \\
& & & & & & \cdot & \cdot \\
& & & & & & & \cdot & \cdot \\
& & & & & & & & \Gamma_{J-1} & I \\
& & & & & & & & & \Gamma_J & I
\end{bmatrix}
\begin{bmatrix}
\mathbf{w}_0 \\
\mathbf{w}_1 \\
\cdot \\
\cdot \\
\mathbf{w}_j \\
\cdot \\
\cdot \\
\mathbf{w}_{J-1} \\
\mathbf{w}_J
\end{bmatrix}
=
\begin{bmatrix}
\mathbf{r}_0 \\
\mathbf{r}_1 \\
\cdot \\
\cdot \\
\mathbf{r}_j \\
\cdot \\
\cdot \\
\mathbf{r}_{J-1} \\
\mathbf{r}_J
\end{bmatrix}
\tag{7A.1.13b}
$$

From Eq. (7A.1.13b) it follows that

$$
\mathbf{w}_0 = \mathbf{r}_0 \tag{7A.1.14a}
$$
$$
\mathbf{w}_j = \mathbf{r}_j - \Gamma_j \mathbf{w}_{j-1} \qquad 1 \leqslant j \leqslant J \tag{7A.1.14b}
$$

We denote the components of the vectors $\mathbf{w}_j$, by

$$
\mathbf{w}_j \equiv \begin{pmatrix} (w_1)_j \\ (w_2)_j \\ (w_3)_j \end{pmatrix} \qquad 0 \leqslant j \leqslant J \tag{7A.1.15}
$$

Then from Eq. (7A.1.14), it follows that for $j = 0$,

$$
(w_1)_0 = (r_1)_0 \qquad (w_2)_0 = (r_2)_0 \qquad (w_3)_0 = (r_3)_0 \tag{7A.1.16a}
$$

and for $1 \leqslant j \leqslant J$,

$$
\begin{aligned}
(w_1)_j &= (r_1)_j - (\gamma_{11})_j (w_1)_{j-1} - (\gamma_{12})_j (w_2)_{j-1} - (\gamma_{13})_j (w_3)_{j-1} \\
(w_2)_j &= (r_2)_j - (\gamma_{21})_j (w_1)_{j-1} - (\gamma_{22})_j (w_2)_{j-1} - (\gamma_{23})_j (w_3)_{j-1} \\
(w_3)_j &= (r_3)_j
\end{aligned} \tag{7A.1.16b}
$$

With $\mathfrak{U}$ defined in Eq. (7A.1.6b) we can write Eq. (7A.1.12) as

$$
\begin{bmatrix}
\Delta_0 & C_0 \\
& \Delta_1 & C_1 \\
& & \cdot & \cdot \\
& & & \cdot & \cdot \\
& & & & \Delta_j & C_j \\
& & & & & \cdot & \cdot \\
& & & & & & \cdot & \cdot \\
& & & & & & & \Delta_{J-1} & C_{J-1} \\
& & & & & & & & \Delta_J
\end{bmatrix}
\begin{bmatrix}
\delta_0 \\
\delta_1 \\
\cdot \\
\cdot \\
\delta_j \\
\cdot \\
\cdot \\
\delta_{J-1} \\
\delta_J
\end{bmatrix}
=
\begin{bmatrix}
\mathbf{w}_0 \\
\mathbf{w}_1 \\
\cdot \\
\cdot \\
\mathbf{w}_j \\
\cdot \\
\cdot \\
\mathbf{w}_{J-1} \\
\mathbf{w}_J
\end{bmatrix}
\tag{7A.1.17}
$$

From Eq. (7A.1.17) it follows that

$$\Delta_J \delta_J = w_J \tag{7A.1.18a}$$

$$\Delta_j \delta_j = w_j - C_j \delta_{j+1} \qquad j = J-1, J-2, \ldots, 0 \tag{7A.1.18b}$$

The unknown vectors, δ_j, can be calculated from Eq. (7A.1.18) by introducing the definitions of Δ_j, w_j and C_j. From Eq. (7A.1.18a) we find the three components of δ_J to be

$$\delta u_J = (w_3)_J \tag{7A.1.19a}$$

$$\delta v_J = \frac{e_2 (\alpha_{11})_J - e_1 (\alpha_{21})_J}{(\alpha_{23})_J (\alpha_{11})_J - (\alpha_{13})_J (\alpha_{21})_J} \tag{7A.1.19b}$$

$$\delta f_J = \frac{e_1 - (\alpha_{13})_J \delta v_J}{(\alpha_{11})_J} \tag{7A.1.19c}$$

where

$$e_1 = (w_1)_J - (\alpha_{12})_J \delta u_J$$

$$e_2 = (w_2)_J - (\alpha_{22})_J \delta u_J$$

The components of δ_j, for $j = J-1, J-2, \ldots, 0$, are found from Eq. (7A.1.18b) to be

$$\delta v_j = \frac{(\alpha_{11})_j [(w_2)_j + e_3 (\alpha_{22})_j] - (\alpha_{21})_j (w_1)_j - e_3 (\alpha_{21})_j (\alpha_{12})_j}{\Delta_2} \tag{7A.1.20a}$$

$$\delta u_j = -\frac{h_{j+1}}{2} \delta v_j - e_3 \tag{7A.1.20b}$$

$$\delta f_j = \frac{(w_1)_j - (\alpha_{12})_j \delta u_j - (\alpha_{13})_j \delta v_j}{(\alpha_{11})_j} \tag{7A.1.20c}$$

where

$$e_3 = (w_3)_j - \delta u_{j+1} + \frac{h_{j+1}}{2} \delta v_{j+1}$$

$$\Delta_2 = (\alpha_{21})_j (\alpha_{12})_j \frac{h_{j+1}}{2} - (\alpha_{21})_j (\alpha_{13})_j - \frac{h_{j+1}}{2} (\alpha_{22})_j (\alpha_{11})_j + (\alpha_{23})_j (\alpha_{11})_j \tag{7A.1.20d}$$

To summarize, one iteration of Newton's method is carried out as follows. The vectors r_j defined in Eq. (7A.1.1) are computed from Eqs. (7.2.14a), (7.2.14b), and (7.2.14c) by using the latest iterate. The matrix elements of A_j, B_j, and C_j defined in Eq. (7A.1.2) are next determined from Eqs. (7.2.14d)–(7.2.14i). Using the relations in Eqs. (7A.1.7) and (7A.1.14), the matrices Γ_j and Δ_j and vectors w_j are calculated. The matrix elements for Γ_j defined in Eq. (7A.1.8a) are determined from Eq. (7A.1.10). For Δ_j defined in Eq. (7A.1.8b) the elements are determined from Eq. (7A.1.9). The components of the vector w_j defined in Eq. (7A.1.15) are determined from Eq. (7A.1.16). The step in which Γ_j, Δ_j, and w_j are calculated is usually referred to as the *forward sweep*. In the so-called *backward sweep* the components of δ_j defined in Eq. (7A.1.1) are computed from Eqs. (7A.1.19) and (7A.1.20).

These calculations are repeated until some convergence criterion is satisfied. In boundary-layer calculations we usually use the wall-shear parameter, $v(0)$, as the convergence criterion. Calculations are stopped when

$$|\delta v_0^{(i)}| < \epsilon_1 \tag{7A.1.21}$$

where ϵ_1 is a prescribed value.

7A.2 FORTRAN PROGRAM FOR THE BLOCK ELIMINATION METHOD

In Appendix 8A we shall describe and present a Fortran program for solving the two-dimensional laminar and turbulent boundary-layer equations by the box method. One of the key subroutines of that program is a subroutine (called SOLV3) that contains the recursion formulas that arise in the block elimination method discussed in Sec. 7A.1. This is a very useful subroutine, since it can be used to solve any third-order ordinary (linear or nonlinear) and/or parabolic PDE as will be described in Appendix 8B.

We present this subroutine here to show how easily the recursion formulas of Sec. 7A.1 can be programed. The Fortran notation for some of the key symbols is also presented.

Fortran name	Symbol
W1, W2, W3	w_1, w_2, w_3
A11, A12, A13, . . .	$\alpha_{11}, \alpha_{12}, \alpha_{13}, \ldots$
G11, G12, G13, . . .	$\gamma_{11}, \gamma_{12}, \gamma_{13}, \ldots$
DELU, DELV, DELF	$\delta u, \delta v, \delta f$

```
      SUBROUTINE SOLV3
      COMMON/BLC0/ NP,NX,NXT,NTR,NFLOW,ETAE,VGP,CNU,DETA(61),A(61),
     1             ETA(61)
      COMMON/BLCP/ DELV(61),F(61,2),U(61,2),V(61,2),B(61,2)
      COMMON/BLC6/ S1(61),S2(61),S3(61),S4(61),S5(61),S6(61),
     1             R1(61),R2(61),R3(61)
      DIMENSION W1(61),W2(61),W3(61),A11(61),A12(61),A13(61),A21(61),
     1           A22(61),A23(61),G11(61),G12(61),G13(61),G21(61),
     2           G22(61),G23(61),DELU(61),DELF(61)
C - - - - - - - - - - - - - - - - - - - - - - - - - - - - - - - -
C  W-ELEMENTS FOR J=1 - SEE (7A.1.16A)
      W1(1) = R1(1)
      W2(1) = R2(1)
      W3(1) = R3(1)
C  ALFA ELEMENTS FOR J=1 - SEE (7A.1.9A)
      A11(1)= 1.0
      A12(1)= 0.0
      A13(1)= 0.0
      A21(1)= 0.0
      A22(1)= 1.0
      A23(1)= 0.0
```

```
C   GAMMA ELEMENTS FOR J=2 - SEE (7A.1.9B)
      G11(2)=-1.0
      G12(2)=-0.5*DETA(1)
      G13(2)= 0.0
      G21(2)= S4(2)
      G23(2)=-2.0*S2(2)/DETA(1)
      G22(2)= G23(2)+S6(2)
C
C   FORWARD SWEEP
C
      DO 500 J=2,NP
      IF(J .EQ. 2) GO TO 100
C   SEE (7A.1.10B)
      DEN    = (A13(J-1)*A21(J-1)-A23(J-1)*A11(J-1)-A(J)*
     1         (A12(J-1)*A21(J-1)-A22(J-1)*A11(J-1)))
      DEN1   = A22(J-1)*A(J)-A23(J-1)
      G11(J)= (A23(J-1)+A(J)*(A(J)*A21(J-1)-A22(J-1)))/DEN
      G12(J)= -(A(J)*A(J)+G11(J)*(A12(J-1)*A(J)-A13(J-1)))/DEN1
      G13(J)= (G11(J)*A13(J-1)+G12(J)*A23(J-1))/A(J)
      G21(J)= (S2(J)*A21(J-1)-S4(J)*A23(J-1)+A(J)*(S4(J)*
     1         A22(J-1)-S6(J)*A21(J-1)))/DEN
      G22(J)= (-S2(J)+S6(J)*A(J)-G21(J)*(A(J)*A12(J-1)-A13(J-1)))/DEN1
      G23(J)= G21(J)*A12(J-1)+G22(J)*A22(J-1)-S6(J)
C   SEE (7A.1.10A)
  100 A11(J)= 1.0
      A12(J)=-A(J)-G13(J)
      A13(J)= A(J)*G13(J)
      A21(J)= S3(J)
      A22(J)= S5(J)-G23(J)
      A23(J)= S1(J)+A(J)*G23(J)
C   SEE (7A.1.16B)
      W1(J)  = R1(J)-G11(J)*W1(J-1)-G12(J)*W2(J-1)-G13(J)*W3(J-1)
      W2(J)  = R2(J)-G21(J)*W1(J-1)-G22(J)*W2(J-1)-G23(J)*W3(J-1)
      W3(J)  = R3(J)
  500 CONTINUE

C
C   BACKWARD SWEEP
C
C   SEE (7A.1.19), (7A.1.20)
      DELU(NP) = W3(NP)
      E1       = W1(NP)-A12(NP)*DELU(NP)
      E2       = W2(NP)-A22(NP)*DELU(NP)
      DELV(NP) = (E2*A11(NP)-E1*A21(NP))/(A23(NP)*A11(NP)-A13(NP)*
     1           A21(NP))
      DELF(NP) = (E1-A13(NP)*DELV(NP))/A11(NP)
      J        = NP
  600 J        = J-1
      E3       = W3(J)-DELU(J+1)+A(J+1)*DELV(J+1)
      DEN2     = A21(J)*A12(J)*A(J+1)-A21(J)*A13(J)-A(J+1)*A22(J)*A11(J)+
     1           A23(J)*A11(J)
      DELV(J)  = (A11(J)*(W2(J)+E3*A22(J))-A21(J)*W1(J)-E3*A21(J)*A12(J)
     1           )/DEN2
      DELU(J)  =-A(J+1)*DELV(J)+E3
      DELF(J)  = (W1(J)-A12(J)*DELU(J)-A13(J)*DELV(J))/A11(J)
      IF(J .GT. 1) GO TO 600
C
      WRITE(6,9100) V(1,2),DELV(1)
      DO 700 J=1,NP
      F(J,2)= F(J,2)+DELF(J)
      U(J,2)= U(J,2)+DELU(J)
  700 V(J,2)= V(J,2)+DELV(J)
      U(1,2)= 0.0
      RETURN
C - - - - - - - - - - - - - - - - - - - - - - - - - - - - - - - - - -
 9100 FORMAT(1H ,5X,8HV(WALL)=,E14.6,5X,6HDELVW=,E14.6)
      END
```

APPENDIX 7B: Newton's Method for Internal Flows

As was discussed in Sec. 7.3, the solution of the governing equations for internal flows requires the pressure drop, β, defined by Eq. (7.3.13a). Since this is not known, we use Newton's method for the iteration procedure. For an assumed value of β we first solve the standard problem. From the solution we can compute ϕ, which represents the error term. It is clear from Eq. (7.3.7) that in order to determine the next value of β, it is necessary to know $\partial\phi/\partial\beta$ at $\eta = \eta_{sp}$. For this reason we differentiate the difference equations with respect to β. This leads to a system of linear difference equations known as the variational equations. When the governing equations are expressed in *transformed coordinates*, the variational equations for Eqs. (7.2.10a), (7.2.10b), and (7.3.12) with f_1, u_1, and v_1 defined by

$$f_1 \equiv \frac{\partial f}{\partial \beta} \qquad u_1 \equiv \frac{\partial u}{\partial \beta} \qquad v_1 \equiv \frac{\partial v}{\partial \beta} \tag{7B.1}$$

are

$$(f_1)_j - (f_1)_{j-1} - \frac{h_j}{2}[(u_1)_j + (u_1)_{j-1}] = 0 \tag{7B.2a}$$

$$(u_1)_j - (u_1)_{j-1} - \frac{h_j}{2}[(v_1)_j + (v_1)_{j-1}] = 0 \tag{7B.2b}$$

$$h_j^{-1}[b_j(v_1)_j - b_{j-1}(v_1)_{j-1}] + \frac{\alpha_1}{2}[f_j(v_1)_j + v_j(f_1)_j + f_{j-1}(v_1)_{j-1} + v_{j-1}(f_1)_{j-1}]$$

$$+ \frac{\alpha}{2}\{v_{j-1/2}^{n-1}[(f_1)_j + (f_1)_{j-1}] - f_{j-1/2}^{n-1}[(v_1)_j + (v_1)_{j-1}]$$

$$- 2u_j(u_1)_j - 2u_{j-1}(u_1)_{j-1}\} = -2\alpha \tag{7B.2c}$$

Note that in Eq. (7B.2c) we have dropped the superscript n for simplicity.

Equation (7B.2c) can also be written as

$$(s_1)_j(v_1)_j + (s_2)_j(v_1)_{j-1} + (s_3)_j(f_1)_j + (s_4)_j(f_1)_{j-1} + (s_5)_j(u_1)_j + (s_6)_j(u_1)_{j-1}$$

$$= -\alpha \tag{7B.2d}$$

with the coefficients $(s_k)_j$ the same as those defined by Eqs. (7.2.14d)–(7.2.14i) with the restrictions imposed by Eq. (7.3.15).

Similarly, the boundary conditions [Eq. (7.3.14)] yield

$$(f_1)_0 = 0 \qquad (u_1)_0 = 0 \qquad (u_1)_J = \frac{1}{u_J} \tag{7B.3}$$

When the governing equations are expressed in physical coordinates, we find the variational equations for Eqs. (7.3.19) and (7.3.21) by differentiating them with respect to β. With the notation

$$F_1 \equiv \frac{\partial F}{\partial \beta} \qquad U_1 = \frac{\partial U}{\partial \beta} \qquad V_1 \equiv \frac{\partial V}{\partial \beta} \tag{7B.4}$$

these equations are

$$(F_1)_j - (F_1)_{j-1} - \frac{h_j}{2}[(U_1)_j + [(U_1)_{j-1}] = 0 \tag{7B.5a}$$

$$(U_1)_j - (U_1)_{j-1} - \frac{h_j}{2}[(V_1)_j + (V_1)_{j-1}] = 0 \tag{7B.5b}$$

$$(s_1)_j(V_1)_j + (s_2)_j(V_1)_{j-1} + (s_3)_j(F_1)_j + (s_4)_j(F_1)_{j-1} + (s_5)_j(U_1)_j + (s_6)_j(U_1)_{j-1}$$
$$= -2\alpha \tag{7B.5c}$$

$$(F_1)_0 = 0 \quad (U_1)_0 = 0 \quad (V_1)_J = \alpha\sqrt{R_L}\,[1 - U_J(U_1)_J] \tag{7B.6}$$

Here the coefficients $(s_k)_j$ are the same as those defined by Eqs. (7.3.23d)–(7.3.23i).

7B.1 SOLUTION OF THE DIFFERENCE EQUATIONS OF THE MOMENTUM EQUATION FOR INTERNAL FLOWS

The block-elimination method discussed in Appendix 7A is a general one for a third-order equation in that it can also be used to solve the difference equations of the momentum equation for internal flows. This can be done by accounting for the differences in the boundary conditions and for the differences in coefficients of the momentum equations. To account for the changes in the boundary conditions, an important point to keep in mind is that the first two rows of the A_0 matrix account for the two wall boundary conditions and the third row of the A_J matrix accounts for the "edge" boundary condition. To clarify this point, let us consider the case when the governing equations are expressed in transformed coordinates. For the standard problem for an assumed value of β, we first solve the system given by Eqs. (7.2.13) and (7.2.15) with the definitions and restrictions imposed by Eqs. (7.3.13)–(7.3.15). Except for the changes in the coefficients $(s_k)_j$ and $(r_2)_j$, the system [Eq. (7A.1.3)] remains unchanged. The boundary conditions, δu_0, δf_0, and δu_J, are the same; therefore the A_0 and A_J matrices remain the same.

For the variational equations there are a few differences. First of all the δ_j vector is defined by

$$\delta_j = \begin{bmatrix} (f_1)_j \\ (u_1)_j \\ (v_1)_j \end{bmatrix} \tag{7B.7}$$

The coefficients $(s_k)_j$ are the *same* as those of the standard problem. The coefficients $(r_1)_j$, $(r_2)_j$, and $(r_3)_j$, however, are different: $(r_1)_j = 0$ for all j, $(r_2)_j = -2\alpha$ for all j, and $(r_3)_j = 0$ for all j except $j = J$. The $(r_3)_J$ term, which was zero before, is now equal to $(r_3)_J = 1/U_J$.

When the governing equations are expressed in physical coordinates, there are also a few minor changes. For the standard problem the δ_j vector is defined by

$$\delta_j = \begin{bmatrix} \delta F_j \\ \delta U_j \\ \delta V_j \end{bmatrix} \qquad (7B.8)$$

and obviously the increments δf_j, δu_j, and δv_j in Eqs. (7A.1.19) and (7A.1.20) are replaced by the new definitions. The coefficients $(s_k)_j$ and (r_j) have new definitions. Since the wall boundary conditions are similar, the A_0 matrix remains unchanged. However, the third row of the A_J matrix changes because the edge boundary condition in this case is different from that given by $\delta U_J = 0$. The new definition of A_J is

$$A_J = \begin{bmatrix} 1 & -h_J/2 & 0 \\ (s_3)_J & (s_5)_J & (s_1)_J \\ 0 & \alpha\sqrt{R_L}\,U_J & 1 \end{bmatrix} \qquad (7B.9)$$

For the variational equations the δ_j vector is defined by

$$\delta_j = \begin{bmatrix} (F_1)_j \\ (U_1)_j \\ (V_1)_j \end{bmatrix} \qquad (7B.10)$$

The coefficients $(s_k)_j$ are the same as those of the standard problem, but the coefficients $(r_1)_j$, $(r_2)_j$, and $(r_3)_j$ are different. The wall boundary conditions are the same; therefore the A_0 matrix remains unaltered. However, the A_J matrix has a slightly different third row due to the edge boundary condition.

Chapter 8

Numerical Solutions of Laminar and Turbulent Boundary Layers

8.1 INTRODUCTION

In Chap. 7 we described an efficient and accurate numerical method, the Box method, to solve the TSL equations. In this chapter we shall use this method to solve the TSL equations for external and internal laminar and turbulent flows. For the turbulent flows we shall use the eddy-viscosity concept to model the Reynolds stresses. This is compatible with any type of turbulence model that predicts the local Reynolds shear stress, $-\rho\overline{u'v'}$, because the eddy viscosity, ϵ_m, is just $-\rho\overline{u'v'}/(\partial u/\partial y)$. For the demonstration calculated below we shall use an algebraic eddy-viscosity formulation of Cebeci and Smith (1974), discussed in Sec. 8.3. In Sec. 8.5 we discuss the Bradshaw et al. (1967) method, one of the simplest of those that use a transport equation for $-\rho\overline{u'v'}$.

Even from a purely computational viewpoint, turbulent boundary layers present a much more difficult problem than do laminar boundary layers. Consider, for example, an incompressible turbulent flow. The skin friction is appreciably greater than that of a laminar flow, yet the boundary layer is much thicker. That means that the dimensionless velocity gradient, $(\delta/u_e)(\partial u/\partial y)$, is greater at the wall. To maintain computational accuracy when $(\partial u/\partial y)$ is large, short steps in y must be taken. Therefore the steps near the wall in a turbulent boundary layer must be shorter than the corresponding steps in a laminar boundary layer under similar

235

conditions. The alternative is to match the calculation to one of the law-of-the-wall formulas (Chap. 6) at a finite distance from the wall, a procedure that may not be easy.

The numerical method described in Chap. 7 is unique in that various kinds of spacings in both the x-direction and the η-direction can be used. In the calculations presented here we use an arbitrary Δx-spacing and a particular kind of $\Delta\eta$-spacing where $\eta = y \sqrt{(u_e/\nu x)}$. The net in the η-direction is a geometric progression having the property that the ratio of lengths of any two adjacent intervals is a constant; that is, $h_j = Kh_{j-1}$. The distance to the jth line is given by the following formula:

$$\eta_j = h_1 \frac{K^j - 1}{K - 1} \quad j = 1, 2, 3, \ldots, J \quad K > 1 \tag{8.1.1}$$

There are two parameters: h_1, the length of the first $\Delta\eta$-step, and K, the ratio of two successive steps. The total number of points, J, can be calculated by the following formula:

$$J = \frac{\ln\left[1 + (K - 1)(\eta_\infty/h_1)\right]}{\ln K} \tag{8.1.2}$$

In our calculations we select the parameters h_1 and K and calculate the η_∞. An idea about the number of points taken across the boundary layer with the variable grid that uses those parameters for different η_∞-values can be obtained from Fig.

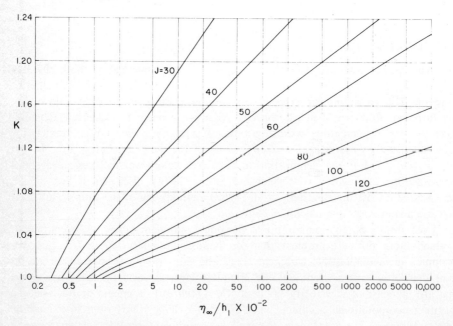

Fig. 8.1 Variation of K with h_1 for different η_∞-values.

8.1. For example, for $h_1 = 0.01$, $K = 1.10$, and $\eta_\infty = 100$, the ratio of η_∞/h_1 is 10,000, and the number of points across the boundary layer is approximately 70. For a uniform spacing ($K = 1$) with $h_1 = 0.01$ and $\eta_\infty = 100$, there would be 10,000 points!

8.2 EXTERNAL LAMINAR BOUNDARY LAYERS

For two-dimensional laminar and turbulent boundary layers the momentum equation with the eddy-viscosity concept is given by Eq. (6.1.4). In terms of transformed variables discussed in Sec. 5.3 that equation can be written as

$$(bf'')' + \frac{m+1}{2} ff'' + m[1 - (f')^2] = \bar{x}\left(f' \frac{\partial f'}{\partial \bar{x}} - f'' \frac{\partial f}{\partial \bar{x}}\right) \tag{8.2.1}$$

Here $\bar{x}$ is the transformed x-variable [see Eq. (3.3.1)], $\epsilon_m^+ = \epsilon_m/\nu$, and

$$b = (1 + t)^{2k}(1 + \epsilon_m^+) \tag{8.2.2}$$

We recall that for two-dimensional flows, $\bar{x} = x$ and $t = 0$. For axisymmetric flows with negligible transverse-curvature effect, $t = 0$. For laminar flows, ϵ_m^+ is equal to zero. Thus Eq. (8.2.1) is applicable to both laminar and turbulent flows. Equation (8.2.1) is subject to the following boundary conditions:

$$f(\bar{x},0) \equiv f_w(\bar{x}) \begin{cases} = 0 & \text{(no mass transfer)} \\ = \dfrac{-1}{(u_e \nu \bar{x})^{1/2}} \displaystyle\int_0^{\bar{x}} v_w \, d\bar{x} \end{cases} \tag{8.2.3a}$$

with transpiration velocity, v_w, and

$$f'(\bar{x},0) = 0 \qquad \lim_{\eta \to \infty} f'(\bar{x},\eta) = 1 \tag{8.2.3b}$$

In this section we shall discuss the application of Eq. (8.2.1) to several two-dimensional and axisymmetric laminar boundary layers. The solutions will be obtained by the numerical method discussed in Chap. 7.

8.2.1 Two–Dimensional Similar Flows

For two-dimensional laminar flows, Eq. (8.2.1) reduces to Eq. (5.3.9), that is,

$$f''' + \frac{m+1}{2} ff'' + m[1 - (f')^2] = x\left(f' \frac{\partial f'}{\partial x} - f'' \frac{\partial f}{\partial x}\right) \tag{5.3.9}$$

We recall that when the external velocity, u_e, is proportional to x^m and the boundary conditions [Eq. (8.3.1)] (with $\bar{x} = x$) are independent of x, then Eq. (5.3.9) reduces to the Falkner-Skan equation:

$$f''' + \frac{m+1}{2} ff'' + m[1 - (f')^2] = 0 \tag{5.4.1}$$

That equation can be solved by the shooting method described in Sec. 4.2.2. It can also be solved by the finite-difference method discussed in Chap. 7. We shall now present a brief description of the use of the latter method to solve Eq. (5.4.1) subject to the boundary conditions

$$\eta = 0 \quad f = f_w \quad f' = 0; \quad \eta = \eta_\infty \quad f' = 1 \tag{8.2.4}$$

Since the linearized form of the equations is being solved, it is necessary to make an initial guess for the functions f, f', and f'' across the boundary layer. These can be obtained by a number of expressions. One possibility is to assume that

$$f' = a + b\eta + c\eta^3 \tag{8.2.5}$$

and obtain the constants a, b, and c in Eq. (8.2.5) by using the boundary conditions [Eq. (8.2.4)] and by using an additional condition $f''(\eta_\infty) = 0$. That procedure shows that

$$f' = \frac{3}{2}\left(\frac{\eta}{\eta_\infty}\right) - \frac{1}{2}\left(\frac{\eta}{\eta_\infty}\right)^3 \tag{8.2.6a}$$

and

$$f = \frac{\eta_\infty}{4}\left(\frac{\eta}{\eta_\infty}\right)^2\left[3 - \frac{1}{2}\left(\frac{\eta}{\eta_\infty}\right)^2\right] \tag{8.2.6b}$$

$$f'' = \frac{3}{2}\frac{1}{\eta_\infty}\left[1 - \left(\frac{\eta}{\eta_\infty}\right)^2\right] \tag{8.2.6c}$$

Recalling that $f' = u$ and $f'' = v$ and using Eq. (8.2.6), we now can determine the quantities $(r_p)_j$ and $(s_k)_j$ from Eq. (7.2.14) for $p = 1$–3 and $k = 1$–6. For our problem all terms superscripted $(n - 1)$ are zero (i.e., x-derivatives at constant η are zero) and $(b)_j = 1$, $\alpha = 0$, $\alpha_1 = (m + 1)/2$, $\alpha_2 = m$, and m is specified. Then the factorization procedure [Eq. (7A.1.3), see Appendix 7A] is carried out, and a solution of the momentum equation is obtained. The inhomogeneous quantities, $(r_p)_j$, and the coefficients, $(s_k)_j$, are calculated with the latest iterates, and the factorization procedure is repeated. The iteration is repeated until some convergence criterion is satisfied. In boundary-layer calculations the greatest error usually appears in the wall shear parameter, f_w''. For that reason we use the wall shear parameter as the convergence criterion. Calculations are stopped when

$$|f_w''^{(i+1)} - f_w''^{(i)}| < \epsilon_1 \tag{8.2.7}$$

where the value of ϵ_1 is prescribed. The results, of course, are the same as those of the shooting method if sufficient grid points are taken across the layer.

8.2.2   Two–Dimensional Nonsimilar Flows

We now apply the numerical method of Chap. 7 to two-dimensional nonsimilar laminar flows. In most practical problems the flow starts as either a flat-plate flow ($m = 0$) or as a stagnation-point flow ($m = 1$). Since the governing momentum equation is similar, Eq. (5.3.9) reduces to Eq. (5.4.1), and its solution can be obtained by the procedure discussed in Sec. 8.2.1. At the next x-location, using the specified velocity distribution, one can calculate m from its definition and then solve the differenced momentum equation by a procedure similar to that of Sec. 8.2.1. For example, in Eq. (7.2.14) all terms superscripted $(n - 1)$ are known and $(b)_j = 1$, $\alpha \neq 0$ (it is calculated from its definition), and α_1 and α_2 are calculated from their definitions. We can use the f, u, and v profiles for $(n - 1)$ to be the new guesses for the terms appearing in the quantities $(r_p)_j$ and $(s_k)_j$ in Eq. (7.2.14). Then the factorization procedure [Eq. (7A.1.3)] is carried out, and a solution of the momentum equation is obtained. The inhomogeneous quantities, $(r_p)_j$, and the coefficients, $(s_k)_j$, are calculated with the latest iterates, and the factorization is repeated until Eq. (8.2.7) is satisfied.

Flow Past a Circular Cylinder

As an example of a two-dimensional nonsimilar laminar flow, let us consider the flow past a circular cylinder of radius r_0, normal to the axis. The inviscid-velocity distribution is ($\bar{x} = x/r_0$)

$$u_e = 2u_\infty \sin \bar{x} \tag{8.2.8}$$

From the definition of m we find that

$$m = \frac{\bar{x} \cos \bar{x}}{\sin \bar{x}} \tag{8.2.9}$$

The flow starts (at $\bar{x} = 0$) as a stagnation point flow:

$$\lim_{\bar{x} \to 0} m = 1$$

Figure 8.2 shows the values of dimensionless wall shear stress,

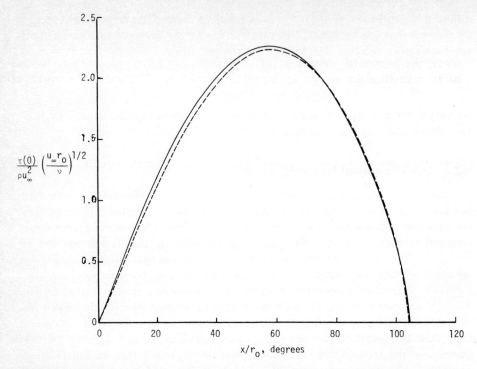

Fig. 8.2 Dimensionless wall shear distribution for a laminar flow past a circular cylinder. The results indicated by the solid line were obtained by the finite-difference method, and those indicated by the dashed line were obtained by Thwaites' method.

$$\frac{\tau(0)}{\rho u_\infty^{\ 2}} \left(\frac{u_\infty r_0}{\nu}\right)^{1/2} \tag{8.2.10}$$

computed by the finite-difference method of Chap. 7 together with Thwaites' method of Sec. 5.6. Both methods indicate separation at $\bar{x} = 105°$. These results, of course, invalidate Eq. (8.2.8), but a realistic calculation would involve iterating an external-flow solution with a solution for the boundary layer and the separated flow region (Chap. 11).

Flow Past a Flat Plate with Uniform Suction

Let us consider a laminar flow past a flat plate with uniform suction ($v_w = $ const). Since, according to Eq. (8.2.3a), f_w is not independent of x as is required by similarity, this is a nonsimilar flow. (To have a similar flow on a flat plate with mass transfer, the mass-transfer velocity must vary with x as $x^{-1/2}$.) Then Eq. (8.2.3a) becomes

$$f_w = -\frac{v_w}{u_e}\sqrt{\text{Re}_x} \tag{8.2.11}$$

and Eq. (8.2.3b) remains the same. Note that $v_w < 0$ for suction. At some distance x from the leading edge the boundary-layer thickness, δ, which in general is a function of x, becomes virtually constant and stays constant with increasing x. As a result, the horizontal component of the velocity, u, varies only with y, that is, $\partial u / \partial x = 0$. Then the momentum equation becomes an ODE and can be written as

$$v_w \frac{du}{dy} = v \frac{d^2 u}{dy^2} \tag{8.2.12}$$

since from the continuity equation, $v(x, y) = v_w = \text{const.}$ The solution of Eq. (8.2.12) is

$$u(y) = u_e \left(1 - \exp \frac{v_w y}{v}\right) \tag{8.2.13}$$

The profile given above, known as the asymptotic suction profile, was briefly mentioned in Sec. 5.5. Using this velocity profile and the definitions of δ^*, θ, and τ_w, we find

$$\delta^* = -\frac{v}{v_w} \qquad \theta = -\frac{1}{2} \frac{v}{v_w} \qquad \tau_w = -\rho v_w u_e \tag{8.2.14a}$$

In dimensionless quantities, Eq. (8.2.14a) is

$$\frac{\delta^*}{x} = -\frac{1}{\bar{v}_w \mathrm{Re}_x} \qquad \frac{\theta}{x} = -\frac{1}{2} \frac{1}{\bar{v}_w \mathrm{Re}_x} \qquad \frac{\tau(0)}{\rho u_e^2} = -\bar{v}_w \tag{8.2.14b}$$

where $\bar{v}_w = v_w / u_e$. From those expressions we see that for the case of asymptotic suction, $\delta^* / \theta = H = 2$ and that the wall shear is a constant independent of the viscosity, μ, of the fluid.

Such an idealized flow does not exist close to the leading edge of the plate, even though uniform suction starts at the leading edge, because the boundary-layer thickness is zero there.

The distance x from the leading edge necessary to develop the asymptotic-suction flow has been determined by Iglisch (1949) to be roughly

$$-\bar{v}_w \mathrm{Re}_x^{1/2} \cong 2 \tag{8.2.15}$$

We now solve Eq. (5.3.9) (with $m = 0$) subject to the boundary conditions [Eqs. (8.2.11) and (8.2.3b)] and examine the way in which the solutions approach the asymptotic value of H. The results are shown in Fig. 8.3. We see that at $\bar{v}_w^2 \mathrm{Re}_x \sim 4$, corresponding to $f_w = 2$, the asymptotic value of H is almost reached.

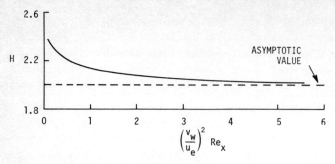

Fig. 8.3 Computed values of H for a flat-plate flow with uniform suction.

8.2.3 Axisymmetric Flows

For axisymmetric flows the general momentum equation for laminar flows is given by Eq. (8.2.1) with $b = (1 + t)^{2k}$. Before we solve that equation subject to the boundary conditions given by Eq. (8.2.3), it is necessary to compute the x-coordinate by using Eq. (3.3.1) for a given body shape. If the body under consideration has significant transverse curvature then the term [see Eq. (5.3.6)]

$$\left(\frac{L}{r_0}\right)^2 2 \cos \phi \left(\frac{\nu \bar{x}}{u_e}\right)^{1/2} \tag{8.2.16}$$

in the definition of t is known from the geometry and from the known external-velocity distribution of the body. Then, as in two-dimensional flows, m is calculated, and the governing momentum equation is solved by a procedure very similar to those discussed in Sec. 8.2.2. Obviously, if the body has negligible transverse curvature, then t is equal to zero.

Flow Past a Sphere

The inviscid-velocity distribution of a flow past a sphere of radius L is given by

$$u_e = \tfrac{3}{2} u_\infty \sin \xi \tag{8.2.17}$$

where $\xi = x/L$. The local cross-sectional radius of the body, r_0, is

$$r_0 = L \sin \xi \tag{8.2.18}$$

The $\bar{x}$-coordinate and the m-distribution of this flow can be obtained from Eqs. (3.3.1) and (5.3.7). First, using Eq. (8.2.18), we find from Eq. (3.3.1) that

$$\bar{x} = L(\xi - \tfrac{1}{4} \sin 2\xi) \tag{8.2.19}$$

Next, using Eqs. (8.2.17) and (8.2.18), we find from Eq. (5.3.7) that

$$m = \frac{[\xi - (1/4) \sin 2\xi] \cos \xi}{\sin^3 \xi} \qquad (8.2.20)$$

At $\bar{x} = 0$ the flow starts as a stagnation-point flow for which the value of m is given by Eq. (8.2.20). That is not obvious from the formula for m, since when this expression is evaluated at $\bar{x} = 0$, m becomes indeterminate. However, by using L'Hopital's rule it can be shown that $m = \frac{1}{3}$. We note that in this case the momentum equation for an axisymmetric flow with negligible transverse curvature is identical to Eq. (5.4.1) with $m = \frac{1}{3}$. Therefore the solution procedure is identical to the procedure described in Sec. 8.2.1.

Away from the stagnation point, Eq. (8.2.1) with $(1 + t)^{2k}$ can be solved at specified values of $\bar{x}$ and m given by Eqs. (8.2.19) and (8.2.20), respectively, again by the procedure described in Sec. 8.2.2.

Figure 8.4 shows the values of dimensionless wall shear

$$\frac{\tau(0)}{\rho u_\infty^2} \left(\frac{u_\infty L}{\nu} \right)^{1/2} \qquad (8.2.21)$$

computed for the boundary conditions given in Eq. (8.2.3) with $f_w = 0$. We note that the flow separation occurs at $107.5°$.

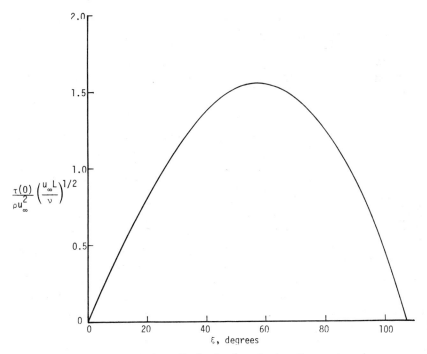

Fig. 8.4 Dimensionless wall shear distribution for a laminar flow past a sphere.

Axial Flow Past a Slender Circular Cylinder

When the radius of a body is of the same order of magnitude as the thickness of the boundary layer, the transverse-curvature effect, which is of second order, becomes quite important and strongly affects the skin friction and heat transfer. A typical example is axial flow past long circular cylinders.

The transverse-curvature effect on slender circular cylinders has been studied by a number of people. Seban and Bond (1951) expanded the stream function in an ascending power series of $2/R$, where R is a transverse-curvature parameter,

$$R = \frac{\mathrm{Re}_x}{4}\left(\frac{r_0}{x}\right)^2 \tag{8.2.22}$$

with $\mathrm{Re}_x = u_e x/\nu$. Kelly (1954) added corrections to their expression by correcting the definition of δ^* for such flows and by correcting numerical coefficients appearing in Eq. (26) of Seban and Bond's paper (1951). With those corrections the Seban-Bond-Kelly results for flow past slender circular cylinders for large R values are given by

$$c_f \mathrm{Re}_x^{1/2} = 0.664\left(1 + \frac{1.050}{R^{1/2}} - \frac{0.480}{R}\right) \tag{8.2.23}$$

The same problem was also studied by Stewartson (1955) analytically for values of R approaching 0. His expression is given by

$$c_f \mathrm{Re}_x^{1/2} = \frac{2}{R^{1/2}}\left\{\frac{1}{\ln\,(1/cR)} - \frac{3.854}{[\ln\,(1/cR)]^3} + \cdots\right\} \tag{8.2.24}$$

where $\ln\,(1/cR) = -\ln R - 0.577$ (c is Euler's constant).

Numerical solutions of this problem were also obtained by Cebeci (1970b), who used an implicit finite-difference method, and by Jaffe and Okamura (1968), who used a shooting method. Later Cebeci (see Cebeci and Smith, 1974) repeated his calculations by using the numerical method described in Chap. 7. The calculations were performed for R from zero to infinity. The numerical calculations for this kind of flow differ from the usual laminar boundary-layer calculations in the sense that although for most laminar flows the transformed boundary-layer thickness is almost constant (i.e., $\eta_\infty \simeq 8$), for this flow, η_∞ is a function of R. For example, for $R \simeq 10^{-3}$, $\eta_\infty \simeq 100$. Furthermore, the wall-shear parameter f_w'' becomes large with decreasing R. In such cases the laminar-velocity profiles look somewhat like turbulent-velocity profiles. In order to maintain computational accuracy it is necessary to take very small $\Delta\eta$-spacing close to the wall. For that reason, Cebeci's calculations were made with the variable grid described in Sec. 8.1.

Figure 8.5 shows the calculated values of $c_f \mathrm{Re}_x^{1/2}$ as a function of R. The results indicate that both Eqs. (8.2.23) and (8.2.24) agree well with the computed

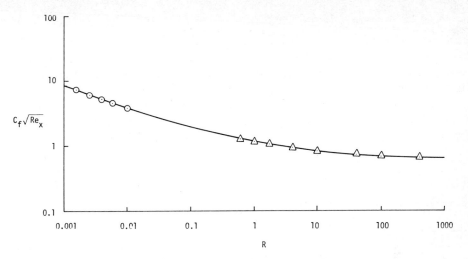

Fig. 8.5 Local skin-friction coefficient for laminar flows over slender circular cylinders. Circles indicate data from Stewartson (1955). Triangles indicate data from Seban-Bond-Kelly (see Seban and Bond, 1951). Solid line denotes numerical solutions of Cebeci (see Cebeci and Smith, 1974).

values within the ranges in which they are valid. It appears that Eq. (8.2.23) is valid for $R > 1$ and that Eq. (8.2.24) is valid for $R < 0.02$.

8.3 EXTERNAL TURBULENT BOUNDARY LAYERS

With the inclusion of ϵ_m^+ in b, Eq. (8.2.1) is applicable to both two-dimensional and axisymmetric external turbulent boundary layers subject to the boundary conditions given by Eq. (8.2.3). In this section we shall discuss the application of that equation to several two-dimensional and axisymmetric turbulent boundary layers. Before Eq. (8.2.1) can be solved, however, it is necessary to express ϵ_m^+ by some suitable formulas. Here we employ the formulation explained and used by Cebeci and Smith (1974). According to that formulation the turbulent boundary layer is treated as a composite layer consisting of inner and outer regions with separate expressions for eddy viscosity in each region. The analytic functions are—in effect—data fits, not theoretical results. In the inner region the eddy-viscosity formula is defined by

$$(\epsilon_m)_i = L^2 \left(\frac{r}{r_0}\right)^k \left|\frac{\partial u}{\partial y}\right| \gamma_{tr} \qquad 0 \leqslant y \leqslant y_c \tag{8.3.1}$$

Here L is defined for two-dimensional flows as the mixing length:

$$L = \kappa y \left[1 - \exp\left(-\frac{y}{A}\right)\right] \tag{8.3.2a}$$

where $\kappa = 0.4$ and A is a damping-length constant (Sec. 6.4 and Fig. 6.14) defined by

$$A = 26 \frac{\nu}{N} u_\tau^{-1} \qquad u_\tau = \left(\frac{\tau_w}{\rho} \right)^{1/2} \tag{8.3.2b}$$

$$N = \left\{ \frac{p^+}{v_w^+} [1 - \exp(11.8 v_w^+)] + \exp(11.8 v_w^+) \right\}^{1/2} \tag{8.3.2c}$$

$$p^+ = \frac{\nu u_e}{u_\tau^3} \frac{du_e}{dx} \qquad v_w^+ = \frac{v_w}{u_\tau} \tag{8.3.2d}$$

For flows with no mass transfer, Eq. (8.3.2c) can be written as

$$N = (1 - 11.8 p^+)^{1/2} \tag{8.3.2e}$$

For axisymmetric flows, L is defined by

$$L = \kappa r_0 \ln\left(\frac{r}{r_0} \right) \left\{ 1 - \exp\left[-\frac{r_0}{A} \ln\left(\frac{r}{r_0} \right) \right] \right\} \tag{8.3.3}$$

As before, the damping length, A, is given by Eq. (8.3.2d).

In Eq. (8.3.1), γ_{tr} is an intermittency factor that accounts for the transitional region that exists between a laminar and turbulent flow (Chap. 6). It is defined by

$$\gamma_{tr} = 1 - \exp\left[-G r_0(x_{tr}) \left(\int_{x_{tr}}^{x} \frac{dx}{r_0} \right) \left(\int_{x_{tr}}^{x} \frac{dx}{u_e} \right) \right] \tag{8.3.4}$$

Here x_{tr} is the location of the start of transition and the empirical factor, G, which has the dimensions of velocity/(length)^2, is given by

$$G = \frac{1}{1200} \frac{u_e^3}{\nu^2} R_{x\,tr}^{-1.34} \tag{8.3.5}$$

The transition Reynolds number is defined as $R_{x_{tr}} = u_e x_{tr}/\nu$. For simple shapes, Eq. (8.3.4) can be simplified considerably. For example, for a straight tube or for a flat plate it becomes, using Eq. (8.3.5),

$$\gamma_{tr} = 1 - \exp\left[-\frac{1}{1200} R_{x\,tr}^{0.66} \left(\frac{x}{x_{tr}} - 1 \right)^2 \right] \tag{8.3.6}$$

for $x > x_{tr}$ only. In this case, taking $R_{x_{tr}} = 10^6$, γ_{tr} reaches 0.99 at $x/x_{tr} = 1.8$.

In the outer region the eddy-viscosity formula is defined by

$$(\epsilon_m)_o = \alpha \left| \int_0^\infty (u_e - u)\, dy \right| \gamma_{tr} \qquad y_c \leqslant y \leqslant \delta \tag{8.3.7}$$

Here α is a universal constant equal to 0.0168 when $R_\theta \geqslant 5000$. For lower values of R_θ, α varies with R_θ according to the empirical formula given by Cebeci (1973a):

$$\alpha = 0.0168 \frac{1.55}{1 + \Pi} \tag{8.3.8}$$

where Π is

$$\Pi = 0.55[1 - \exp(-0.243 z_1^{1/2} - 0.298 z_1)] \tag{8.3.9}$$

with $z_1 = (R_\theta/425 - 1)$. For an ordinary boundary layer, Eq. (8.3.7) reduces to $(\epsilon_m)_o = 0.0168 u_e \delta^*$.

The condition used to define the inner and outer regions is the continuity of the eddy viscosity; from the wall outward the expression for the inner eddy viscosity is applied until $(\epsilon_m)_i = (\epsilon_m)_o$, which defines y_c.

In most practical boundary-layer calculations, it is necessary to calculate a complete boundary-layer flow. That is, for a given external-velocity distribution and for a given (natural) transition point it is necessary to calculate laminar, transitional, and turbulent boundary layers by starting the calculations at the leading edge or at the forward stagnation point of the body. In most boundary-layer prediction methods, however, the calculation of transitional boundary layers is avoided by assuming the transitional region to be a switching point between laminar and turbulent regions. In general, especially at low Reynolds numbers, that is not a good procedure, and it can lead to substantial errors. With the intermittency factor defined by Eq. (8.3.4) the transitional region can be accounted for rather more satisfactorily. For incompressible two-dimensional flows the start of transition can be satisfactorily calculated by using the empirical correlation curve given by Eq. (6.1.1). Sometimes, however, the laminar-flow calculations indicate flow separation before the transition point can be calculated, say, by Eq. (6.1.1). In those cases the wall shear becomes negative and prevents the solutions from converging. In situations like those, one may assume the laminar separation point to be the transition point and start the turbulent flow calculations at that point.

If the above procedure is used to calculate a complete flow field, then starting calculations at $\bar{x} = 0$ is identical to the process described in Sec. 8.2.1. The only difference occurs when the flow is turbulent at the first x-station. Before the momentum equation can be solved, it is necessary to establish the inner and outer regions for the eddy-viscosity formulas. Since the eddy-viscosity formulas contain terms such as $\partial u/\partial y$ (which is proportional to f'') and $\int_0^\infty (u_e - u)\, dy$, these two regions are not known until a solution of the momentum equation is generated. Thus an iteration process is necessary. For the first iteration, $(\epsilon_m^+)_i$ and $(\epsilon_m^+)_o$ are obtained from the solution at $\bar{x} = x_{n-1}$, and the inner and outer regions are established by the continuity of the eddy-viscosity expression. Thereafter, the solution of Eq. (8.2.1) for turbulent flows is similar to that for laminar flows. For a more detailed discussion see Cebeci and Smith (1974).

The calculations can also be started at any $\bar{x} > 0$, by $\bar{x} = x_n$ and by inputting the initial profiles for f, f', and f''. When the procedure is used, then the turbulent-flow calculation at $\bar{x} = x_{n+1}$ is the same as the procedure described above.

8.3.1 Two-Dimensional Flows

For two-dimensional turbulent flows, Eq. (8.2.1) reduces to

$$[(1 + \epsilon^+)f'']' + \frac{m + 1}{2} ff'' + m[1 - (f')^2] = x \left(f' \frac{\partial f'}{\partial x} - f'' \frac{\partial f}{\partial x} \right) \tag{8.3.10}$$

The solution of Eq. (8.3.10) subject to Eq. (8.2.3) is similar to the solution of laminar flows discussed in Sec. 8.2. Here we show examples of various flows computed by Cebeci and Smith (1974). We also show a comparison of calculated results with the experiment to show the "accuracy" of the turbulence model used for the Reynolds shear stress term.

Figure 8.6 shows the variation of local skin-friction coefficient, c_f, with Reynolds number, Re_x, for a flat-plate flow. In this example, laminar flow starts at the leading edge, and transition was assumed to be very close to the leading edge.

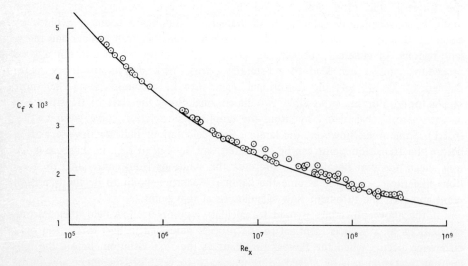

Fig. 8.6 Comparison of calculated local skin-friction coefficient with experiment for zero pressure gradient incompressible turbulent flow. The symbols denote the experiment, and the solid lines denote the numerical solutions of Cebeci and Smith (1974).

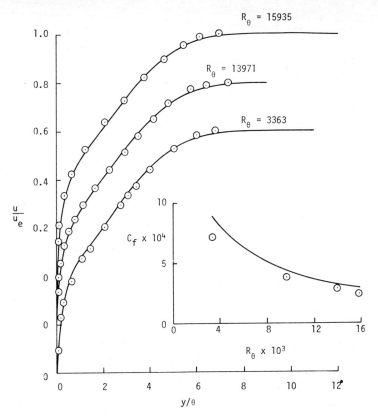

Fig. 8.7 Comparison of calculated velocity profiles for the flow measured by Simpson et al. (1967), where $v_w/u_e = 0.00784$. Symbols denote experimental data, and the solid lines denote the numerical solutions of Cebeci and Smith (1974).

Correlations on R_x cannot be general, since $R_{x_{tr}}$ varies from case to case; in general, comparisons should be made at given Re rather than given R_x. Shown in the figure are the experimental data and the predictions of Eq. (6.7.3). It is seen that the numerical results agree quite well with experiment.

Figure 8.7 shows the results for a porous flat plate. Here the mass-transfer parameter is $v_w/u_e = 0.00784$. The experimental data are those of Simpson et al. (1967). In addition to the velocity profiles, Fig. 8.7 also shows a comparison between the calculated local skin-friction values and the values given by Simpson et al.

Figure 8.8 shows a comparison of calculated and experimental skin-friction values for a flow past an ellipse, which has a minor axis $d = 3.97$ in. The experimental velocity distribution obtained by Schubauer (1939) was given for a free-stream velocity of $u_\infty = 60$ ft/s, corresponding to a Reynolds number $R_d = u_\infty d/v = 1.18 \times 10^5$. For this experiment there were three distinct regions: laminar, transitional, and turbulent. The transitional region extended from $x/d = 1.25$ to $x/d = 2.27$, and separation was indicated at $x/d = 2.91$. The numerical calculations were made for the given experimental external-velocity distribution and transition point. The

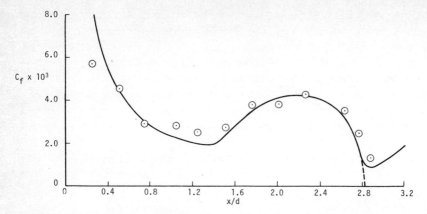

Fig. 8.8 Comparison of calculated laminar, transitional, and turbulent local skin-friction coefficients with the experiment for Schubauer's ellipse. The circles indicate data from the experiment, $R_d = 1.18 \times 10^5$. Solid and dashed lines denote numerical solutions (see Cebeci and Smith, 1974) for experimental and extrapolated u_e/u_∞, respectively.

calculations were started as laminar at the stagnation point ($m = 1$), and in the subsequent stations the m-distribution was calculated from the external-velocity distribution and the momentum equation was solved for nonsimilar flows. At the specified transition point ($x/d = 1.25$) the eddy-viscosity formulas, including γ_{tr} were activated, and the turbulent flow calculations were started. Note from the results shown in Fig. 8.8 that when the experimental velocity distribution is used, the local skin-friction coefficient begins to increase near separation because the real pressure gradient decreases as the boundary layer displacement thickness grows rapidly (Chap. 11). Therefore a calculation in which the pressure distribution is taken as known, instead of being calculated by a viscous-inviscid interaction method, will either predict separation at, or upstream of, the separation point or fail to predict it at all. However, when the calculations are repeated by using an extrapolated velocity distribution like that obtained by an inviscid method, the skin friction goes to zero at $x/d = 2.82$, a value that agrees satisfactorily with the experimental data.

An important class of boundary-layer flows comprises those that have been perturbed in some manner and then allowed to recover to some equilibrium condition. As an example of that class, we present the results for an equilibrium boundary-layer in moderate positive pressure gradient measured by Bradshaw and Ferriss (1965). From $x = 1.92$ to 6.92 ft, the flow has an external-velocity distribution in which $u_e \sim x^{-0.255}$. Figure 8.9 shows computed and experimental results. The calculations were started at $x = 1.92$ ft. It can be seen that although the agreement in velocity profiles is fairly good except close to the wall, the agreement in skin friction is poor.

8.3.2 Axisymmetric Flows

For axisymmetric flows the general momentum equation for laminar and turbulent flows is given by Eq. (8.2.1). As in laminar flows, before we solve that equation

subject to the boundary conditions given by Eq. (8.2.3), we first compute the
$\bar{x}$-coordinate by using Eq. (3.3.1) for a given body shape. Figure 8.10 shows the
results for a 285-ft-long airship with fineness ratio of 4.2. The experimental data are
due to Cornish and Boatwright (1960); the measurement was along the top, where
the flow should be nearly axisymmetric with negligible transverse curvature. The
external-velocity distribution and boundary-layer measurements were made in flight
at speeds from 35 to 70 mph. No transition data were given, but from the configuration of the airship it was inferred that the boundary layer was tripped at approximately $x/L = 0.05$.

As in laminar flows the transverse-curvature effect also plays an important role
in turbulent flows. As an example we consider turbulent flows past slender circular

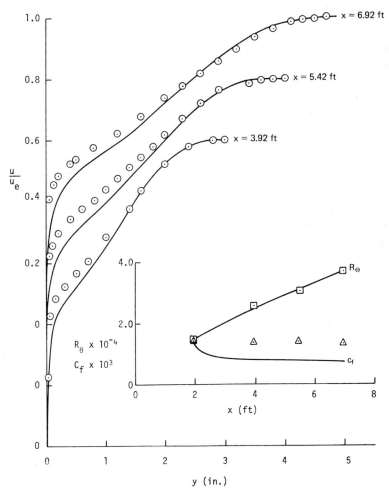

Fig. 8.9 Comparison of calculated results for the equilibrium flow of Bradshaw and Ferriss
(1965). The symbols denote experimental data, and the solid lines denote numerical solutions
of Cebeci and Smith (1974).

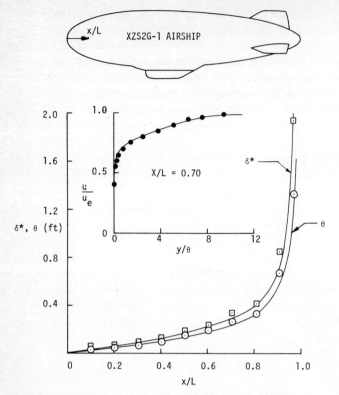

Fig. 8.10 Comparison of calculated and experimental results for the XZS2G-1 airship measured by Cornish and Boatwright (1960) at $R_L = 188 \times 10^6$ corresponding to 70 mph. The symbols denote the experimental data, and the solid lines denote the numerical solutions of Cebeci and Smith (1974).

cylinders. Table 8.1 shows the deviation of the local skin-friction coefficient from that of a flat plate on two slender cylinders of diameters $d = 1$ in. and 0.024 in. The cylinder c_f-values were calculated by Cebeci (1973b) by using the eddy-viscosity formulation presented in Sec. 8.3. The flat-plate c_f-values ($c_{f_{fp}}$) were obtained by using Eqs. (6.7.10) and (6.7.13).

Table 8.1   Transverse-Curvature Effect on Local Skin-Friction Coefficient of Slender Circular Cylinders $(R_{\theta 3-d} = u_e \theta_{3-d}/\nu)^a$

d, in.	$R_{\theta 3-d}$	δ/r_0	$c_f/c_{f_{fp}}$
0.024	2100	75	2.15
1.00	8750	2	1.40

^{a}See Eq. (1.9.15) in Cebeci and Smith (1974) for the definition of θ_{3-d}.

The local skin-friction coefficient of slender circular cylinders can be obtained by using the empirical formulas given by White (1972).

According to the studies conducted by Cebeci (1973b), the eddy-viscosity formulation presented above can also be used to compute turbulent boundary layers successfully with transverse-curvature effect. Figure 8.11 shows the computed velocity profiles for the two cylinders mentioned above together with the experimental data of Richmond (1957). For further comparisons, see Cebeci and Smith (1974).

8.4 INTERNAL TURBULENT FLOWS

We now apply the numerical method discussed in Sec. 7.3 to compute internal flows. In Secs. 5.7.1.1 and 5.7.1.2 we have presented results obtained by this method for the entrance-region laminar flow in a pipe and for the entrance-region laminar flow between parallel plates. We shall now present results for turbulent flows. We use the eddy-viscosity formulation of Sec. 8.3 except that the edge velocity, u_e, in the outer eddy-viscosity formula [Eq. (8.3.7)] is replaced by u_{max}.

The internal-flow method discussed in Sec. 5.7 and 7.3, like the external-flow method, applies to both laminar and turbulent flows. The calculations can be started as laminar at the duct entry ($\bar{x} = 0$) and can be continued as turbulent by specifying the transition location. As in external flows, the calculations can also be

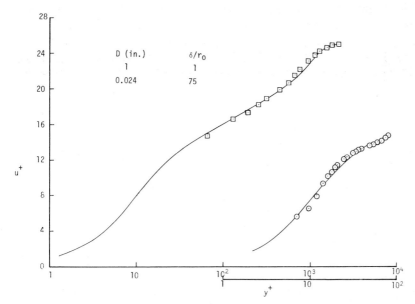

Fig. 8.11 Comparison of calculated and experimental dimensionless velocity profiles for two cylinders. The experimental u^+ values (denoted by symbols) were obtained by normalizing the measured u values by the *calculated* friction velocity, u_τ. The lower y^+ side refers to the 0.024-in. cylinder. The solid lines denote numerical solutions of Cebeci and Smith (1974). The data are from Richmond (1957).

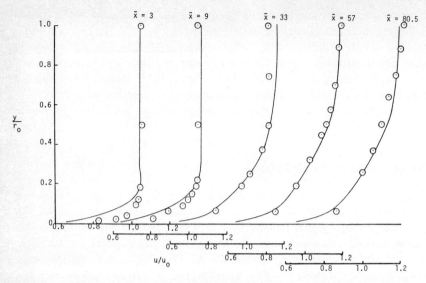

Fig. 8.12 Comparison of calculated and experimental velocity profiles in the entrance region of a pipe for turbulent flow at $R_d = 194,000$. The symbols denote the data of Barbin and Jones (1963), and the solid lines denote the numerical solutions of Cebeci and Chang (1977).

started as turbulent by inputting the initial-velocity profiles. The results shown in Figs. 8.12–8.14 were obtained using the latter procedure. They are for entrance-region turbulent flow in a pipe with uniform entry velocity, u_0. The experimental data in Figs. 8.12 and 8.13 are from Barbin and Jones (1963), and those in Fig. 8.14 are from Richman and Azad (1974). For further comparisons see Cebeci and Chang (1977), and for some recent mean-flow data see Reichert and Azad (1976).

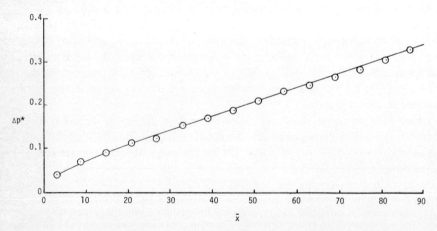

Fig. 8.13 Comparison of calculated and experimental pressure drop in the entrance region of a pipe for turbulent flow at $R_d = 194,000$. The symbols denote the data of Barbin and Jones (1963), and the solid lines denote the numerical solutions of Cebeci and Chang (1977).

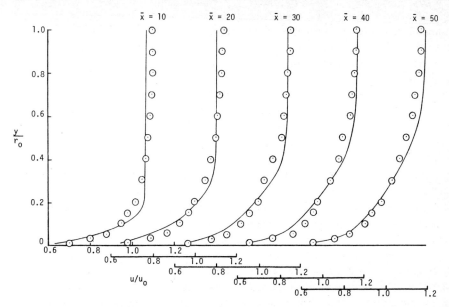

Fig. 8.14 Comparison of calculated and experimental velocity profiles in the entrance region of a pipe for turbulent flow at $R_d = 100,000$. The symbols denote the data of Richman and Azad (1974), and the solid lines denote the numerical solutions of Cebeci and Chang (1977).

8.5 A TRANSPORT-EQUATION MODEL OF TURBULENCE

An exact equation for the shear stress, $-\rho\overline{u'v'}$, or for any of the other Reynolds stresses, can be obtained from the NS equations for the instantaneous velocity components $u + u'$, $v + v'$, and $w + w'$. We add v' times the $(u + u')$ equation to u' times the $(v + v')$ equation and take the time mean. The result is an equation giving the rate of change of $-\rho\overline{u'v'}$ along a mean streamline, that is, $D(-\rho\overline{u'v'})/Dt$, in the notation of Chap. 2, in terms of the local mean-velocity gradients, such as $\partial U/\partial y$, and some turbulence quantities. The turbulence quantities appearing in this transport equation for $-\rho\overline{u'v'}$ include some or all of the other Reynolds stresses and also more complicated quantities such as $\overline{p'u'}$ (where p' is the pressure fluctuation) and $\overline{u'v'^2}$. No matter how many transport equations we use, we always have more unknowns than equations.

Physically, the transport equations relate the rate of change of Reynolds stress along a mean streamline (or equivalently the net rate of addition of Reynolds stress to fluid as it is convected through a unit CV by the mean flow) to the net sum of generation or destruction of Reynolds stress within the CV and the transport of Reynolds stress into the CV by the turbulent fluctuations themselves. The concept of a stress being convected is a difficult one; it is easier here to think of $-\rho\overline{u'v'}$ as a measure of the strength of turbulent eddies rather than a stress. The quantity $\frac{1}{2}\rho(\overline{u'^2} + \overline{v'^2} + \overline{w'^2})$, $-\frac{1}{2}$ times the sum of the Reynolds normal stresses, is a plausible definition of the kinetic energy of the turbulence per unit volume, the mean-flow kinetic energy being $\frac{1}{2}\rho(u^2 + v^2 + w^2)$. Transport of this quantity is easier to visualize. Its transport equation is just $-\frac{1}{2}$ times the sum of the transport equations for $-\rho\overline{u'^2}$, $-\rho\overline{v'^2}$, and $-\rho\overline{w'^2}$ separately.

Although the transport equations for Reynolds stress are not soluble as they stand, they have long been used as a guide to turbulence processes. In particular, they tell us that Reynolds stresses are not directly linked to the local mean-velocity gradient as is assumed in the eddy-viscosity and mixing-length formulas. The Reynolds stress at a given point depends on the whole history of the turbulence passing that point: the Reynolds-stress transport equation is a DE for Reynolds stress, whereas the eddy-viscosity formula is an algebraic equation. However, in self-similar or slowly changing flows, the history of the turbulence is in some sense uniform, so that history can be related to local conditions, and the eddy viscosity, defined as $-\overline{u'v'}/(\partial u/\partial y)$, is simply behaved. For more demanding cases we try to find empirical correlations, not for eddy viscosity, but for the unknown turbulence quantities on the right-hand sides of the Reynolds-stress transport equations. This process is called "modeling." For example, the quantity $\overline{u'v'^2}$, mentioned above, could be assumed to be proportional to $(-\overline{u'v'})^{3/2}$, where the factor of proportionality might be a function of y/δ and possibly of some other variables.

The most obvious way of empirically correlating the behavior of the unknown turbulence quantities is to use experimental data. Unfortunately, these quantities are difficult to measure, and virtually no reliable measurement of pressure fluctuations is available. As luck would have it, the most important unknown term in the $\overline{u'v'}$ transport equation contains pressure fluctuations. Considerable progress has been made by a combination of physicomathematical argument and trial-and-error adjustment of empirical constants to optimize the predictions of $\overline{u'v'}$, but this is obviously inferior to checking direct measurements. Another basic difficulty is that a turbulence length scale is needed, since the dimensions of the terms in the $D(-\rho\overline{u'v'})/Dt$ equation are all $\rho \times$ (velocity)^3/length. Transport equations for suitably defined length scales can be written but are not well understood. A good impression of the achievements and difficulties of advanced transport-equation modeling is given in the papers of Hanjalic and Launder (1972) and Launder et al. (1975). Here we describe an older, simpler method that has been applied to a large variety of shear-layer problems, including rapidly changing boundary layers. It is sophisticated enough to be a worthwhile improvement on eddy viscosity for the more demanding problems while simple enough for its limits of validity to be fairly easily assessed. The main feature of interest is that a modeled transport equation for $\overline{u'v'}$ is deduced from the turbulent-energy equation, whose main terms have been measured in most types of shear layer.

The turbulent-energy equation for a two-dimensional TSL can be written, using symbol q^2 for $u'^2 + v'^2 + w'^2$, as

$$\frac{D}{Dt}(\tfrac{1}{2}\bar{q}^2) = -\overline{u'v'}\,\frac{\partial u}{\partial y} - \frac{\partial}{\partial y}(\overline{p'v'} + \tfrac{1}{2}\overline{q^2 v'}) - (\text{dissipation rate}) \qquad (8.5.1)$$

In order, the terms on the right-hand side are the production of turbulent kinetic energy by work done by the rate of shear strain, $\partial u/\partial y$, against the shear stress,

$-\rho\overline{u'v'}$; the diffusion of turbulent energy by turbulence processes; and the viscous term, which is closely equal to the rate of dissipation of turbulent energy into heat. For more details see Bradshaw (1971b). If we now define quantities a_1 (dimensionless), L (a length), and G (dimensionless) by

$$a_1 = \frac{-\overline{u'v'}}{\overline{q^2}} \tag{8.5.2a}$$

$$L = \frac{(\overline{u'v'})^{3/2}}{(\text{dissipation rate})} \tag{8.5.2b}$$

$$G = \frac{(\overline{p'v'} + 1/2\,\overline{q^2 v'})}{(-\overline{u'v'})(-\overline{u'v'})_{\text{max}}^{1/2}} \tag{8.5.2c}$$

we can rewrite Eq. (8.5.1) as

$$\frac{D}{Dt}\left(\frac{-\overline{u'v'}}{2a_1}\right) = (-\overline{u'v'})\frac{\partial u}{\partial y} - \frac{\partial}{\partial y}[G(-\overline{u'v'})(-\overline{u'v'})_{\text{max}}^{1/2}] - \frac{(-\overline{u'v'})^{3/2}}{L} \tag{8.5.3}$$

The use of $(\overline{u'v'})_{\text{max}}^{1/2}$ is suggested by physical arguments about the large eddies that effect most of the diffusion of turbulent energy. Now Eq. (8.5.3) together with the mean-momentum equation for a two-dimensional TSL,

$$\frac{Du}{Dt} = -\frac{1}{\rho}\frac{dp}{dx} - \frac{\partial\overline{u'v'}}{\partial y} \tag{8.5.4}$$

and the continuity equation,

$$\frac{\partial u}{\partial x} + \frac{\partial v}{\partial y} = 0 \tag{8.5.5}$$

form a group of three PDEs in the three unknowns u, v, and $-\overline{u'v'}$, if a_1, L, and G can be correlated empirically. Bradshaw et al. (1967) found that good results in boundary layers were obtained by taking $a_1 = 0.15$ and L/δ and G as functions of y/δ. Various extensions have been made since that date (see Bradshaw and Ferriss, 1972), but the basic functions remain unchanged.

In the inner layer of a wall flow the transport terms become small in comparison to production and dissipation, and the modeled turbulent energy equation reduces to

$$0 = (-\overline{u'v'})\frac{\partial u}{\partial y} - \frac{(-\overline{u'v'})^{3/2}}{L}$$

which is equivalent to the mixing-length formula [Eq. (6.3.3)] if $L = l$. In the inner layer, $l = \kappa y \approx 0.41y$. In the outer layer, transport terms are *not* negligible, l is *not* equal to L, and it is likely that L, being a ratio of two turbulence quantities, is less dependent on flow history than l, which is the ratio of a turbulence quantity $(-\overline{u'v'})^{1/2}$ to a mean-flow quantity $\partial u/\partial y$. L is the most important of the three empirical functions of Eq. (8.5.2); a_1 is not very critical and a constant value is sufficient (though not necessary mathematically), while G is important only near the outer edge of a boundary layer where the growth of the layer is fed by energy diffusion from below.

As they are written, Eqs. (8.5.2) and (8.5.3) apply to shear layers with $\overline{u'v'} < 0$. Layers with a change of sign of $\overline{u'v'}$, such as wall jets, are really pairs of interacting shear layers and are best treated as such, the shear stress near the point of reversal being the net sum of contributions from the "positive" and "negative" sides of the layer. Asymmetrical interacting layers have $\partial u/\partial y$ and $-\overline{u'v'}$ going to zero at different points, so that a calculation method that used a smooth function for eddy viscosity, $-\overline{u'v'}/(\partial u/\partial y)$, would not be acceptable.

Note that Eq. (8.5.3) could equally well be thought of as a directly modeled version of the exact $\overline{u'v'}$ transport equation, which has terms whose effect, though not their cause, is similar to that of the terms in Eq. (8.5.1). The advantage of the roundabout procedure used here is that a_1, L, and G can all be measured, except for the $\overline{p'v'}$ term in G, which seems to be small. Although currently available measurements of turbulent quantities are not as accurate as the predictions of $\overline{u'v'}$ must be, they are much better than nothing. They define the error band within which a_1, L, and G can be arbitrarily adjusted and, what is even more important, they give advance warning of breakdown of the correlations in difficult cases.

APPENDIX 8A: Fortran Program for Solving TSL Equations for Two-Dimensional Laminar and Turbulent Boundary Layers

8A.1   GENERAL DESCRIPTION OF THE METHOD

In this section we present a Fortran program for solving the TSL equations for two-dimensional laminar and turbulent boundary layers. We use the numerical method discussed in Chap. 7. It is programed in FORTRAN IV for the IBM 370/165; it can be used on other computers with minor changes. It can also be used to solve third-order parabolic PDEs as will be described later.

The program consists of a MAIN routine, which contains the logic of the computations, and eight subroutines: INPUT, GRID, IVPL, GROWTH, EDDY, CMOM, SOLV3, and OUTPUT. The function of each subroutine is as follows:

INPUT   We read ν and η_∞ at $x = 0$ and u_e, and m as functions of x.

GRID   Generates the grid across the boundary layer for a uniform or non-uniform net.

IVPL   Generates the initial-velocity profile for a laminar flow.

GROWTH Controls the growth of the boundary layer.

EDDY   Contains the eddy-viscosity formulas.

CMOM   Contains the coefficients of the differenced momentum equation.

SOLV3   Contains the recursion formulas that arise in the block elimination method (see Sec. 7A.2).

OUTPUT   Prints out the desired boundary-layer parameters and profiles.

The following sections present a brief description of each routine together with the Fortran notation used in each routine. Units are dimensionless unless noted.

8A.1.1 MAIN

This routine contains the logic of the computations. It also checks the convergence of the iterations. For laminar flows, it uses Eq. (7A.1.21) for convergence with

```
       COMMON/BLC0/ NP,NX,NXT,NTR,NFLOW,ETAE,VGP,CNU,DETA(61),A(61),
      1            ETA(61)
       COMMON/BLCC/ X(60),UE(60),P1(60),P2(60),CEL(60),RX(60),CF,P1P,P2P,
      1            RTHETA(60)
       COMMON/BLCP/ DELV(61),F(61,2),U(61,2),V(61,2),B(61,2)
C - - - - - - - - - - - - - - - - - - - - - - - - - - - - - - - - - -
       ITMAX = 6
       CEL(1)= 0.0
       NX    = 1
       CALL INPUT
       CALL GRID
       CALL IVPL
C
    25 WRITE(6,9100) NX,X(NX)
       RX(NX)= UE(NX)*X(NX)/CNU
       IF(NX .GT. 1) CEL(NX) = 0.5*(X(NX)+X(NX-1))/(X(NX)-X(NX-1))
       IT    = 0
       P1P   = P1(NX)+CEL(NX)
       P2P   = P2(NX)+CEL(NX)
       IF(NX .LT. NTR) GO TO 60
       CALL EDDY
    60 IT    = IT+1
       IF(IT .LE. ITMAX) GO TO 70
       WRITE(6,2500)
       GO TO 90
C
    70 CALL CMOM
       CALL SOLV3
C
C   CHECK FOR CONVERGENCE
    61 IF(NX .GE. NTR) GO TO 62
C---LAMINAR FLOW
       IF(ABS(DELV(1)) .GT. 1.0E-05) GO TO 60
       GO TO 75
C---TURBULENT FLOW
    62 IF(ABS(DELV(1)/(V(1,2)+0.5*DELV(1))) .GT. 0.02) GO TO 60
C
C   CHECK FOR GROWTH
    75 IF(NX .EQ. 1) GO TO 90
       IF(NP .EQ. 61) GO TO 90
       IF(ABS(V(NP,2)) .LE. 1.0E-03) GO TO 90
       CALL GROWTH
       IT    = 0
       GO TO 60
    90 CALL OUTPUT
       GO TO 25
C - - - - - - - - - - - - - - - - - - - - - - - - - - - - - - - - - -
  2500 FORMAT(1H0,16X,25HITERATIONS EXCEEDED ITMAX)
  9100 FORMAT(1H0,4HNX =,I3,5X,3HX =,F10.3)
       END
```

$\epsilon_1 = 10^{-5}$ to give about four-figure accuracy for most predicted quantities. For turbulent flows the convergence criterion is expressed on a percentage basis. After the OUTPUT subroutine is called, the profiles F, U, V, and B, which represent the variables f_j, u_j, v_j, and b_j, are shifted.

Fortran name	Symbol
ITMAX	Iteration count
CEL, P1P, P2P	α, α_1, α_2 [see Eq. (7.2.7a)]
DELV(1)	δv [see Eq. (7A.1.21)]

8A.1.2 Subroutine INPUT η_∞

[handwritten: $m = \dfrac{x}{U_\infty}\dfrac{dU_\infty}{dx}$]

This subroutine defines v and specifies initial η_∞ at $x = 0$. If the boundary layer wants to grow (see GROWTH), the program computes its own η_∞ at the subsequent stations. Here we read u_e and m as functions of x. Parameter m is either input or is computed from the given external-velocity distribution, $u_e(x)$, and from the

[handwritten left margin: modify to use for dh/dx]

```
      SUBROUTINE INPUT
      COMMON/BLCO/ NP,NX,NXT,NTR,NFLOW,ETAE,VGP,CNU,DETA(61),A(61),
     1             ETA(61)
      COMMON/BLCC/ X(60),UE(60),P1(60),P2(60),CEL(60),RX(60),CF,P1P,P2P,
     1             RTHETA(60)
C - - - - - - - - - - - - - - - - - - - - - - - - - - - - - - - - - - -
      ETAE = 8.0
      CNU  = 1.6E-04
      READ(5,8000) NXT,NTR,NP2,DETA(1),VGP
C  IF P2 IS CALCULATED (NP2=1), READ P2(1)--P2 FOR 2,NXT MAY BE BLANK
      READ(5,8100) (X(I),UE(I),P2(I),I=1,NXT)
      WRITE(6,9000) NXT,NTR,NP2,ETAE,DETA(1),VGP,CNU
      GO TO (50,100), NP2
C  CALCULATION OF PRESSURE-GRADIENT PARAMETER P=P2
   50 DO 80 I=2,NXT
      IF(I .EQ. NXT) GO TO 60
      A1   = (X(I)-X(I-1))*(X(I+1)-X(I-1))
      A2   = (X(I)-X(I-1))*(X(I+1)-X(I))
      A3   = (X(I+1)-X(I))*(X(I+1)-X(I-1))
      DUDS = -(X(I+1)-X(I))/A1*UE(I-1) + (X(I+1)-2.0*X(I)+X(I-1))/
     1       A2*UE(I) + (X(I)-X(I-1))/A3*UE(I+1)
      GO TO 70
   60 A1   = (X(I-1)-X(I-2))*(X(I)-X(I-2))
      A2   = (X(I-1)-X(I-2))*(X(I)-X(I-1))
      A3   = (X(I)-X(I-1))*(X(I)-X(I-2))
      DUDS = (X(I)-X(I-1))/A1*UE(I-2) - (X(I)-X(I-2))/A2*UE(I-1) +
     1       (2.0*X(I)-X(I-2)-X(I-1))/A3*UE(I)
   70 P2(I) = X(I)/UE(I)*DUDS
   80 CONTINUE
  100 DO 90 I=1,NXT
   90 P1(I) = 0.5*(P2(I)+1.0)
      RETURN
C - - - - - - - - - - - - - - - - - - - - - - - - - - - - - - - - - - -
 8000 FORMAT(3I3,2F10.0)
 8100 FORMAT(3F10.0)
 9000 FORMAT(1H0,6HNXT  =,I3,14X,6HNTR  =,I3,14X,6HNP2  =,I3/
     1       1H ,6HETAE =,E14.6,3X,6HDETA1=,E14.6,3X,6HVGP  =,E14.6,3X,
     2       6HNU   =,E14.6/)
      END
```

definition of m. The derivative of du_e/dx is obtained by using three-point Lagrange interpolation formulas given by $(n < N)$:

$$\left(\frac{du_e}{dx}\right)_n = -\frac{u_e^{n-1}}{A_1}(x_{n+1} - x_n) + \frac{u_e^{n}}{A_2}(x_{n+1} - 2x_n + x_{n-1}) + \frac{u_e^{n+1}}{A_3}(x_n - x_{n-1})$$

$$(8A.1.1)$$

Here N refers to the last x^n station and

$A_1 = (x_n - x_{n-1})(x_{n+1} - x_{n-1})$
$A_2 = (x_n - x_{n-1})(x_{n+1} - x_n)$
$A_3 = (x_{n+1} - x_n)(x_{n+1} - x_{n-1})$

The derivative of du_e/dx at the end point $n = N$ is given by

$$\left(\frac{du_e}{dx}\right)_N = -\frac{u_e^{N-2}}{A_1}(x_N - v_{N-1}) - \frac{u_e^{N-1}}{A_2}(x_N - x_{N-2})$$

$$+ \frac{u_e^{N}}{A_3}(2x_N - x_{N-2} - x_{N-1}) \quad (8A.1.2)$$

where now

$A_1 = (x_{N-1} - x_{N-2})(x_N - x_{N-2})$
$A_2 = (x_{N-1} - x_{N-2})(x_N - x_{N-1})$
$A_3 = (x_N - x_{N-1})(x_N - x_{N-2})$

Fortran name	Symbol	Units
CNU	ν	ft^2/s, $\text{m}^2\,\text{s}^{-1}$
ETAE	η_∞	
NXT	Total number of x-stations	
NTR	x_{tr}	ft, m
DETA(1), VGP	h_1, K	
UE	$u_e(x)$	ft/s, m s^{-1}
P2	m	
DUDS	du_e/dx	s^{-1}

8A.1.3 Subroutine GRID

This subroutine generates the grid normal to the flow for the grid described in Sec. 8.1. It requires the first initial $\Delta\eta$-spacing (denoted by h_1 in Sec. 8.1) and the variable grid parameter K.

Fortran name	Symbol
NP	J (total number of j-points)
DETA, VGP	h_j, K [see Eqs. (8.1.1) and (8.1.2)]

```
      SUBROUTINE GRID
      COMMON/BLC0/ NP,NX,NXT,NTR,NFLOW,ETAE,VGP,CNU,DETA(61),A(61),
     1             ETA(61)
C - - - - - - - - - - - - - - - - - - - - - - - - - - - - - - - - - -
      IF((VGP-1.0) .LE. 0.001) GO TO 5
      NP   = ALOG((ETAE/DETA(1))*(VGP-1.0)+1.0)/ALOG(VGP) + 1.0001
      GO TO 10
    5 NP   = ETAE/DETA(1) + 1.0001
   10 IF(NP .LE. 61) GO TO 15
      WRITE(6,9000)
      STOP
   15 ETA(1)= 0.0
      DO 20 J=2,61
      DETA(J)=VGP*DETA(J-1)
      A(J)   = 0.5*DETA(J-1)
   20 ETA(J)= ETA(J-1)+DETA(J-1)
      RETURN
C - - - - - - - - - - - - - - - - - - - - - - - - - - - - - - - - - -
 9000 FORMAT(1H0,36HNP EXCEEDED 61 -- PROGRAM TERMINATED)
      END
```

8A.1.4 Subroutine IVPL

This subroutine generates initial-velocity profiles for laminar boundary layers at $x = 0$. The initial profiles for f_j, u_j, and v_j are given by Eqs. (8.2.6b), (8.2.6a), and (8.2.6c), respectively.

```
      SUBROUTINE IVPL
      COMMON/BLC0/ NP,NX,NXT,NTR,NFLOW,ETAE,VGP,CNU,DETA(61),A(61),
     1             ETA(61)
      COMMON/BLCP/ DELV(61),F(61,2),U(61,2),V(61,2),B(61,2)
C - - - - - - - - - - - - - - - - - - - - - - - - - - - - - - - - - -
      ETANPQ= 0.25*ETA(NP)
      ETAU15= 1.5/ETA(NP)
      DO 30 J=1,NP
      ETAB  = ETA(J)/ETA(NP)
      ETAB2 = ETAB**2
      F(J,2)= ETANPQ*ETAB2*(3.0-0.5*ETAB2)
      U(J,2)= 0.5*ETAB*(3.0-ETAB2)
      V(J,2)= ETAU15*(1.0-ETAB2)
      B(J,2)= 1.0
   30 CONTINUE
      RETURN
      END
```

8A.1.5 Subroutine GROWTH

For most laminar-boundary-layer flows the transformed boundary-layer thickness $\eta_\infty(x)$ is almost constant. A value of $\eta_\infty = 8$ is sufficient. However, for turbulent

```
      SUBROUTINE GROWTH
      COMMON/BLCO/ NP,NX,NXT,NTR,NFLOW,ETAE,VGP,CNU,DETA(61),A(61),
     1            ETA(61)
      COMMON/BLCP/ DELV(61),F(61,2),U(61,2),V(61,2),B(61,2)
C - - - - - - - - - - - - - - - - - - - - - - - - - - - - - - - - - -
      NPO    = NP
      NP1    = NP+1
      NP     = NP+1
      IF(NX .EO. NTR) NP     = NP+3
      IF(NP .GT. 61) NP = 61
C
C   DEFINITION OF PROFILES FOR NEW NP
      DO 35 J=NP1,NP
      F(J,1) = U(NPO,1)*(ETA(J)-ETA(NPO))+F(NPO,1)
      U(J,1) = U(NPO,1)
      V(J,1) = 0.0
      B(J,1) = B(NPO,1)
      F(J,2) = U(NPO,2)*(ETA(J)-ETA(NPO))+F(NPO,2)
      U(J,2) = U(NPO,2)
      V(J,2) = V(J,1)
      B(J,2) = B(NPO,2)
   35 CONTINUE
      NNP    = NP-(NP1-1)
      WRITE(6,6000) NNP
      RETURN
C - - - - - - - - - - - - - - - - - - - - - - - - - - - - - - - - - -
 6000 FORMAT(1H0,5X,13HETAE GROWTH -,I3,14H -POINTS ADDED)
      END
```

boundary layers, $\eta_\infty(x)$ generally increases with increasing x. We determine an estimate of $\eta_\infty(x)$ by the following procedure.

We always require that $\eta_\infty(x_n) \geq \eta_\infty(x_{n-1})$, and in fact the calculations start with $\eta_\infty(0) = \eta_\infty(x_1)$. When the computations on $x = x_n$ (for any $n \geq 1$) have been completed, we test to see if $|v_J{}^n| \leq \epsilon_v$, where $\eta_J = \eta_\infty(x_n)$ and, say, $\epsilon_v = 10^{-3}$. This test is done in MAIN. If this test is satisfied, we set $\eta_\infty(x_{n+1}) = \eta_\infty(x_n)$. Otherwise, we call GROWTH and set $J_{new} = J_{old} + t$, where t is a number of points, say $t = 3$. In this case we also specify values of $(f_j{}^n, u_j{}^n, v_j{}^n, b_j{}^n)$ for the new η_j points. We take the values of $u_j = 1$, $v_j{}^n = v_J{}^n$, $f_j{}^n = (\eta_j - \eta_\infty) u_J{}^n + f_J{}^n$, and $b_j{}^n = b_J{}^n$. This is also done for the values of $f_j{}^{n-1}$, $u_j{}^{n-1}$, $v_j{}^{n-1}$, and $b_j{}^{n-1}$.

8A.1.6 Subroutine EDDY

For simplicity we use an eddy-viscosity formulation that does not include the low Reynolds number effect and the mass transfer effect. These capabilities, if desired, can easily be incorporated into the formulas defined in the subroutine. The formulas for inner and outer eddy-viscosity expressions are given by

$$(\epsilon_m)_i = L^2 \frac{\partial u}{\partial y} \gamma_{tr} \qquad (\epsilon_m)_i \leq (\epsilon_m)_o$$

$$(\epsilon_m)_o = 0.0168 \left| \int_0^\infty (u_e - u)\, dy \right| \gamma_{tr} \qquad (\epsilon_m)_o \geq (\epsilon_m)_i$$

where

$$L = 0.4y \left[1 - \exp\left(-\frac{y}{A} \right) \right]$$

$$A = 26 \frac{\nu}{N} u_\tau^{-1} \qquad u_\tau = \left(\frac{\tau_w}{\rho} \right)^{1/2}$$

$$N = (1 - 11.8p^+)^{1/2} \qquad p^+ = \frac{\nu u_e}{u_\tau^3} \frac{du_e}{dx}$$

$$\gamma_{tr} = 1 - \exp\left[-G(x - x_{tr}) \int_{x_{tr}}^{x} \frac{dx}{u_e} \right]$$

$$G = 8.35 \times 10^{-4} \left(\frac{u_e^3}{\nu^2} \right) \mathrm{Re}_x^{-1.34}$$

```
      SUBROUTINE EDDY
      COMMON/BLCO/ NP,NX,NXT,NTR,NFLOW,ETAE,VGP,CNU,DETA(61),A(61),
     1            ETA(61)
      COMMON/BLCP/ DELV(61),F(61,2),U(61,2),V(61,2),B(61,2)
      COMMON/BLCC/ X(60),UE(60),P1(60),P2(60),CFL(60),RX(60),CF,P1P,P2P,
     1            RTHETA(60)
C - - - - - - - - - - - - - - - - - - - - - - - - - - - - - - - - - -
      GAMTR = 1.0
      UEINTG= 0.0
      U1    = 1.0/UE(NTR-1)
      DO 10 I=NTR,NX
      U2    = 1.0/UE(I)
      UEINTG= UEINTG+(U1+U2)*(X(I)-X(I-1))*0.5
   10 U1    = U2
      GG    = 8.35E-04*UE(NX)**3/(RX(NTR-1)**1.34*CNU**2)
      EXPTM = GG*(X(NX)-X(NTR-1))*UEINTG
      IF(EXPTM .LE. 10.0) GO TO 15
      WRITE(6,9100) GG,UEINTG,EXPTM
      GO TO 20
   15 GAMTR = 1.0-EXP(-EXPTM)
   20 CONTINUE
      IFLGD = 0
      RX2   = SQRT(RX(NX))
      RX4   = SQRT(RX2)
      PPLUS = P2(NX)/(RX4*V(1,2)**1.5)
      RX216 = RX2*0.16
      CN    = SQRT(1.0-11.8*PPLUS)
      CRSQV = CN*RX4*SQRT(V(1,2))/26.0
      J = 1
      EDVO  = 0.0168*RX2*(ETA(NP)-F(NP,2)+F(1,2))*GAMTR
   50 IF(IFLGD .EQ. 1) GO TO 100
      YOA   = CRSQV*ETA(J)
      EDVI  = RX216*ETA(J)**2*V(J,2)*(1.0-EXP(-YOA))**2*GAMTR
      IF(EDVI   .LT. EDVO  ) GO TO 200
C----
      IFLGD = 1
  100 EDV   = EDVO
      GO TO 300
  200 EDV   = EDVI
  300 B(J,2)= 1.0+EDV
      J     = J+1
      IF(J .LE. NP) GO TO 50
      RETURN
 9100 FORMAT(1H0,2X,3HGG=,E14.6,3X,7HUEINTG=,E14.6,2X,6HEXPTM=,E14.6)
      END
```

In terms of transformed variables these formulas become (noting that $v \propto \partial u / \partial y$)

$$(\epsilon_m^+)_i = 0.16 \mathrm{Re}_x^{1/2} \left[1 - \exp \left(-\frac{y}{A} \right) \right]^2 \eta^2 v \gamma_{tr}$$

$$(\epsilon_m^+)_o = 0.0168 \mathrm{Re}_x^{1/2} [\eta_\infty - f(\eta_\infty)] \gamma_{tr}$$

$$\frac{y}{A} = \frac{N}{26} \mathrm{Re}_x^{1/4} v_w^{1/2} \eta \qquad p^+ = m \mathrm{Re}_x^{-1/4} (v_w)^{-3/2} \qquad \mathrm{Re}_x = \frac{u_e x}{v}$$

$$\epsilon_m^+ = \frac{\epsilon_m}{v}$$

Fortran name	Symbol
RX	Re_x
PPLUS	p^+
EDVO	$(\epsilon_m^+)_o$
YOA	y/A
CN	N
EDVI	$(\epsilon_m^+)_i$
GAMTR	γ_{tr}

8A.1.7 Subroutine CMOM

This is one of the most important subroutines of the computer program. It defines the coefficients of the linearized momentum equation given by Eqs. (7.2.14a)–(7.2.14i).

Fortran name	Symbol
FB, UB, VB	$f_{j-1/2}^n, u_{j-1/2}^n, v_{j-1/2}^n$
USB	$(u^2)_{j-1/2}^n$
FVB	$(fv)_{j-1/2}^n$
DERBV	$[(bv)']_{j-1/2}^n$
CFB, CVB, CFVB, CUSB	$f_{j-1/2}^{n-1}, v_{j-1/2}^{n-1}, (fv)_{j-1/2}^{n-1}, (u^2)_{j-1/2}^{n-1}$
S1, S2, S3, S4, S5, S6	$S_1, S_2, S_3, S_4, S_5, S_6$ [see Eqs. (7.2.14d)–(7.2.14i)]
R1, R3, R2	r_1, r_3, r_2 [see Eqs. (7.2.14a)–(7.2.14c)]
CDERBV	$[(bv)']_{j-1/2}^{n-1}$
CRB	$R_{j-1/2}^{n-1}$ [see Eq. (7.2.8)]
CLB	$L_{j-1/2}^{n-1}$

```
      SUBROUTINE CMOM
      COMMON/BLC0/ NP,NX,NXT,NTR,NFLOW,ETAF,VGP,CNU,DETA(61),A(61),
     1             ETA(61)
      COMMON/BLCC/ X(60),UE(60),P1(60),P2(60),CEL(60),RX(60),CF,P1P,P2P,
     1             RTHETA(60)
      COMMON/BLCP/ DELV(61),F(61,2),U(61,2),V(61,2),B(61,2)
      COMMON/BLC6/ S1(61),S2(61),S3(61),S4(61),S5(61),S6(61),
     1             R1(61),R2(61),R3(61)
C - - - - - - - - - - - - - - - - - - - - - - - - - - - - - - - - -
      DO 100 J=2,NP
C  PRESENT STATION
      USB   = 0.5*(U(J,2)**2+U(J-1,2)**2)
      FVB   = 0.5*(F(J,2)*V(J,2)+F(J-1,2)*V(J-1,2))
      FB    = 0.5*(F(J,2)+F(J-1,2))
      UB    = 0.5*(U(J,2)+U(J-1,2))
      VB    = 0.5*(V(J,2)+V(J-1,2))
      DERBV = (B(J,2)*V(J,2)-B(J-1,2)*V(J-1,2))/DETA(J-1)
      IF(NX .GT. 1) GO TO 10
C  PREVIOUS STATION
      CFB   = 0.0
      CVB   = 0.0
      CFVB  = 0.0
      CUSB  = 0.0
      GO TO 20
   10 CFB   = 0.5*(F(J,1)+F(J-1,1))
      CVB   = 0.5*(V(J,1)+V(J-1,1))
      CFVB  = 0.5*(F(J,1)*V(J,1)+F(J-1,1)*V(J-1,1))
      CUSB  = 0.5*(U(J,1)**2+U(J-1,1)**2)
      CDERBV= (B(J,1)*V(J,1)-B(J-1,1)*V(J-1,1))/DETA(J-1)
C
C  COEFFICIENTS OF THE DIFFERENCED MOMENTUM EQ.
   20 S1(J) = B(J,2)/DETA(J-1)+(P1P*F(J,2)-CEL(NX)*CFB)*0.5
      S2(J) =-B(J-1,2)/DETA(J-1)+(P1P*F(J-1,2)-CEL(NX)*CFB)*0.5
      S3(J) = 0.5*(P1P*V(J,2)+CEL(NX)*CVB)
      S4(J) = 0.5*(P1P*V(J-1,2)+CEL(NX)*CVB)
      S5(J) =-P2P*U(J,2)
      S6(J) =-P2P*U(J-1,2)
C
C  DEFINITIONS OF RJ
      IF(NX .EQ. 1) GO TO 30
      CLB   = CDERBV+P1(NX-1)*CFVB+P2(NX-1)*(1.0-CUSB)
      CRB   = -P2(NX)+CEL(NX)*(CFVB-CUSB)-CLB
      GO TO 35
   30 CRB   = -P2(NX)
   35 R2(J) = CRB-(DERBV+P1P*FVB-P2P*USB-CEL(NX)*(CFB*VB-CVB*FB))
      R1(J) = F(J-1,2)-F(J,2)+DETA(J-1)*UB
      R3(J-1)=U(J-1,2)-U(J,2)+DETA(J-1)*VB
  100 CONTINUE
      R1(1) = 0.0
      R2(1) = 0.0
      R3(NP)= 0.0
      RETURN
      END
```

8A.1.8 Subroutine OUTPUT

This subroutine prints out the desired profiles such as f_j, u_j, v_j, and b_j as functions of η_j. It also computes the boundary-layer parameters, θ, δ^*, H, c_f, R_θ, $R_{\delta}*$, and Re_x. The integration is made by the trapezoidal rule.

Fortran name	Symbol
THETA1	θ_1
THETA	θ
DELS	δ^*
H	H
CF	c_f
RTHETA	R_θ
RX	Re_x

```
      SUBROUTINE OUTPUT
      COMMON/BLC0/ NP,NX,NXT,NTR,NFLOW,ETAE,VGP,CNU,DETA(61),A(61),
     1             ETA(61)
      COMMON/BLCC/ X(60),UE(60),P1(60),P2(60),CEL(60),RX(60),CF,P1P,P2P,
     1             RTHETA(60)
      COMMON/BLCP/ DELV(61),F(61,2),U(61,2),V(61,2),B(61,2)
C - - - - - - - - - - - - - - - - - - - - - - - - - - - - - - - - - - -
      WRITE(6,4400)
      WRITE(6,4500) (J,ETA(J),F(J,2),U(J,2),V(J,2),B(J,2),J=1,NP,4)
      WRITE(6,4500) NP,ETA(NP),F(NP,2),U(NP,2),V(NP,2),B(NP,2)
      IF(NX .EQ. 1) GO TO 210
      F1     = 0.
      THETA1= 0.
      DO 150 J=2,NP
      F2     = U(J,2)*(1.-U(J,2))
      THETA1= THETA1+(F1+F2)*0.5*DETA(J-1)
  150 F1     = F2
      THETA = THETA1*X(NX)/SQRT(RX(NX))
      DELS  = (ETA(NP)-F(NP,2))*X(NX)/SQRT(RX(NX))
      H     = DELS/THETA
      CF    = 2.0*V(1,2)/SQRT(RX(NX))
      RTHETA(NX) = UE(NX)*THETA/CNU
      RDELS = UE(NX)*DELS/CNU
      WRITE(6,9000) X(NX),THETA,DELS,H,CF,RX(NX),RTHETA(NX),RDELS,
     1              UE(NX),P2(NX)
C
  210 NX     = NX+1
      IF(NX .GT. NXT) STOP
C
C   SHIFT PROFILES
      DO 250 J=1,NP
      F(J,1)= F(J,2)
      U(J,1)= U(J,2)
      V(J,1)= V(J,2)
  250 B(J,1)= B(J,2)
      RETURN
C - - - - - - - - - - - - - - - - - - - - - - - - - - - - - - - - - - -
 4400 FORMAT(1H0,2X,1HJ,4X,3HETA,9X,1HF,13X,1HU,13X,1HV,13X,1HB)
 4500 FORMAT(1H ,I3,F10.3,4E14.6)
 9000 FORMAT(1H0,6X,1HX,11X,5HTHETA,10X,4HDELS,11X,1HH,13X,2HCF/
     1            1H ,6X,2HRX,10X,6HRTHETA,8X,5HRDELS,11X,2HUE,12X,2HP2/
     2            1H0,5E14.6/1H ,5E14.6)
      END
```

(handwritten annotation:) calc h, δ_R, PGRAD, HGRAD

8A.2 DESCRIPTION OF INPUT

The input to the Fortran program of Sec. 8A.1 is described below.

Card 1 The first four variables are integers with field length of 3; the remaining variables are F10.0 format:

NXT   Total number of x-stations, not to exceed 60.

NTR   NX-station where transition occurs.

NP2   Determines pressure-gradient parameter. If NP2 = 1, P_2 calculated, must read $P_2(1)$; remainder may be blank. If NP2 = 2, P_2 input.

DETA(1) $\Delta\eta$-initial step size of the variable grid system. Use $\Delta\eta = 0.01$ for turbulent flows. If desired, it may be changed.

VGP K is the variable-grid parameter. Use $K = 1.0$ for laminar flow and $K = 1.14$ for turbulent flow. For a flow consisting of both laminar and turbulent flows, use $K = 1.14$.

1 2 3	4 5 6	7 8 9	10 11 12 13 14 15 16 17 18 19	20 21 22 23 24 25 26 27 28 29
NXT	NTR	NP2	DETA(1)	VGP

Load sheet for card 1.

Cards 2 to NXT+1 These variables are punched in 3F10.0 format:

X Surface distance, feet or meters.

u_e Velocity, feet per second or meters per second.

P_2 Dimensionless pressure-gradient parameter. If NP2 = 1, read $P_2(1)$ only; remainder are blank.

1 2 3 4 5 6 7 8 9 10	11 12 13 14 15 16 17 18 19 20	21 22 23 24 25 26 27 28 29 30
x	u_e	P_2

Load sheet for cards 2 to NXT+1.

EXAMPLE 8A.1

For a two-dimensional incompressible laminar flow, for which the external velocity is given by

$$u_e^* = 1 - \tfrac{1}{8}x^* \tag{E8A.1.1}$$

where $u_e^* = u_e/u_\infty$ and $x^* = x/L$, compute the boundary-layer parameters with the computer program described in Sec. 8A.1 for $0 \leqslant x^* \leqslant 0.55$.

Solution From Eq. (E8A.1.1) and from the definition of P2, it follows that

$$P2 = \frac{x^*}{u_e^*}\left(-\frac{1}{8}\right)$$

Let us choose $\Delta x^* = 0.05$. Then NXT = 12 and ETAE = 8.0. Since the flow is laminar, we choose NTR to be a number greater than NXT, say 99. For illustration purposes we also choose DETA(1) to be 0.20 with VGP = 1.0. That choice gives us 41 points across the boundary layer. If higher accuracy is desired, then a smaller value of DETA(1) should be used. In general, 41 points [DETA(1) = 0.2] is sufficient. With these selections the load sheet for card 1 becomes

1 2 3	4 5 6	7 8 9	10 11 12 13 14 15 16 17 18 19 20	21 22 23 24 25 26 27 28 29
NXT	NTR	NP2	DETA(1)	VGP
1 2	9 9	2	0.20	1.0

Similarly, the first three cards of the last sheet for cards 2 to NXT+1 become

1 2 3 4 5 6 7 8 9 10 11 12 13 14 15 16 17 18 19 20	21 22 23 24 25 26 27 28 29 30	
x	u_e	P_2
0.	1.0	0.
0.05	0.99375	-0.0006289
0.10	0.9875	-0.012658

The rest of the cards are similar to those given above.

The output in terms of velocity profiles and the usual boundary layer parameters are presented below for the first two stations.

```
NXT   = 12              NTR   = 99           NP2   = 2
ETAE  = 0.800000E+01    DETA1= 0.200000E+00  VGP   = 0.100000E+01   NU  = 0.160000E-03

NX =  1      X =      0.0
       V(WALL)=  0.187500E+00      DELVW=  0.198672E+00
       V(WALL)=  0.386172E+00      DELVW= -0.487825E-01
       V(WALL)=  0.337389E+00      DELVW= -0.533280E-02
       V(WALL)=  0.332057E+00      DELVW= -0.687143E-04
       V(WALL)=  0.331988E+00      DELVW= -0.242560E-07

  J     ETA        F                 U               V               B
  1    0.0       0.0               0.0             0.331099E+00    0.100000E+01
  5    0.800     0.106039E+00      0.264539E+00    0.327184E+00    0.100000E+01
  9    1.600     0.419983E+00      0.516251E+00    0.296438E+00    0.100000E+01
 13    2.400     0.921102E+00      0.728224E+00    0.228076E+00    0.100000E+01
 17    3.200     0.155701E+01      0.875407E+00    0.139362E+00    0.100000E+01
 21    4.000     0.230300E+01      0.955155E+00    0.644665E-01    0.100000E+01
 25    4.800     0.309226E+01      0.987404E+00    0.219446E-01    0.100000E+01
 29    5.600     0.387715E+01      0.997476E+00    0.516174E-02    0.100000E+01
 33    6.400     0.467620E+01      0.999621E+00    0.961500E-03    0.100000E+01
 37    7.200     0.547608E+01      0.999961E+00    0.121839E-03    0.100000E+01
 41    8.000     0.627607E+01      0.100000E+01    0.109555E-04    0.100000E+01
 41    8.000     0.627607E+01      0.100000E+01    0.109555E-04    0.100000E+01

NX =  2      X =      0.050
       V(WALL)=  0.331988E+00      DELVW=  0.962110E-02
       V(WALL)=  0.341609E+00      DELVW= -0.401776E-04
       V(WALL)=  0.341569E+00      DELVW=  0.583146E-07

  J     ETA        F                 U               V               B
  1    0.0       0.0               0.0             0.341569E+00    0.100000E+01
  5    0.800     0.108555E+00      0.270214E+00    0.331838E+00    0.100000E+01
  9    1.600     0.427033E+00      0.523870E+00    0.295801E+00    0.100000E+01
 13    2.400     0.935005E+00      0.734883E+00    0.225611E+00    0.100000E+01
 17    3.200     0.159532E+01      0.876673E+00    0.136177E+00    0.100000E+01
 21    4.000     0.233737E+01      0.957184E+00    0.621958E-01    0.100000E+01
 25    4.800     0.310407E+01      0.988303E+00    0.209320E-01    0.100000E+01
 29    5.600     0.389926E+01      0.997651E+00    0.508041E-02    0.100000E+01
 33    6.400     0.469830E+01      0.999652E+00    0.887021E-03    0.100000E+01
 37    7.200     0.549827E+01      0.999965E+00    0.110625E-03    0.100000E+01
 41    8.000     0.629926E+01      0.100000E+01    0.967713E-05    0.100000E+01
 41    8.000     0.629826E+01      0.100000E+01    0.967713E-05    0.100000E+01

          X            THETA           DELS            H            CF
          RX           RTHETA          RDELS           UE           P2

    0.500000E-01   0.186906E-02    0.482833E-02    0.258205E+01   0.387654E-01
    0.310547E+03   0.116142E+02    0.299884E+02    0.993750E+00   0.528300E-02
```

EXAMPLE 8A.2

For a two-dimensional flow past a NACA 0012 airfoil the velocity distribution for zero angle of attack is given in Fig. 8A.1 as a function of chordwise distance. Compute the laminar and turbulent boundary-layer, say δ^* and c_f, distribution with the computer program described in Sec. 8A.1 for $0 \leqslant x/c \leqslant 1.0$. Assume transition to be at $x/c = 0.20$, with the chord Reynolds number, $u_\infty c/\nu$, at 10^6.

Solution In general, approximately 25 x-stations are sufficient to compute the complete velocity field on an airfoil. We choose the Δx-spacing according to the external-velocity distribution. The points on Fig. 8A.1 show our choice of the x-stations. A tabulation of u_e is given on p. 144.

Since the flow is laminar initially and becomes turbulent at $x/c = 0.20$, which corresponds to NTR = 13, we use a variable grid with DETA(1) of 0.05 and with VGP = 1.2. With these selections the load sheet for card 1 becomes

1 2 3 4 5 6 7 8 9 10 11 12 13 14 15 16 17 18 19 20 21 22 23 24 25 26 27 28 29			
NXT	NTR	NP2	DETA(1)	VGP
25	13	1	0.05	1.2

Similarly, the first five cards of the last load sheet for cards 2 to NXT+1 become

1 2 3 4 5 6 7 8 9 10 11 12 13 14 15 16 17 18 19 20 21 22 23 24 25 26 27 28 29 30		
X	u_e	P_2
0.	0.	1.0
0.005	1 60.	
0.013	1 68.	
0.025	1 79.2	
0.05	1 85.6	

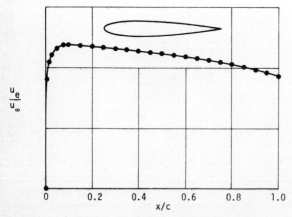

Fig. 8A.1 External-velocity distribution for NACA 0012 airfoil at zero angle of attack. The symbols denote the input at (x/c)-stations.

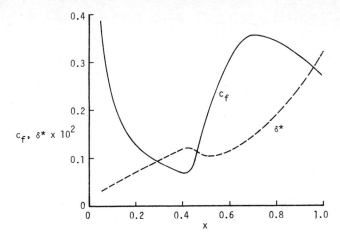

Fig. 8A.2 Distribution of δ^* and c_f for Ex. 8A.2. The x denotes the surface distance.

Note that in this case P2 is calculated from the specified external-velocity distribution and that at NX = 1, P2 = 1.

Figure 8A.2 shows the computed values of δ^* and c_f distributions.

APPENDIX 8B: Solution of Third–Order Ordinary or Parabolic Partial Differential Equations

The Fortran program presented in Appendix 8A is a general computer program for solving a third-order nonlinear parabolic PDE. It can also be used to solve other third-order linear or nonlinear ODEs or PDEs. To illustrate its use, we present three examples.

EXAMPLE 8B.1

Solve the following equations

$$f''' + 2ff'' - (f')^2 + (g')^2 = 0 \qquad\qquad \text{(E8B.1.1)}$$
$$g''' + 2fg'' - 2f'g' = 0 \qquad\qquad \text{(E8B.1.2)}$$

subject to the boundary conditions

$$f(0) = f'(0) = g(0) = 0 \quad g'(0) = 1 \quad f'(\eta_\infty) = 0 \quad g'(\eta_\infty) = 0 \qquad \text{(E8B.1.3)}$$

Solution It would be ideal to solve the system under discussion by expressing it as a system of six first-order equations and then solving it by an algorithm similar to SOLV3. However, that procedure will require a new algorithm; at the expense of a less efficient code it is best and sufficient to solve the required system by making use of our already existing SOLV3 algorithm. In this case we solve Eqs. (E8B.1.1) and (E8B.1.2) separately rather than together.

We first consider Eq. (E8B.1.1) and write it in terms of new variables $u(\eta)$ and $v(\eta)$ as

$$f' = u \tag{E8B.1.4a}$$
$$u' = v \tag{E8B.1.4b}$$
$$v' + 2fv - u^2 = -w^2 \tag{E8B.1.4c}$$

with $g' = w$. This system is subject to the boundary conditions

$$f(0) = 0 \quad u(0) = 0 \quad u(\eta_\infty) = 0 \tag{E8B.1.5}$$

The finite-difference approximations of Eqs. (E8B.1.4a) and (E8B.1.4b) are the same as those given by Eqs. (7.2.5a) and (7.2.5b). For Eq. (E8B.1.4c) they are

$$h_j^{-1}(v_j - v_{j-1}) + 2(fv)_{j-1/2} - (u^2)_{j-1/2} = -(w^2)_{j-1/2} \tag{E8B.1.6}$$

The linearized equations of Eqs. (E8B.1.4a) and (E8B.1.4b) are the same as those given by Eqs. (7.2.13a) and (7.2.13b). For Eq. (E8B.1.6) they are of the form given by Eq. (7.2.13c) except that now

$$(s_1)_j = \frac{1}{h_j} + f_j \quad (s_2)_j = -\frac{1}{h_j} + f_{j-1} \quad (s_3)_j = v_j \quad (s_4)_j = v_{j-1}$$

$$(s_5)_j = -u_j \quad (s_6)_j = -u_{j-1} \tag{E8B.1.7}$$

$$(r_2)_j = -(w^2)_{j-1/2} - [h_j^{-1}(v_j - v_{j-1}) + 2(fv)_{j-1/2} - (u^2)_{j-1/2}] \tag{E8B.1.8}$$

In terms of perturbation quantities the boundary conditions [Eq. (E8B.1.5)] become

$$\delta f_0 = 0 \quad \delta u_0 = 0 \quad \delta u_J = 0 \tag{E8B.1.9}$$

Since the boundary conditions in Eq. (E8B.1.9) are the same as those given by Eq. (7.2.15), no significant changes in SOLV3 are needed.

Let us now consider Eq. (E8B.1.7) and write it in terms of new variables, $w(\eta)$ and $t(\eta)$, as

$$g' = w \tag{E8B.1.10a}$$
$$w' = t \tag{E8B.1.10b}$$
$$t' + 2ft - 2uw = 0 \tag{E8B.1.10c}$$

The boundary conditions are

$$g(0) = 0 \quad w(0) = 1 \quad w(\eta_\infty) = 0 \tag{E8B.1.11}$$

The finite-difference approximations of Eqs. (E8B.1.10a) and (E8B.1.10b) are similar to those given by Eqs. (7.2.5a) and (7.2.5b), that is,

$$\frac{g_j - g_{j-1}}{h_j} = w_{j-1/2} \tag{E8B.1.12a}$$

$$\frac{w_j - w_{j-1}}{h_j} = t_{j-1/2} \tag{E8B.1.12b}$$

The finite-difference approximation of Eq. (E8B.1.10c) is

$$h_j^{-1}(t_j - t_{j-1}) + 2f_{j-1/2}t_{j-1/2} - 2u_{j-1/2}w_{j-1/2} = 0 \tag{E8B.1.12c}$$

With Newton's method we can write Eqs. (E8B.1.12a) and (E8B.1.12b) as

$$\delta g_j - \delta g_{j-1} - \frac{h_j}{2}(\delta w_j + \delta w_{j-1}) = (r_1)_j \tag{E8B.1.13a}$$

$$\delta w_j - \delta w_{j-1} - \frac{h_j}{2}(\delta t_j + \delta t_{j-1}) = (r_3)_j \tag{E8B.1.13b}$$

and Eq. (E8B.1.10c) in the form

$$(s_1)_j \delta t_j + (s_2)_j \delta t_{j-1} + (s_3)_j \delta g_j + (s_4)_j \delta g_{j-1} + (s_5)_j \delta w_j + (s_6)_j \delta w_{j-1} = (r_2)_j \tag{E8B.1.13c}$$

Here

$$(r_1)_j = g_{j-1} - g_j + h_j w_{j-1/2} \tag{E8B.1.14a}$$

$$(r_2)_j = -[h_j^{-1}(t_j - t_{j-1}) + 2f_{j-1/2}t_{j-1/2} - 2u_{j-1/2}w_{j-1/2}] \tag{E8B.1.14b}$$

$$(r_3)_j = w_{j-1} - w_j + h_j t_{j-1/2} \tag{E8B.1.14c}$$

$$(s_1)_j = \frac{1}{h_j} + f_{j-1/2} \quad (s_2)_j = -\frac{1}{h_j} + f_{j-1/2} \quad (s_3)_j = (s_4)_j = 0$$

$$(s_5)_j = (s_6)_j = -u_{j-1/2} \tag{E8B.1.15}$$

In terms of perturbation quantities the boundary conditions [Eq. (E8B.1.11)] become

$$\delta g_0 = 0 \quad \delta w_0 = 0 \quad \delta w_J = 0 \tag{E8B.1.16}$$

We alter five subroutines of our Fortran program. In subroutine IVPL we define new initial-velocity profiles that satisfy the boundary conditions given by Eqs. (E8B.1.5) and (E8B.1.11). Subroutine CMOM contains the coefficients of the two momentum equations and their boundary conditions. Programing changes are made to MAIN, SOLV3, and OUTPUT.

The initial profiles for the system [Eq. (E8B.1.4)] and [Eq. (E8B.1.5)] can be satisfied by assuming $u(\eta)$ to be

$$u = \frac{\eta}{\eta_\infty}\left(1 - \frac{\eta}{\eta_\infty}\right) \tag{E8B.1.17a}$$

From this it follows that

$$f = \eta\frac{\eta}{\eta_\infty}\left(\frac{1}{2} - \frac{1}{3}\frac{\eta}{\eta_\infty}\right) \tag{E8B.1.17b}$$

$$v = \frac{1}{\eta_\infty}\left(1 - 2\frac{\eta}{\eta_\infty}\right) \tag{E8B.1.17c}$$

The initial profiles for the system Eqs. (E8B.1.10) and (E8B.1.11) can be satisfied by assuming $w(\eta)$ to be

$$w = 1 - \frac{\eta}{\eta_\infty} \tag{E8B.1.18a}$$

From this it follows that

$$g = \eta\left(1 - \frac{1}{2}\frac{\eta}{\eta_\infty}\right) \tag{E8B.1.18b}$$

$$t = -\frac{1}{\eta_\infty} \tag{E8B.1.18c}$$

A sample calculation made with $\eta_\infty = 8$, $h_1 = 0.2$, and $K = 1$ is shown below. The solutions of the two equations were obtained in succession, that is, for the given initial-velocity profiles; at first Eq. (E8B.1.1) was solved, then Eq. (E8B.1.2) was solved. The procedure was repeated until convergence, which was specified for $|\delta v_w| \leqslant 10^{-5}$. The results below also show the iterations. It is seen from them that the convergence rate is very slow. Although we have used Newton's method to linearize the governing equations, we cheated in linearizing Eq. (E8B.1.4c). Rather than solving Eqs. (E8B.1.1)–(E8B.1.3) as a system of six first-order equations, we treated w as known in Eq. (E8B.1.1) so that we can use the existing SOLV3 algorithm.

```
NXT   =   1            NTR   =  99           NP2   =   2
ETAE  =  0.800000E+01   DETA1=  0.200000E+00  VGP   =  0.100000E+01   NU   =  0.160000E-03

NX =   1     X =      0.0
      V(WALL)=   0.125000E+00    DELVW=   0.957217E+00
      T(WALL)=  -0.125000E+00    DELTW=  -0.866291E+00
      V(WALL)=   0.108222E+01    DELVW=  -0.391470E+00
      T(WALL)=  -0.991291E+00    DELTW=   0.175636E+00
      V(WALL)=   0.690746E+00    DELVW=  -0.148577E+00
      T(WALL)=  -0.815654E+00    DELTW=   0.122844E+00
      V(WALL)=   0.542170E+00    DELVW=  -0.412911E-01
      T(WALL)=  -0.692810E+00    DELTW=   0.653338E-01
      V(WALL)=   0.500878E+00    DELVW=   0.524172E-02
      T(WALL)=  -0.627476E+00    DELTW=   0.119227E-01
      V(WALL)=   0.506120E+00    DELVW=   0.460060E-02
      T(WALL)=  -0.615553E+00    DELTW=  -0.176018E-02
      V(WALL)=   0.510721E+00    DELVW=  -0.332446E-03
      T(WALL)=  -0.617314E+00    DELTW=   0.383484E-03
      V(WALL)=   0.510388E+00    DELVW=   0.164712E-03
      T(WALL)=  -0.616930E+00    DELTW=  -0.146649E-03
      V(WALL)=   0.510555E+00    DELVW=  -0.545627E-04
      T(WALL)=  -0.617077E+00    DELTW=   0.487718E-04
      V(WALL)=   0.510498E+00    DELVW=   0.183542E-04
      T(WALL)=  -0.617028E+00    DELTW=  -0.170852E-04
```

ITERATIONS EXCEEDED ITMAX

J	ETA	F	U	V	B
1	0.0	0.0	0.0	0.510517E+00	0.100000E+01
5	0.800	0.959972E-01	0.180631E+00	0.300887E-01	0.100000E+01
9	1.600	0.233285E+00	0.149914E+00	-0.745390E-01	0.100000E+01
13	2.400	0.329245E+00	0.912795E-01	-0.652717E-01	0.100000E+01
17	3.200	0.384203E+00	0.492877E-01	-0.401143E-01	0.100000E+01
21	4.000	0.413031E+00	0.250668E-01	-0.218126E-01	0.100000E+01
25	4.800	0.427427E+00	0.122407E-01	-0.112512E-01	0.100000E+01
29	5.600	0.434327E+00	0.570386E-02	-0.566488E-02	0.100000E+01
33	6.400	0.437440E+00	0.243071E-02	-0.282070E-02	0.100000E+01
37	7.200	0.438663E+00	0.804882E-03	-0.139771E-02	0.100000E+01
41	8.000	0.438950E+00	0.0	-0.691390E-03	0.100000E+01
41	8.000	0.438950E+00	0.0	-0.691390E-03	0.100000E+01

J	ETA	G	W	T	
1	0.0	0.0	0.100000E+01	-0.617045E+00	
5	0.800	0.615318E+00	0.560323E+00	-0.448768E+00	
9	1.600	0.943952E+00	0.286573E+00	-0.247111E+00	
13	2.400	0.110939E+01	0.142171E+00	-0.125523E+00	
17	3.200	0.119100E+01	0.697184E-01	-0.622979E-01	
21	4.000	0.123098E+01	0.338893E-01	-0.307460E-01	
25	4.800	0.125014E+01	0.161899E-01	-0.151639E-01	
29	5.600	0.125923E+01	0.748269E-02	-0.748313E-02	
33	6.400	0.126330E+01	0.317453E-02	-0.369578E-02	
37	7.200	0.126489E+01	0.104130E-02	-0.182664E-02	
41	8.000	0.126526E+01	0.0	-0.903257E-03	
41	8.000	0.126526E+01	0.0	-0.903257E-03	

EXAMPLE 8B.2

Solve the following equation

$$f''' - 18\eta^2 f'' - 36(\eta f' - f) = -1.8330 \qquad \text{(E8B.2.1)}$$

subject to the boundary conditions

$$f(0) = f'(0) = 0 \qquad f''(\eta_\infty) = 1.0 \qquad \text{(E8B.2.2)}$$

Solution To solve the above system we need to use a new MAIN, CMOM, and SOLV3. Since Eq. (E8B.2.1) is a linear equation, we do not need the IVPL subroutine.

For subroutine CMOM we need to define new coefficients for $(s_k)_j$ $(k = 1\text{-}6)$ and $(r_2)_j$. So we let

$$f' = u \qquad \text{(E8B.2.3a)}$$

$$u' = v \qquad \text{(E8B.2.3b)}$$

$$v' - 18\eta^2 v - 36(\eta u - f) = -1.8330 \qquad \text{(E8B.2.3c)}$$

The finite-difference approximations of Eqs. (E8B.2.3a) and (E8B.2.3b) are the same as those given by Eqs. (7.2.5a) and (7.2.5b). For Eq. (E8B.2.3c) they are of the form given by Eq. (7.2.13c) except that since Eq. (E8B.2.3c) is linear, the perturbation quantities δf_j, δu_j, and δv_j are replaced by f_j, u_j, and v_j, that is,

$$(s_1)_j v_j + (s_2)_j v_{j-1} + (s_3)_j f_j + (s_4)_j f_{j-1} + (s_5)_j u_j + (s_6)_j u_{j-1} = (r_2)_j \qquad \text{(E8B.2.4)}$$

Here

$$(s_1)_j = \frac{1}{h_j} - 9\eta_{j-1/2}^2 \qquad (s_2)_j = -\frac{1}{h_j} - 9\eta_{j-1/2}^2 \qquad (s_3)_j = 18 \qquad (s_4)_j = 18$$

$$(s_5)_j = (s_6)_j = -18\eta_{j-1/2} \qquad \text{(E8B.2.5a)}$$

$$(r_2)_j = -1.833 \qquad \text{(E8B.2.5b)}$$

The boundary conditions [Eq. (E8B.2.2)] become

$$f_0 = 0 \qquad u_0 = 0 \qquad v_J = 1 \tag{E8B.2.6}$$

To account for the different edge boundary condition, we consider A_J. The only change occurs in the last row. From Eq. (7A.1.2) we write A_J as

$$A_J = \begin{pmatrix} 1 & -h_J/2 & 0 \\ (s_3)_J & (s_5)_J & (s_1)_J \\ 0 & 0 & 1 \end{pmatrix} \tag{E8B.2.7}$$

From Eq. (7A.1.7c), for $j = J$, we write Δ_J as

$$\Delta_J = \begin{pmatrix} (\alpha_{11})_J & (\alpha_{12})_J & (\alpha_{13})_J \\ (\alpha_{21})_J & (\alpha_{22})_J & (\alpha_{23})_J \\ 0 & 0 & 1 \end{pmatrix} \tag{E8B.2.8}$$

Here the first two rows of Δ_J are the same as before. From Eq. (7A.1.18a), remembering the definition of δ_j,

$$\begin{pmatrix} (\alpha_{11})_J & (\alpha_{12})_J & (\alpha_{13})_J \\ (\alpha_{21})_J & (\alpha_{22})_J & (\alpha_{23})_J \\ 0 & 0 & 1 \end{pmatrix} \begin{pmatrix} f_J \\ u_J \\ v_J \end{pmatrix} = \begin{pmatrix} (w_1)_J \\ (w_2)_J \\ (w_3)_J \end{pmatrix} \tag{E8B.2.9}$$

Here $(w_1)_J$ and $(w_2)_J$ are the same as before, but $(w_3)_J = (r_3)_J$ with $(r_3)_J = 1$. Solving Eq. (E8B.2.9), we get

$$v_J = (w_3)_J$$

$$u_J = \frac{(w_2)_J(\alpha_{11})_J - (w_1)_J(\alpha_{21})_J}{(\alpha_{22})_J(\alpha_{12})_J - (\alpha_{12})_J(\alpha_{21})_J}$$

$$f_J = \frac{(w_1)_J - (\alpha_{12})_J u_J}{(\alpha_{11})_J}$$

In summary, we redefine the coefficients $(s_k)_j$, $(r_2)_j$, and $(r_3)_j$ in subroutine CMOM and replace the formulas for the components of δ_J with the new ones to account for the new boundary conditions at $\eta = \eta_\infty$ in subroutine SOLV3. Since we are solving a linear equation, no iteration is required; the solutions in SOLV3 obtained as DELF(J), DELU(J), and DELV(J) are F(J,2), U(J,2), and V(J,2).

A sample calculation obtained with $\eta_\infty = 8$, $h_1 = 0.2$, and $K = 1.0$ is shown below.

```
NXT  =  1              NTR   = 99           NP2  =  2
ETAE =  0.800000E+01   DETA1= 0.200000E+00  VGP  =  0.100000E+01   NU  =  0.160000E-03

NX  =  1      X  =    0.0
 J     ETA          F              U               V               B
 1    0.0         0.0            0.0           0.174171E+01    0.100000E+01
 5    0.800       0.478282E+00   0.114838E+01   0.131566E+01    0.130000E+01
 9    1.600       0.181486E+01   0.219009E+01   0.127392E+01    0.100000E+01
13    2.400       0.398135E+01   0.322482E+01   0.123085E+01    0.100000E+01
17    3.200       0.697507E+01   0.425834E+01   0.118607E+01    0.100000E+01
21    4.000       0.107954E+02   0.529148E+01   0.114480E+01    0.100000E+01
25    4.800       0.154421E+02   0.632446E+01   0.110839E+01    0.100000E+01
29    5.600       0.209150E+02   0.735735E+01   0.107679E+01    0.100000E+01
33    6.400       0.272142E+02   0.839019E+01   0.104946E+01    0.100000E+01
37    7.200       0.343397E+02   0.942303E+01   0.102589E+01    0.100000E+01
41    8.000       0.422926E+02   0.104553E+02   0.100000E+01    0.100000E+01
41    8.000       0.422926E+02   0.104553E+02   0.100000E+01    0.100000E+01
```

EXAMPLE 8B.3

Solve the laminar boundary-layer equations for two-dimensional flows in physical coordinates.

Solution In this case the governing equations are given by Eqs. (5.1.1)–(5.1.3). To solve this system by the Fortran program described in Appendix 8A, it would appear to be logical to reduce Eqs. (5.1.1) and (5.1.2) to a first-order system by defining new variables $w(x, y)$ and $t(x, y)$ and write them as

$$u' = w$$

$$\frac{\partial u}{\partial x} + v' = 0$$

$$u \frac{\partial u}{\partial x} + vw = u_e \frac{du_e}{dx} + vw'$$

However, this is not the best use of the SOLV3 subroutine, since the second equation above is a PDE rather than an ODE as in Sec. 7.2, Eq. (7.2.1b). Solution of the system given above will require some modifications in the SOLV3 subroutine. For this reason, to avoid extra work we shall use the stream function concept and write the system [Eqs. (5.1.1) and (5.1.2)] as

$$f''' = f' \frac{\partial f'}{\partial \xi} - f'' \frac{\partial f}{\partial \xi} - \bar{u}_e \frac{d\bar{u}_e}{d\xi} \tag{E8B.3.1}$$

Here $\psi(\xi, y) = \sqrt{u_0 \nu L}\, f(\xi, Y)$, $Y = \sqrt{u_0/\nu L}\, y$, $\bar{u}_e = u_e/u_0$, and $\xi = x/L$ with u_0 denoting a reference velocity and with primes denoting differentiation with respect to Y. Equation (E8B.3.1) is subject to the boundary conditions

$$Y = 0 \quad f = 0 \quad f' = 0; \quad Y = Y_\infty \quad f' = \bar{u}_e \tag{E8B.3.2}$$

With the introduction of new variables $u(\xi, Y)$ and $v(\xi, Y)$, Eqs. (E8B.3.1) and (E8B.3.2) can be written as

$$f' = u \tag{E8B.3.3a}$$

$$u' = v \tag{E8B.3.3b}$$

$$v' = u \frac{\partial u}{\partial \xi} - v \frac{\partial f}{\partial \xi} - \bar{u}_e \frac{d\bar{u}_e}{d\xi} \tag{E8B.3.3c}$$

The finite-difference approximations of Eqs. (E8B.3.3a) and (E8B.3.3b) are the same as those given by Eqs. (7.2.5a) and (7.2.5b). For Eq. (E8B.3.3c) they are

$$h_j^{-1}(v_j{}^n - v_{j-1}^n) + \alpha[(u^2)_{j-1/2}^n - (fv)_{j-1/2}^n + f_{j-1/2}^{n-1} v_{j-1/2}^n - v_{j-1/2}^{n-1} f_{j-1/2}^n] = R_{j-1/2}^{n-1} \tag{E8B.3.4}$$

where

$$R_{j-1/2}^{n-1} = -h_j^{-1}(v_j{}^{n-1} - v_{j-1}^{n-1}) + \alpha[-(u^2)_{j-1/2}^{n-1} + (fv)_{j-1/2}^{n-1} - (\bar{u}_e{}^2)^n + (\bar{u}_e{}^2)^{n-1}] \tag{E8B.3.5a}$$

$$\alpha = \frac{1}{k_n} \tag{E8B.3.5b}$$

The linearized equations of Eqs. (E8B.3.3a) and (E8B.3.3b) are the same as those given by Eqs. (7.2.13a) and (7.2.13b). For Eq. (E8B.3.4) they are of the form given by Eq. (7.2.13c) except that now

$$(s_1)_j = \frac{1}{h_j} + \frac{\alpha}{2}(f_j{}^n - f_{j-1/2}^{n-1}) \qquad (s_2)_j = -\frac{1}{h_j} + \frac{\alpha}{2}(f_{j-1}^n - f_{j-1/2}^{n-1})$$

$$(s_3)_j = \frac{\alpha}{2}(v_j{}^n + v_{j-1/2}^{n-1}) \qquad (s_4)_j = \frac{\alpha}{2}(v_{j-1}^n + v_{j-1/2}^{n-1}) \qquad (s_5)_j = -\alpha u_j{}^n$$

$$(s_6)_j = -\alpha u_{j-1}^n$$

$$(r_2)_j = R_{j-1/2}^{n-1} - \{h_j^{-1}(v_j{}^n - v_{j-1}^n) + \alpha[-(u^2)_{j-1/2}^n + (fv)_{j-1/2}^n - f_{j-1/2}^{n-1} v_{j-1/2}^n$$
$$+ v_{j-1/2}^{n-1} f_{j-1/2}^n]\} \tag{E8B.3.6}$$

The boundary conditions [Eq. (E8B.3.2)] become

$$\delta f_0 = \delta u_0 = 0 \qquad \delta u_J = 0$$

Thus the system [Eq. (E8B.3.1)] and [Eq. E8B.3.2)] can be solved by changing the coefficients $(s_k)_j$ ($k = 1$–6) and $(r_2)_j$ in the CMOM subroutine. The other change occurs in the starting profiles, since the boundary layer equations in physical coordinates are singular at $x = 0$. That does not cause much difficulty since these initial profiles can be obtained by using the existing program and the solution procedure can be switched to the physical coordinates at the desired x-location.

As an example, let us compute the flow for which the external-velocity distribution is given by Eq. (E8A.1.1) with $u_e^* = u_e/u_0$ and $\xi = x^*$, and let us switch to the solution of Eq. (E8B.3.1) at $\xi = 0.10$. We can start the solutions by using the existing program, with P2 computed as in Ex. 8.1 and with $\Delta\xi = 0.05$. As a result, with $\eta_\infty = 8$, $h_1 = 0.40$, and $K = 1.0$, for three ξ-stations, namely, $\xi = 0$, 0.05, and 0.10, we perform the calculations in transformed variables as shown below. At $\xi = 0.10$, after we obtain a converged solution, we define a new grid and new dependent variables by

$$Y = \sqrt{\xi/\bar{u}_e}\,\eta$$
$$f_{pky} = \sqrt{\xi\bar{u}_e}\,f_{tr}$$
$$u_{pky} = \bar{u}_e u_{tr}$$
$$v_{pky} = (\bar{u}_e)^{3/2}\frac{v_{tr}}{\sqrt{\xi}}$$

(E8B.3.7)

and perform the calculations in physical variables at $\xi = 0.11$, 0.12, and 0.13, as shown below.

```
NXT =  8             NTR = 99
ETAE =  0.800000E+01  DETA1= 0.40000E+00   VGP  =  0.100000E+01   NU  =  0.160000E-03

NX =  1      X =   0.0
       V(WALL)= -0.187500E+00        DELVW=  0.197418E+00
       V(WALL)=  0.384918E+00        DELVW= -0.479463E-01
       V(WALL)=  0.336971E+00        DELVW= -0.513705E-02
       V(WALL)=  0.331834E+00        DELVW= -0.635397E-04
       V(WALL)=  0.331771E+00        DELVW=  0.963620E-09

   J     ETA        F             U             V             B
   1     0.0       0.0           0.0           0.331771E+00  0.100000E+01
   5     1.600     0.418554E+00  0.514736E+00  0.295759E+00  0.100000E+01
   9     3.200     0.156079E+01  0.873374E+00  0.140056E+00  0.100000E+01
  13     4.800     0.307308E+01  0.987392E+00  0.221712E-01  0.100000E+01
  17     6.400     0.466671E+01  0.999638E+00  0.902141E-03  0.100000E+01
  21     8.000     0.626657E+01  0.100000E+01  0.739541E-05  0.100000E+01
  21     8.000     0.626657E+01  0.100000E+01  0.739541E-05  0.100000E+01

NX =  2      X =   0.050
       V(WALL)=  0.331771E+00        DELVW= -0.963403E-02
       V(WALL)=  0.322137E+00        DELVW= -0.423889E-04
       V(WALL)=  0.322094E+00        DELVW=  0.255563E-07

   J     ETA        F             U             V             B
   1     0.0       0.0           0.0           0.322094E+00  0.100000E+01
   5     1.600     0.410453E+00  0.506929E+00  0.295291E+00  0.100000E+01
   9     3.200     0.154206E+01  0.868835E+00  0.143256E+00  0.100000E+01
  13     4.800     0.305053E+01  0.986606E+00  0.233345E-01  0.100000E+01
  17     6.400     0.464372E+01  0.999602E+00  0.983256E-03  0.100000E+01
  21     8.000     0.624357E+01  0.100000E+01  0.853068E-05  0.100000E+01
  21     8.000     0.624357E+01  0.100000E+01  0.853068E-05  0.100000E+01

      X              THETA          DELS          H             CF
      RX             RTHETA         RDELS         UE            P2

   0.500000E-01   0.189715E-02  0.498351E-02  0.263634E+01  -0.265552E-01
   0.310547E+03   0.117831E+02  0.309521E+02  0.893750E+00  -0.629630E-02

NX =  3      X =   0.100
       V(WALL)=  0.322094E+00        DELVW= -0.988920E-02
       V(WALL)=  0.312205E+00        DELVW= -0.631782E-04
       V(WALL)=  0.312142E+00        DELVW=  0.222305E-07

   J     ETA        F             U             V             B
   1     0.0       0.0           0.0           0.968629E+00  0.100000E+01
   5     0.509     0.126347E+00  0.492572E+00  0.914592E+00  0.100000E+01
   9     1.018     0.478447E+00  0.853226E+00  0.455316E+00  0.100000E+01
  13     1.527     0.951198E+00  0.973432E+00  0.762862E-01  0.100000E+01
  17     2.037     0.145170E+01  0.987068E+00  0.333020E-02  0.100000E+01
  21     2.546     0.195444E+01  0.987500E+00  0.303282E-04  0.100000E+01
  21     2.546     0.195444E+01  0.987500E+00  0.303282E-04  0.100000E+01

      X              THETA          DELS          H             CF
      RX             RTHETA         RDELS         UE            P2

   0.100000E+00   0.210057E-02  0.228030E-02  0.253283E+01  -0.172732E-01
   0.617187E+03   0.380191E+01  0.128930E+02  0.987500E+00  -0.126582E-01

NX =  4      X =   0.110
       V(WALL)=  0.968629E+00        DELVW= -0.517191E-01
       V(WALL)=  0.916910E+00        DELVW= -0.140796E-02
       V(WALL)=  0.915502E+00        DELVW= -0.161832E-05

   J     ETA        F             U             V             B
   1     0.0       0.0           0.0           0.915501E+00  0.100000E+01
   5     0.509     0.119812E+00  0.468396E+00  0.881240E+00  0.100000E+01
   9     1.018     0.458091E+00  0.828259E+00  0.484173E+00  0.100000E+01
  13     1.527     0.922531E+00  0.965362E+00  0.101702E+00  0.100000E+01
  17     2.037     0.142098E+01  0.985293E+00  0.656356E-02  0.100000E+01
  21     2.546     0.192300E+01  0.986250E+00  0.250298E-03  0.100000E+01
  21     2.546     0.192300E+01  0.986250E+00  0.250258E-03  0.100000E+01

      X              THETA          DELS          H             CF
      RX             RTHETA         RDELS         UE            P2

   0.110000E+00   0.103788E-02  0.263085E-02  0.253483E+01  -0.703167E-01
   0.678047E+03   0.639754E+01  0.162167E+02  0.986250E+00  -0.139417E-01
```

```
NX =   5    X  =    0.120
       V(WALL)=  0.915501E+00      DELVW= -0.452527E-01
       V(WALL)=  0.870248E+00      DELVW= -0.114413E-02
       V(WALL)=  0.869104E+00      DELVW= -0.814130E-06

   Y     ETA          F                   U                   V            B
   1     0.0      0.0            0.0            0.869103E+00   0.100000E+01
   5     0.509    0.114052E+00   0.446387E+00   0.849755E+00   0.100000E+01
   9     1.018    0.439585E+00   0.804065E+00   0.505707E+00   0.100000E+01
  13     1.527    0.895431E+00   0.956090E+00   0.127647E+00   0.100000E+01
  17     2.037    0.139142E+01   0.983190E+00   0.112225E-01   0.100000E+01
  21     2.546    0.189268E+01   0.985000E+00   0.721192E-03   0.100000E+01
  21     2.546    0.189268E+01   0.985000E+00   0.721192E-03   0.100000E+01

     X        THETA        RDELS         H           FE
     RX       RTHETA       RDELS         UE          P2
  0.120000E+00  0.113601E-02  0.288346E-02  0.253834E+01 -0.639517E-01
  0.738750E+03  0.699356E+01  0.177513E+02  0.985000E+00 -0.153244E-01

NX =   6    X  =    0.130
       V(WALL)=  0.869103E+00      DELVW= -0.403144E-01
       V(WALL)=  0.828788E+00      DELVW= -0.954093E-03
       V(WALL)=  0.827834E+00      DELVW= -0.472788E-06

       ETAE GROWTH - 1 -POINTS ADDED
       V(WALL)=  0.827834E+00      DELVW=  0.224249E-06

   Y     ETA          F                   U                   V            B
   1     0.0      0.0            0.0            0.827834E+00   0.100000E+01
   5     0.509    0.108906E+00   0.427551E+00   0.820225E+00   0.100000E+01
   9     1.018    0.422642E+00   0.780769E+00   0.521387E+00   0.100000E+01
  13     1.527    0.869739E+00   0.945781E+00   0.153224E+00   0.100000E+01
  17     2.037    0.136287E+01   0.980688E+00   0.173505E-01   0.100000E+01
  21     2.546    0.186327E+01   0.983674E+00   0.656795E-03   0.100000E+01
  22     2.673    0.198849E+01   0.983750E+00   0.540666E-03   0.100000E+01

     X        THETA        RDELS         H           FE
     RX       RTHETA       RDELS         UE          P2
  0.130000E+00  0.122380E-02  0.314358E-02  0.253985E+01 -0.585024E-01
  0.738750E+03  0.762420E+01  0.193572E+02  0.983750E+00 -0.153184E-01
```

Chapter 9

Stability and Transition

9.1 INTRODUCTION

At sufficiently high Reynolds numbers, most flows are turbulent rather than laminar, and this transition to turbulence (Sec. 1.3) has been the object of many studies utilizing several approaches. The most popular approach is small-disturbance stability theory, in which a small sinusoidal disturbance is imposed on a given steady laminar flow to see whether the disturbance will amplify or decay in time. If the disturbance decays, the flow will stay laminar; if the disturbance amplifies sufficiently, the flow must change in some way, probably to become turbulent. The small-disturbance theory does not predict the details of the nonlinear process by which the flow changes from laminar to turbulent. It establishes which shapes of velocity profiles are unstable, identifies those frequencies that amplify fastest, and indicates how the parameters governing the flow can be changed to delay transition. Although a direct theory linking stability and transition is lacking, several empirical theories leading to a satisfactory prediction of transition exist for two-dimensional and axisymmetric incompressible flows (Sec. 9.6.2), and the more refined ones make direct use of small-disturbance calculations. The small-disturbance theory is therefore quantitatively useful, and a fairly detailed treatment is justified. We shall first discuss the derivation of the small-disturbance equations for two-dimensional and three-dimensional incompressible TSLs.

9.2 SMALL-DISTURBANCE EQUATIONS

We consider the x-component NS equation, Eq. (2.3.9). As in the discussion of turbulence in Sec. 2.4, we add fluctuating parts of u', v', w', and p' to the mean values of u, v, w, and p. If we neglect the body force, f_x, Eq. (2.3.9) becomes identical to Eq. (2.4.2). If u', v', w', and p' are small enough for second-order products like $u'v'$ to be negligible, the mean velocity and pressure satisfy the three-dimensional steady equations of motion. For simplicity we assume that the mean flow is a TSL so that u, v, w, and p satisfy Eqs. (3.2.1), (3.2.2), (3.2.3), and (2.2.2) with the Reynolds stress terms $-\rho\overline{u'v'}$ and $-\rho\overline{u'w'}$ set equal to zero. Subtracting Eq. (3.2.1) from Eq. (2.3.9) and retaining only the linear terms in u', v', and w', that is, neglecting the Reynolds stresses produced by the disturbance, we get

$$\frac{\partial u'}{\partial t} + u'\frac{\partial u}{\partial x} + u\frac{\partial u'}{\partial x} + v'\frac{\partial u}{\partial y} + v\frac{\partial u'}{\partial y} + w'\frac{\partial u}{\partial z} + w\frac{\partial u'}{\partial z} = -\frac{1}{\rho}\frac{\partial p'}{\partial x} + \nu\nabla^2 u' \quad (9.2.1)$$

If this procedure is repeated for the y-component NS equation, Eq. (2.3.11), and for the z-component, which in the absence of body forces is

$$\frac{dw}{dt} = -\frac{1}{\rho}\frac{\partial p}{\partial z} + \nu\nabla^2 w \qquad (9.2.2)$$

we get

$$\frac{\partial v'}{\partial t} + u'\frac{\partial v}{\partial x} + u\frac{\partial v'}{\partial x} + v'\frac{\partial v}{\partial y} + v\frac{\partial v'}{\partial y} + w'\frac{\partial v}{\partial z} + w\frac{\partial v'}{\partial z} = -\frac{1}{\rho}\frac{\partial p'}{\partial y} + \nu\nabla^2 v' \quad (9.2.3)$$

and

$$\frac{\partial w'}{\partial t} + u'\frac{\partial w}{\partial x} + u\frac{\partial w'}{\partial x} + v'\frac{\partial w}{\partial y} + v\frac{\partial w'}{\partial y} + w'\frac{\partial w}{\partial z} + w\frac{\partial w'}{\partial z} = -\frac{1}{\rho}\frac{\partial p'}{\partial z} + \nu\nabla^2 w'$$

$$(9.2.4)$$

The instantaneous continuity equation for three-dimensional flow is

$$\frac{\partial u'}{\partial x} + \frac{\partial v'}{\partial y} + \frac{\partial w'}{\partial z} = 0 \qquad (9.2.5)$$

The above equations are still complicated and can be simplified further. We shall see below that disturbances of practical interest have streamwise wavelengths of only a few boundary-layer thicknesses. Therefore x-wise gradients of u' and v' are of the same order as y-wise gradients. It follows from Eq. (9.2.5) that u' and v' are of the same order. We can then see that in Eq. (9.2.1), $v\,\partial u'/\partial y$ is of order (v/u) times $u\,\partial u'/\partial x$ and is therefore negligible. Similarly, in Eq. (9.2.3), $v\,\partial v'/\partial y$ is negligible. The result of neglecting these terms is the same as putting $v = 0$ and is therefore called the *parallel-flow approximation*. Unfortunately, the streamwise distance traveled by a disturbance while it amplifies by a factor of two, say, is often so long that the shear layer grows appreciably in the meantime. Therefore the parallel-flow solutions are not so accurate as one would expect from simple order-of-magnitude analysis of the equations but are adequate for discussion purposes. With the parallel-flow approximation, Eqs. (9.2.1), (9.2.3), and (9.2.4) become

$$\frac{\partial u'}{\partial t} + u \frac{\partial u'}{\partial x} + v' \frac{du}{dy} + w \frac{\partial u'}{\partial z} = -\frac{1}{\rho} \frac{\partial p'}{\partial x} + \nu \nabla^2 u' \tag{9.2.6}$$

$$\frac{\partial v'}{\partial t} + u \frac{\partial v'}{\partial x} + w \frac{\partial v'}{\partial z} = -\frac{1}{\rho} \frac{\partial p'}{\partial y} + \nu \nabla^2 v' \tag{9.2.7}$$

$$\frac{\partial w'}{\partial t} + u \frac{\partial w'}{\partial x} + v' \frac{dw}{dy} + w \frac{\partial w'}{\partial z} = -\frac{1}{\rho} \frac{\partial p'}{\partial z} + \nu \nabla^2 w' \tag{9.2.8}$$

For two-dimensional mean flows with two-dimensional disturbances, Eqs. (9.2.5)–(9.2.8) become

$$\frac{\partial u'}{\partial x} + \frac{\partial v'}{\partial y} = 0 \tag{9.2.9}$$

$$\frac{\partial u'}{\partial t} + u \frac{\partial u'}{\partial x} + v' \frac{du}{dy} = -\frac{1}{\rho} \frac{\partial p'}{\partial x} + \nu \left(\frac{\partial^2 u'}{\partial x^2} + \frac{\partial^2 u'}{\partial y^2} \right) \tag{9.2.10}$$

$$\frac{\partial v'}{\partial t} + u \frac{\partial v'}{\partial x} = -\frac{1}{\rho} \frac{\partial p'}{\partial y} + \nu \left(\frac{\partial^2 v'}{\partial x^2} + \frac{\partial^2 v'}{\partial y^2} \right) \tag{9.2.11}$$

We assume (p. 11) that the small disturbance is a sinusoidal traveling plane wave (a Tollmien-Schlichting wave; more complicated disturbances can be represented by a Fourier series) and write a three-dimensional disturbance as

$$g'(x,y,z,t) = g(y) \exp \left[i(\alpha x + \beta z - \omega t) \right] \tag{9.2.12}$$

Here $g(y)$ is the complex amplitude function of a typical real flow variable g'; α and β are complex wave numbers $2\pi/\lambda_x$ and $2\pi/\lambda_z$, where λ_x and λ_z are the wavelengths in the x- and z-direction, respectively; and ω is the radian (circular) frequency of the disturbance.

Here g', g, α, β, and ω are all in general *complex*. As in alternating-current theory the real and imaginary parts of g', g'_r, and ig'_i, say, are used for the "in-phase" and "quadrature" components with respect to an arbitrary line of zero phase. The magnitude of g' is $(g'^2_r + g'^2_i)^{1/2}$, and its relative phase angle is $\tan^{-1} (g'_i/g'_r)$. The real part of the exponential term represents a growth of disturbance amplitude in x or t, while the imaginary part, $\exp (i\theta)$ say, can be rewritten as $\cos \theta + i \sin \theta$ and therefore represents the sinusoidal oscillation in x or t itself. If we write $\alpha = \alpha_r + i\alpha_i$ and $\omega = \omega_r + i\omega_i$, then the real part of the exponential for $\beta = 0$ is

$\exp\left(-\alpha_i x + \omega_i t\right)$

and the imaginary part is

$\exp\left[i(\alpha_r x - \omega_r t)\right]$

The ratio g_i/g_r represents the variation of phase angle with y. At a given time the magnitude of g' is constant along the line $\alpha x + \beta z = \text{const}$, which is thus a line of constant phase in the xz-plane. When $\beta = 0$, the direction of propagation of the wave (i.e., the direction normal to the constant-phase line in the xz-plane) is in the x-direction. If $\beta \neq 0$, the wave-propagation direction is inclined at the angle γ with respect to the free stream; it is given by

$$\gamma = \tan^{-1}\frac{\beta}{\alpha} \tag{9.2.13}$$

Even if the mean flow is two-dimensional, the oblique disturbances are three-dimensional.

Inspection of Eqs. (9.2.5)–(9.2.8) reveals that w' occurs only as $\partial w'/\partial z$ and w only as the coefficient of $\partial(\ )'/\partial z$ terms except in the z-momentum equation. Consequently, if the disturbances have the form of Eq. (9.2.12), and the x- and z-axes are rotated about the y-axis so that the new z-axis is parallel to the line of constant phase, the z-derivatives in the new coordinate system, $\tilde{x}$ and $\tilde{z}$ (see Fig. 9.1), must be zero, and the w' and w terms will drop out of Eqs. (9.2.5), (9.2.6), and (9.2.7). Specifically, the rotated coordinates yield the new variables

$$\tilde{u} = u + w \tan\gamma \qquad \tilde{w} = -u \tan\gamma + w$$
$$\tilde{u}' = u' + w' \tan\gamma \qquad \tilde{w}' = -u' \tan\gamma + w'$$
$$\tilde{v}' = \frac{v'}{\cos\gamma} \qquad \tilde{p} = \frac{p'}{\cos^2\gamma} \tag{9.2.14}$$
$$\tilde{x} = x \cos\gamma + z \sin\gamma \qquad \tilde{z} = -x \sin\gamma + z \cos\gamma$$

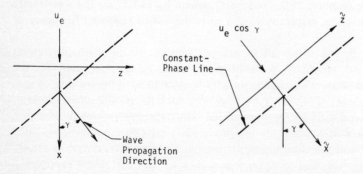

Fig. 9.1 The rotated-coordinate system.

With this transformation, Dunn and Lin (1955) showed that the resulting equations for Eqs. (9.2.5), (9.2.6), and (9.2.7) are identical to those equations for two-dimensional disturbance in the shear layer, $\tilde{u}(y)$. That is, if we change the variables in Eqs. (9.2.5)–(9.2.8) by means of Eq. (9.2.14), the first three equations become identical to those in Eqs. (9.2.9)–(9.2.11) with the old variables replaced by the new ones. For the new $\tilde{z}$-momentum equation we get

$$\frac{\partial \tilde{w}'}{\partial t} + \tilde{u} \frac{\partial \tilde{w}'}{\partial \tilde{x}} + \tilde{v} \frac{d\tilde{w}}{dy} = \nu \left(\frac{\partial^2 \tilde{w}'}{\partial \tilde{x}^2} + \frac{\partial^2 \tilde{w}'}{\partial y^2} \right) \qquad (9.2.15)$$

As a result of these important observations we may take $\beta = 0$ in Eq. (9.2.12) and use the resulting two-dimensional disturbance in Eqs. (9.2.9)–(9.2.11). The resulting stability equations are also valid for three-dimensional disturbances provided that we simply adjoin Eq. (9.2.15). Note, however, that this $\tilde{z}$-momentum disturbance equation is *uncoupled* from the others. Thus the two-dimensional stability theory can be used for both two- and three-dimensional stability studies, as long as the mean flow is two-dimensional. This useful result is known as Squire's theorem. We therefore restrict our attention to the case of the two-dimensional disturbances.

9.3 ORR–SOMMERFELD EQUATION

For two-dimensional incompressible flows the small-disturbance equations are given by Eqs. (9.2.9)–(9.2.11). Eliminating pressure from these equations[*] and introducing a stream function, $\psi(x, y, t)$, such that

$$u' = \frac{\partial \psi}{\partial y} \qquad v' = -\frac{\partial \psi}{\partial x} \qquad (9.3.1)$$

we can write Eqs. (9.2.10) and (9.2.11) as

$$\frac{\partial}{\partial t} \nabla^2 \psi + u \frac{\partial}{\partial x} (\nabla^2 \psi) - \frac{\partial \psi}{\partial x} \frac{d^2 u}{dy^2} = \nu \nabla^4 \psi \qquad (9.3.2)$$

where

$$\nabla^4 = \nabla^2 \nabla^2$$

and it can be seen that $-\nabla^2 \psi$ is the fluctuating z-component of vorticity, $\partial v'/\partial x - \partial u'/\partial y$, so that Eq. (9.3.2) is an equation for the rate of change of fluctuating vorticity following the fluid along a mean streamline. Compare the derivation of the vorticity equation in Sec. 2.5.

---

[*]This is done by differentiating Eq. (9.2.10) with respect to y and Eq. (9.2.11) with respect to x and by subtracting one of the resulting expressions from the other.

We write the disturbance stream function in complex notation in a form similar to that of Eq. (9.2.12), with $\beta = 0$, as

$$\psi(x,y,t) = \phi(y) \exp [i(\alpha x - \omega t)] \tag{9.3.3}$$

Here $\phi(y)$ is the complex amplitude of the disturbance stream function ($\phi = \phi_r + i\phi_i$), α is the wave number of the disturbance related to the wavelength of the disturbance, λ, by $\lambda = 2\pi/\alpha$. If α is complex ($\equiv \alpha_r + i\alpha_i$), then the amplitude will vary with x as $\exp(-\alpha_i x)$. The analysis for this case is referred to as the *spatial-amplification theory*. If ω is complex ($\equiv \omega_r + i\omega_i$), the amplitude will vary with time as $\exp(-\omega_i t)$. The analysis for this case is referred to as the *temporal-amplification theory*. If α and ω are both real, then the disturbance propagates through the parallel mean flow with constant amplitude $|\phi(y)|$. If α and ω are both complex, the disturbance amplitude will vary in both time and space.

Introducing Eq. (9.3.3) into Eq. (9.3.2), we obtain the following fourth-order ODE for the amplitude $\phi(y)$:

$$\left(u - \frac{\omega}{\alpha}\right)(\phi'' - \alpha^2\phi) - u''\phi = -\frac{i\nu}{\alpha}(\phi^{IV} - 2\alpha^2\phi'' + \alpha^4\phi) \tag{9.3.4}$$

This is known as the Orr-Sommerfeld equation and is the fundamental equation for incompressible stability theory. Here primes denote differentiation with respect to y. For convenience we introduce dimensionless variables by dividing all velocities by a reference velocity, u_0, and all lengths by a reference length, l. We also introduce a dimensionless time ($\equiv tu_0/l$) and write Eq. (9.3.4) in dimensionless form as

$$(\bar{u} - \bar{c})(\phi'' - \alpha_l^2\phi) - \bar{u}''\phi = -\frac{i}{\alpha_l R}(\phi^{IV} - 2\alpha_l^2\phi'' + \alpha_l^4\phi) \tag{9.3.5}$$

Now the primes denote differentiation with respect to $\bar{y}(\equiv y/l)$, and

$$\bar{c} = \frac{c}{u_0} \qquad c = \frac{\omega}{\alpha} \qquad \bar{u} = \frac{u}{u_0} \qquad R = \frac{u_0 l}{\nu} \qquad \alpha_l = \alpha l \tag{9.3.6}$$

The quantity c ($\equiv c_r + ic_i$) is the complex propagation velocity of the disturbance.

9.3.1 Boundary Conditions

The solution of the stability equation or equations requires conditions to be imposed on the bounding surfaces of the flow. There are many flow configurations to which the stability theory can be applied, but we shall restrict our discussion to internal flows and boundary-layer flows.

On any rigid boundary surface the perturbation velocities must vanish. Thus for two-dimensional flows with a rigid wall at $y = y_0$ we must have $u' = v' = 0$, or from Eqs. (9.3.3) and (9.3.1),

$$\phi(y_0) = 0 \quad \phi'(y_0) = 0 \tag{9.3.7}$$

Many internal flows are symmetric about some plane $y = y_1$. In most cases the most unstable disturbance is antisymmetric. The appropriate boundary conditions on the disturbance at $y = y_1$ are

$$\frac{d}{dy} u'(y_1) = v'(y_1) = 0 \tag{9.3.8a}$$

for the disturbance symmetric about $y = y_1$, and

$$u'(y_1) = \frac{d^2}{dy^2} u'(y_1) = 0 \tag{9.3.8b}$$

for the disturbance antisymmetric about $y = y_1$. Thus in terms of the perturbation stream function, we have

$$\phi(y_1) = 0 \quad \phi''(y_1) = 0 \tag{9.3.9a}$$

symmetric about $y = y_1$, or

$$\phi'(y_1) = 0 \quad \phi'''(y_1) = 0 \tag{9.3.9b}$$

antisymmetric about $y = y_1$.

Finally, if the bounding surface is to represent the undisturbed free stream, we must ensure that the perturbation velocities decay as the edge of the boundary layer is approached. In such cases the appropriate conditions can be obtained (Keller, 1976) by considering the Orr-Sommerfeld equation in the neighborhood of the edge of the boundary layer, $y = \delta$ or $\bar{y} = 1$. Since $\bar{u}(1) = 1$ and $\bar{u}''(1) = 0$, Eq. (9.3.5) becomes, as $\bar{y} \to 1$,

$$\phi^{IV} - (\alpha_l^2 + \xi^2)\phi'' + \alpha_l^2 \xi^2 \phi = 0 \tag{9.3.10}$$

We use this form for $\bar{y}$ near 1. Here $\xi^2 = \alpha_l^2 + i\alpha_l R(1 - \bar{c})$. The general solution of this reduced Orr-Sommerfeld equation is

$$\phi = a_1 \exp(\alpha_l \bar{y}) + a_2 \exp(-\alpha_l \bar{y}) + a_3 \exp(\xi \bar{y}) + a_4 \exp(-\xi \bar{y}) \tag{9.3.11}$$

We require that real parts of α_l and ξ are greater than zero. In order that the boundary-layer disturbances decay near the edge of the boundary layer, it is necessary that $a_1 = a_3 = 0$. This can be assured by imposing the following boundary conditions, with $D \equiv d/d\bar{y}$:

$$(D + \alpha)(D^2 - \xi^2)\phi(1) = 0 \tag{9.3.12a}$$

$$(D + \xi)(D^2 - \alpha^2)\phi(1) = 0 \tag{9.3.12b}$$

9.4 PROPERTIES OF THE ORR–SOMMERFELD EQUATION

Since the Orr-Sommerfeld equation and the boundary conditions are homogeneous, the trivial solution, $\phi(y) \equiv 0$, is valid for all values of five scalars, namely, the Reynolds number, R (always real), and the real and imaginary parts of $\bar{c}$ and α_l. Nontrivial solutions, $\phi(y) \not\equiv 0$, may also exist. The simplest and commonest way of posing the problem is to seek the solution that yields the largest rate of amplification (i.e., the largest ω_i or α_i) for chosen values of all the other parameters. For a discussion of the other solutions see Mack (1976). Posed in this simple way, the problem becomes a more-or-less typical eigenvalue-eigenfunction problem. If we take the case of spatial amplification ($\omega_i = 0$) and choose the Reynolds number, R, and wave number, α_r, the solution provides an eigenfunction, $\phi(y)$, which is in general complex, and a complex eigenvalue for the propagation velocity, $c_r + ic_i$. Note that for $\omega_i = 0$ and given α_r a complex value of $c_r + ic_i$ implies values of the real frequency, ω_r, and the spatial-amplification rate $-\alpha_i$.

The eigenvalues of the Orr-Sommerfeld equation for the spatial-amplification case are often presented in an $\alpha_l R$-diagram in which each point of the plane corresponds to a pair of values of c_r and c_i (see Fig. 9.2, for example). For a given Reynolds number the disturbance may be in one of three states: damped, neutral, or amplified. The locus, $c_i = 0$, called the *curve of neutral* stability, separates the damped (stable) region from the amplified (unstable) region. The point on this curve at which R has its smallest value is of special interest because at values of R less than this all (infinitesimal and traveling-wave) disturbances are stable. This smallest Reynolds number is known as the *critical Reynolds number.*

A slightly different kind of stability theory is based on the solution of the Orr-Sommerfeld equation when $\alpha_l R \to \infty$. It is known as the inviscid stability theory. In this case the effect of viscosity on the disturbances is neglected, and the solutions are obtained from

$$(\bar{u} - \bar{c})(\phi'' - \alpha_l^2\phi) - \bar{u}''\phi = 0 \tag{9.4.1}$$

This is known as the Rayleigh equation. Since Eq. (9.4.1) is only of second order, only two boundary conditions are required. For boundary-layer flows we take, on the wall,

$$\bar{y} = 0 \quad \phi(0) = 0 \tag{9.4.2a}$$

and at the edge, by an argument similar to that leading to Eq. (9.3.12) with real part of α_l greater than zero,

$$\bar{y} = 1 \quad (D + \alpha)\phi(1) = 0 \tag{9.4.2b}$$

Earlier studies on stability theory were conducted mostly on the system Eqs. (9.4.1) and (9.4.2) rather than on the system given by Eqs. (9.3.5), (9.3.7), and (9.3.12). This was due to the difficulties of solving the full equation without the help of high-speed computers. A good review of the work done on the inviscid stability theory is given by Drazin and Howard (1966). Although this theory is not applicable exactly to flows at finite Reynolds number, it has been useful in deriving several important general theorems concerning the stability of laminar velocity profiles. Here we present two important theorems for *inviscid* instability due to Rayleigh (see Schlichting, 1968).

1. In order to have amplified disturbances, it is necessary that the velocity profile, $\bar{u}(y)$, has a point of inflection, $(\bar{u}'' = 0)$, within the flow. This theorem was first proved by Rayleigh. Later Tollmien (see Schlichting, 1968) showed that the existence of a point of inflection is also a *sufficient* condition for the amplification of disturbances. As a result of this theorem we can conclude that the velocity profiles with a point of inflection are unstable at $R = \infty$.
2. If $\bar{u}'' < 0$, there is for neutral disturbances with $c_i = 0$ at least one point, $y = y_c$, where $\bar{u} - \bar{c} = 0$. That is, at some point inside the flow, the wave velocity is equal to the mean velocity. The plane $y = y_c$, where $\bar{u} = \bar{c}$, is called the *critical layer* of the mean flow.

The second theorem states the rather obvious fact that any traveling-wave disturbance that is to avoid rapid decay must move at roughly the speed of the flow. The first theorem implies that profiles with a point of inflection will be unstable at quite low Reynolds numbers in real life. It is, of course, natural to suppose that large viscosity will tend to damp out disturbances even if the flow is unstable at infinite Reynolds number, so the αR-diagram for this case will look much the same as Fig. 9.2 except that the upper branches of the curves have positive slopes everywhere (i.e., the unstable range of α spreads in both directions as R increases). For jets and wakes, R_{crit} is typically of order 10, but the parallel-flow assumption is so poor at such low Reynolds numbers that its quantitative predictions are not reliable. Figure 9.10 shows how rapidly the instability Reynolds number of a boundary layer decreases when du_e/dx is less than zero and an inflection occurs.

Now the behavior just described—maximum instability range at infinite Reynolds number—is plausible and easy to understand. However, it applies only to profiles with an inflection. If a profile without an inflection is unstable, it is by virtue of amplification of disturbances *by viscous effects*. This is an unlikely sounding mechanism, and its prediction by small-disturbance theory was not generally accepted until it was found experimentally. There is no simple but satisfying explanation; it just happens that the viscous-stress fluctuations set up by a traveling-wave disturbance can sometimes shift the phase of u' with respect to v' so that the Reynolds shear stress $-\rho\overline{u'v'}$ is nonzero. The Reynolds stress interacts with the mean-velocity gradient, $\partial u/\partial y$, to extract kinetic energy, $\frac{1}{2}(u^2 + v^2 + w^2)$ per unit mass, from the mean flow and feed it to the disturbance energy, $\frac{1}{2}(\overline{u'^2} + \overline{v'^2} + \overline{w'^2})$. The total (mean plus disturbance) kinetic energy decreases by viscous action, as is required by the second law of thermodynamics.

Over the years a number of numerical methods have been developed to solve the Orr-Sommerfeld equation at any Reynolds number. In Sec. 9.5 we shall describe a recent one. This method for ODEs uses the same techniques as does the box scheme for PDEs (see Chap. 7) coupled with appropriate iteration and extrapolation techniques that yield an extremely stable, accurate, efficient, and flexible method. It has been developed by Keller and Cebeci (1977).

In this method the eigenvalue problem is formulated as a first-order system of ODEs, and the box scheme is employed with nonuniform spacing in the y-direction. The eigenvector is normalized at one of the endpoints, and one of the boundary conditions is used as an equation to determine the eigenvalues. The roots of this equation are computed by Newton's method, which converges quadratically. The Newton iterates are computed by using the linearized form of the difference equations. As a result we can easily compute the rates of change (i.e., influence coefficients) of the eigenvalues with respect to Reynolds number.

9.5 NUMERICAL SOLUTION OF THE ORR-SOMMERFELD EQUATION

To formulate the numerical scheme, we must first replace the Orr-Sommerfeld equation and the appropriate boundary conditions by an equivalent first-order system. We do this in two stages. First, we define

$$\psi \equiv \phi'' - \alpha_I^2 \phi \tag{9.5.1}$$

and the Orr-Sommerfeld equation (9.3.5) becomes

$$\psi'' - \alpha_I^2 \psi = i\alpha_I R\left[(\bar{u} - \bar{c})\psi - \bar{u}''\phi\right] \tag{9.5.2}$$

Recall that ϕ and ψ are both complex so that Eqs. (9.5.1) and (9.5.2) are a coupled system of four second-order ODEs. We write them as

$$\psi \equiv f(y) + ie(y) \qquad \phi \equiv h(y) + ig(y) \tag{9.5.3}$$

and introduce the derivatives

$$f' = u \tag{9.5.4a}$$
$$e' = v \tag{9.5.4b}$$
$$h' = w \tag{9.5.4c}$$
$$g' = z \tag{9.5.4d}$$

Now Eqs. (9.5.1) and (9.5.2) become, by separating real and imaginary parts,

$$u' = -a_1 e - a_2 f + \bar{u}'' R(\alpha_i h + \alpha_r g) \tag{9.5.4e}$$
$$v' = -a_3 e - a_4 f - \bar{u}'' R(\alpha_r h - \alpha_i g) \tag{9.5.4f}$$

$$w' = (\alpha_r^2 - \alpha_i^2)h - 2\alpha_r\alpha_i g + f \qquad (9.5.4g)$$

$$z' = 2\alpha_i\alpha_r h + (\alpha_r^2 - \alpha_i^2)g + e \qquad (9.5.4h)$$

Here we have introduced the coefficients

$$
\begin{aligned}
a_1 &= 2\alpha_i\alpha_r + \alpha_i RC_i + R\alpha_r(\bar{u} - c_r) \\
a_2 &= \alpha_i^2 - \alpha_r^2 + \alpha_i R(\bar{u} - c_r) - R\alpha_r c_i \\
a_3 &= -(\alpha_r^2 - \alpha_i^2) - \alpha_r c_i R + \alpha_i R(\bar{u} - c_r) \\
a_4 &= -2\alpha_i\alpha_r - \alpha_r R(\bar{u} - c_r) - R\alpha_i c_i
\end{aligned}
\qquad (9.5.5)
$$

In stability calculations we consider the antisymmetric velocity disturbances, since they determine the stability of the flow. By using the quantities Eqs. (9.5.1), (9.5.3), and (9.5.4a)–(9.5.4d), the various boundary conditions discussed in Sec. 9.3.1 can be written as follows:

$$u(0) = v(0) = w(0) = z(0) = 0 \qquad (9.5.6a)$$

antisymmetric about $y = 0$,

$$g(\bar{y}_0) = h(\bar{y}_0) = w(\bar{y}_0) = z(\bar{y}_0) = 0 \qquad (9.5.6b)$$

for the rigid wall at $\bar{y} = \bar{y}_0$, and

$$u - a_5 w + a_6 z + \alpha_r f - \alpha_i e - a_8 h + a_7 g = 0 \qquad (9.5.6c)$$

$$v - a_5 z - a_6 w + \alpha_i f + \alpha_r e - a_7 h - a_8 g = 0 \qquad (9.5.6d)$$

$$u + \xi_r f - \xi_i e = 0 \qquad (9.5.6e)$$

$$v + \xi_r e + \xi_i f = 0 \qquad (9.5.6f)$$

for the free stream at $\bar{y} = 1$. These conditions follow from Eqs. (9.3.12a) and (9.3.12b) using Eqs. (9.5.3) and (9.5.4). In any given problem, boundary conditions are applied at two points only. The coefficients $a_5, \ldots, a_8$ are

$$a_5 = [\alpha_r c_i - \alpha_i(1 - c_r)]R \qquad (9.5.7a)$$

$$a_6 = [\alpha_i c_i + \alpha_r(1 - c_r)]R \qquad (9.5.7b)$$

$$a_7 = [2\alpha_i\alpha_r c_i + (1 - c_r)(\alpha_r^2 - \alpha_i^2)]R \qquad (9.5.7c)$$

$$a_8 = [c_i(\alpha_r^2 - \alpha_i^2) - 2\alpha_i\alpha_r(1 - c_r)]R \qquad (9.5.7d)$$

Here ξ_r and ξ_i are the real and imaginary parts of ξ computed as

$$\xi_r = \left(\frac{\gamma_r + \sqrt{\gamma_r^2 + \gamma_i^2}}{2}\right)^{1/2} \qquad (9.5.7e)$$

$$\xi_i = \frac{\gamma_i}{2\xi_r} \tag{9.5.7f}$$

where $\xi^2 = \gamma_r + i\gamma_i$ with

$$\gamma_r = \alpha_r^2 - \alpha_i^2 + \alpha_r Rc_i - \alpha_i R(1 - c_r) \tag{9.5.7g}$$
$$\gamma_i = 2\alpha_r\alpha_i + \alpha_r R(1 - c_r) + \alpha_i c_i R \tag{9.5.7h}$$

The difference equations for the system [Eq. (9.5.4)] at points in the interval $y = (0, 1)$,

$$\bar{y}_0 = 0 \quad \bar{y}_j = \bar{y}_{j-1} + \Delta\bar{y}_j \quad 1 \leqslant j \leqslant J \quad \bar{y}_J = 1 \tag{9.5.8}$$

are taken to be, via the box scheme,

$$f_j - f_{j-1} - p_j(u_j + u_{j-1}) = (r_5)_j = 0 \tag{9.5.9a}$$
$$e_j - e_{j-1} - p_j(v_j + v_{j+1}) = (r_6)_j = 0 \tag{9.5.9b}$$
$$h_j - h_{j-1} - p_j(w_j + w_{j-1}) = (r_7)_j = 0 \tag{9.5.9c}$$
$$g_j - g_{j-1} - p_j(z_j + z_{j-1}) = (r_8)_j = 0 \tag{9.5.9d}$$
$$u_j - u_{j-1} + (s_1)_j(e_j + e_{j-1}) + (s_2)_j(f_j + f_{j-1}) + (s_3)_j(h_j + h_{j-1})$$
$$+ (s_4)_j(g_j + g_{j-1}) = (r_1)_{j+1} = 0 \tag{9.5.9e}$$
$$v_j - v_{j-1} + (s_2)_j(e_j + e_{j-1}) - (s_1)_j(f_j + f_{j-1}) - (s_4)_j(h_j + h_{j-1})$$
$$+ (s_3)_j(g_j + g_{j-1}) = (r_2)_{j+1} = 0 \tag{9.5.9f}$$
$$w_j - w_{j-1} + (s_5)_j(h_j + h_{j-1}) + (s_6)_j(g_j + g_{j-1}) - p_j(f_j + f_{j-1}) = (r_3)_{j+1} = 0 \tag{9.5.9g}$$
$$z_j - z_{j-1} - (s_6)_j(h_j + h_{j-1}) + (s_5)_j(g_j + g_{j-1}) - p_j(e_j + e_{j-1}) = (r_4)_{j+1} = 0 \tag{9.5.9h}$$

Here $p_j = \Delta\bar{y}_j/2$ and

$$(s_1)_j = p_j[2\alpha_i\alpha_r + \alpha_i Rc_i + R\alpha_r(\bar{u}_{j-1/2} - c_r)] \tag{9.5.10a}$$
$$(s_2)_j = p_j[\alpha_i^2 - \alpha_r^2 + \alpha_i R(\bar{u}_{j-1/2} - c_r) - R\alpha_r c_i] \tag{9.5.10b}$$
$$(s_3)_j = -p_j(\bar{u}'')_{j-1/2}R\alpha_i \tag{9.5.10c}$$
$$(s_4)_j = -p_j(\bar{u}'')_{j-1/2}R\alpha_r \tag{9.5.10d}$$
$$(s_5)_j = -p_j(\alpha_r^2 - \alpha_i^2) \tag{9.5.10e}$$
$$(s_6)_j = 2p_j\alpha_i\alpha_r \tag{9.5.10f}$$

Equations (9.5.9a)–(9.5.9h) are imposed for $j = 1, 2, \ldots, J - 1$. Additional conditions at $j = 0$ and $j = J$ are obtained from the appropriate boundary conditions. These difference equations have trivial solutions, $u_j = v_j = f_j = e_j = g_j = w_j = z_j = h_j = 0$ for all j. To find the special parameter values for which nontrivial solutions exist, we use special iteration procedures described in the next section.

9.5.1 Computation of the Eigenvalues and Eigenfunctions

For the case of an internal flow with rigid wall at $\bar{y} = \pm 1$, we must use the boundary conditions given by Eqs. (9.5.6a) and (9.5.6b). Since the Orr-Sommerfeld equation and the boundary conditions are homogeneous, the trivial solution $\phi(y) = 0$ is valid for all values of $\bar{c}$ and α_l. For this reason, to compute the eigenvalues and eigenfunctions, we first drop the boundary condition $\phi'(0) = 0$ (that is $w_0 = z_0 = 0$) from Eq. (9.5.6a) and replace it by the condition $\phi(0) = 1$, that is, $h_0 = 1$ and $g_0 = 0$. Now the difference equations will have a nontrivial solution, since $\phi(0) \neq 0$. We seek to adjust or to determine parameter values such that the dropped boundary conditions are satisfied. We do this by an iteration scheme using Newton's method. Specifically, we use in the computations the boundary conditions

$$h_0 = (r_1)_1 \equiv 1 \qquad g_0 = (r_2)_1 = 0 \qquad u_0 = (r_3)_1 = 0 \qquad v_0 = (r_4)_1 = 0 \qquad (9.5.11a)$$

$$h_J = (r_5)_J = 0 \qquad g_J = (r_6)_J = 0 \qquad w_J = (r_7)_J = 0 \qquad z_J = (r_8)_J = 0 \qquad (9.5.11b)$$

Then the complete system of difference equations (9.5.9) and altered boundary conditions (9.5.11) can be written as

$$\mathcal{Q}\,\Delta = R \qquad (9.5.12)$$

Here $\mathcal{Q}$, Δ, and R have forms similar to those described in Appendix 7A, and the solution of Eq. (9.5.12) can be obtained by the block-elimination method (see Appendix 7A). For further details, the reader is referred to Keller and Cebeci (1977).

The solution of the difference equations (9.5.12) depends upon the five scalars R, c_r, c_i, α_r, and α_i. We can denote this dependence by writing

$$\Delta = \Delta(R, c_r, c_i, \alpha_r, \alpha_i) \qquad (9.5.13)$$

With any three of these scalars fixed we seek the remaining two scalars such that the missing boundary condition, $\phi'(0) = 0$, is satisfied. In our finite-difference notation this corresponds to the two conditions $w_0 = z_0 = 0$. With the above notation these are

$$w_0(R, c_r, c_i, \alpha_r, \alpha_i) = 0 \qquad (9.5.14a)$$

$$z_0(R, c_r, c_i, \alpha_r, \alpha_i) = 0 \qquad (9.5.14b)$$

For example, temporal-amplification theory requires $\alpha_i = 0$ with, say, R and c_i specified; to find the contour of zero amplification in the $R\alpha_r$-plane, we would choose $c_i = 0$. Then Eq. (9.5.14) represents two equations in the two unknowns (c_r, α_r). We solve these equations by means of Newton's method. Specifically, if $(c_r^{\nu}, \alpha_r^{\nu})$ are the νth iterates, then the $(\nu + 1)$st iterates are determined by using $c_r^{\nu+1} = c_r^{\nu} + \delta c_r^{\nu}$ and $\alpha_r^{\nu+1} = \alpha_r^{\nu} + \delta\alpha_r^{\nu}$ in Eq. (9.5.14), expanding about $\delta c_r^{\nu} = \delta\alpha_r^{\nu} = 0$, and retaining at most linear terms in the expansion. This gives the linear system

$$(w_0)^{\nu} + \left(\frac{\partial w_0}{\partial \alpha_r}\right)^{\nu} \delta\alpha_r^{\nu} + \left(\frac{\partial w_0}{\partial c_r}\right)^{\nu} \delta c_r^{\nu} = 0 \tag{9.5.15a}$$

$$(z_0)^{\nu} + \left(\frac{\partial z_0}{\partial \alpha_r}\right)^{\nu} \delta\alpha_r^{\nu} + \left(\frac{\partial z_0}{\partial c_r}\right)^{\nu} \delta c_r^{\nu} = 0 \tag{9.5.15b}$$

Here we have used the notation

$$(w_0)^{\nu} \equiv w_0(R, c_r^{\nu}, c_i, \alpha_r^{\nu}, \alpha_i)$$

etc. The solution of Eqs. (9.5.15a) and (9.5.15b) is obviously

$$\delta c_r^{\nu} = \frac{w_0^{\nu}(\partial z_0/\partial \alpha_r)^{\nu} - z_0^{\nu}(\partial w_0/\partial \alpha_r)^{\nu}}{(\partial w_0/\partial \alpha_r)^{\nu}(\partial z_0/\partial c_r)^{\nu} - (\partial z_0/\partial \alpha_r)^{\nu}(\partial w_0/\partial c_r)^{\nu}} \tag{9.5.16a}$$

$$\delta\alpha_r^{\nu} = -\frac{w_0^{\nu} + \delta c_r^{\nu}(\partial w_0/\partial c_r)^{\nu}}{(\partial w_0/\partial \alpha_r)^{\nu}} \tag{9.5.16b}$$

To evaluate the derivatives of w_0 and z_0, we need only differentiate Eq. (9.5.12). Since the vector $\mathbf{R}$ is independent of α_r and c_r, we get

$$\mathcal{C}\left(\frac{\partial \Delta}{\partial \alpha_r}\right)^{\nu} = -\left(\frac{\partial \mathcal{C}}{\partial \alpha_r}\right)^{\nu}(\Delta)^{\nu} \tag{9.5.17a}$$

$$\mathcal{C}\left(\frac{\partial \Delta}{\partial c_r}\right)^{\nu} = -\left(\frac{\partial \mathcal{C}}{\partial c_r}\right)^{\nu}(\Delta)^{\nu} \tag{9.5.17b}$$

Thus to solve for all of the required derivatives, we need only solve two linear systems with the *same* coefficient matrix, $\mathcal{C}$, already computed and factored to solve Eq. (9.5.12). The right-hand-side vectors in Eq. (9.5.17) are determined from Eqs. (9.5.9) and (9.5.11). They are

$$(r_1)_j = (r_2)_j = (r_3)_j = (r_4)_j = 0 \tag{9.5.18}$$

for $1 \leqslant j \leqslant J$,

$$(r_5)_j = -2p_j\{e_{j-1/2}[2\alpha_i + R(\bar{u}_{j-1/2} - c_r)] - f_{j-1/2}(2\alpha_r + Rc_i) - g_{j-1/2}R\bar{u}''_{j-1/2}\}$$

$$(r_6)_j = -2p_j\{-e_{j-1/2}(2\alpha_r + Rc_i) - f_{j-1/2}[2\alpha_i + R(\bar{u}_{j-1/2} - c_r)] + h_{j-1/2}\bar{u}''_{j-1/2}R\}$$

$$(r_7)_j = -2p_j(-2\alpha_r h_{j-1/2} + 2g_{j-1/2}\alpha_i) \tag{9.5.19}$$

$$(r_8)_j = -2p_j(-2\alpha_i h_{j-1/2} - 2\alpha_r g_{j-1/2})$$

for $1 \leqslant j \leqslant J - 1$, and

$$(r_5)_J = (r_6)_J = (r_7)_J = (r_8)_J = 0 \tag{9.5.20}$$

for $j = J$. Note that for generality we have included the α_i and c_i terms.
For Eq. (9.5.17b) the right-hand-side vectors are

$$(r_1)_j = (r_2)_j = (r_3)_j = (r_4)_j = (r_7)_j = (r_8)_j = 0 \tag{9.5.21}$$

for $1 \leqslant j \leqslant J$,

$$(r_5)_j = -2p_j R(-e_{j-1/2}\alpha_r - f_{j-1/2}\alpha_i)$$
$$(r_6)_j = -2p_j R(-e_{j-1/2}\alpha_i + f_{j-1/2}\alpha_r) \tag{9.5.22}$$

for $1 \leqslant j \leqslant J - 1$, and

$$(r_5)_J = (r_6)_J = 0 \tag{9.5.23}$$

for $j = J$.
For the case of an external boundary-layer flow with free stream at $\bar{y} = 1$ and rigid wall at $\bar{y} = 0$, we must use the boundary conditions, Eqs. (9.5.6b)–(9.5.6f). As in the internal-flow problem we again drop one wall boundary condition, $\phi'(0) = 0$, and replace it by the condition $\psi'(0) = 1$, that is, $u_0 = 1$ and $v_0 = 0$. As a result, the altered boundary conditions become

$$h_0 = (r_1)_1 = 0 \quad g_0 = (r_2)_1 = 0 \quad u_0 = (r_3)_1 = 1 \quad v_0 = (r_4)_1 = 0 \tag{9.5.24a}$$
$$u_J - a_5 w_J + a_6 z_J + \alpha_r f_J - \alpha_i e_J - a_8 h_J + a_7 g_J = (r_5)_J = 0 \tag{9.5.24b}$$
$$v_J - a_5 z_J - a_6 w_J + \alpha_i f_J + \alpha_r e_J - a_7 h_J - a_8 g_J = (r_6)_J = 0 \tag{9.5.24c}$$
$$u_J + \xi_r f_J - \xi_i e_J = (r_7)_J = 0 \tag{9.5.24d}$$
$$v_J + \xi_r e_J + \xi_i f_J = (r_8)_J = 0 \tag{9.5.24e}$$

Equations similar to those given by Eqs. (9.5.18)–(9.5.23) can also be written. For details see Keller and Cebeci (1977).

To illustrate the numerical method for internal and external flows, we now present two sample calculations, one for plane Poiseuille flow and one for flat-plate flow.

9.5.2 Results for Plane Poiseuille Flow

In the case of plane Poiseuille flow between stationary parallel plates (see Sec. 5.7) the velocity profile is given by Eq. (5.7.32). It can be shown that when the plates are located at $\bar{y} = y/h = 1$ and $\bar{y} = -1$ so that the maximum dimensionless velocity $\bar{u}(\equiv u/u_0)$ at $\bar{y} = 0$ is unity, Eq. (5.7.32) can be written (see Prob. 5.22) as

$$\bar{u} = 1 - \bar{y}^2 \tag{9.5.25}$$

The rigid wall boundary conditions are $\phi = \phi' = 0$ at $\bar{y} = \pm 1$. If we consider only the half channel, $\bar{y} = 1$ to $\bar{y} = 0$, then the boundary conditions are those given by Eqs. (9.5.6a) and (9.5.6b), and the method described in the previous sections can be used to solve the Orr-Sommerfeld equation for the velocity profile given by Eq. (9.5.25).

The stability computer program calculates and prints out the real and imaginary parts of ϕ and their derivatives (eight dependent variables altogether) as well as the corresponding eigenvalues. The calculations can be done either for temporal-amplification theory (α_I real) or for spatial-amplification theory (ω complex).

Figures 9.2 and 9.3 show two stability diagrams obtained by using the temporal-amplification theory for three values of c_i, namely, 0, 0.004, and 0.0075. Figure 9.2 shows the variation of $\alpha_I(\equiv \alpha_r)$ with Reynolds number, and Fig. 9.3 shows the variation of c_r with Reynolds number. These calculations were started by assuming α_I and c_r values for fixed values of c_i and R_I. Once a solution was obtained, the Reynolds number was incremented while the value of c_i was kept constant; new values of c_r and α_I are calculated at the new Reynolds number. Around the minimum critical Reynolds number, special care must be taken to compute the eigenvalues c_r and α_I for a given R and c_i. For details the reader is referred to Keller and Cebeci (1977). In this way one stability loop was generated for a given value of c_i; this procedure was repeated to obtain another stability loop for another fixed value of c_i. The results agree quite well with those obtained by

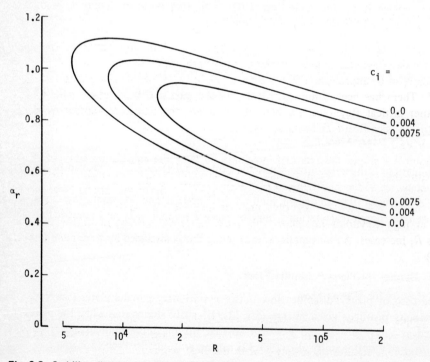

Fig. 9.2 Stability diagram for plane Poiseuille flow.

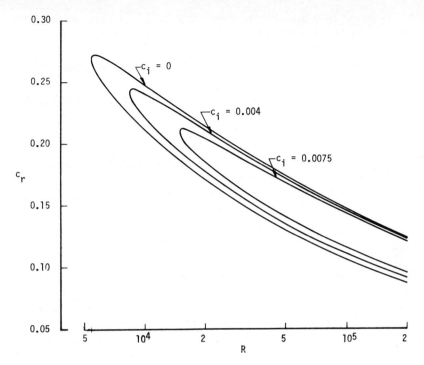

Fig. 9.3 Stability diagram for plane Poiseuille flow.

other numerical methods (see Radbill and McCue, 1970, for example). Note that the minimum Reynolds number attained varies with each value of c_i; it is 5772.22 ± 0.01 for $c_i = 0$, approximately equal to 8500 for $c_i = 0.004$, and 15,000 for $c_i = 0.0075$. There has been some controversy in the past about the exact value of this and other critical Reynolds numbers. To evaluate the Reynolds number in an air flow to six significant figures, we would need to know the temperature to an accuracy of 2×10^{-4} °C.

Figure 9.4 shows the real and imaginary parts of the eigenfunction, ϕ, for $R_l = 10^4$. Note that ϕ_r is a smooth function but ϕ_i shows a rapid variation between the wall and the region just beyond the critical layer, $\bar{y}_c$. Since the critical layer moves closer to the wall as the Reynolds number increases, the variation of ϕ_i becomes worse at high Reynolds numbers. This requires denser spacing of points near the wall as R_l increases. An automatic way of doing this is discussed by Keller and Cebeci (1977).

9.5.3 Results for Blasius Flow

As a sample test case for an external boundary-layer flow we consider the Blasius flow discussed in Sec. 5.5. In this case the dimensionless velocity profile, $\bar{u}$, in the Orr-Sommerfeld equation is simply f' with the edge velocity, u_e, chosen to be u_0. A convenient length scale is δ^*, so that $\bar{y} = y/\delta^*$. The second derivative of the

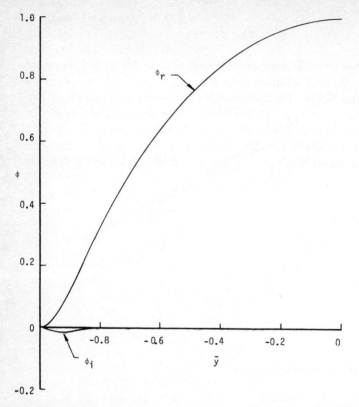

Fig. 9.4 The real and imaginary parts of the eigenfunction of ϕ corresponding to the lower branch of neutral curve at $R_h = 10^4$ for plane Poiseuille flow ($\bar{y} = 0$ at center line; $\bar{y} = -1$ on lower wall).

dimensionless velocity, $\bar{u}''$, can be calculated from

$$\bar{u}'' = f'''(\eta_\infty - f_\infty)^2 \qquad\qquad (9.5.26)$$

Here f', f''', and f_∞ are obtained by solving the Blasius equation. We can use either the shooting method discussed in Sec. 4.2 or the finite-difference (box) method discussed in Chap. 7 to solve the Blasius equation.

Figure 9.5 shows the real and imaginary parts of the eigenfunction, ϕ, corresponding to the lower branch of the neutral curve at $R_{\delta^*} = 902$. This particular solution was chosen because it corresponds to the experimental velocity fluctuation data of Schubauer and Skramstad (1947) and also because it provides us with an opportunity to check our numerical solutions with those obtained by Radbill and McCue (1970). Note that the behavior of ϕ_r and ϕ_i for Blasius flow across the layer is similar to that for internal flow; ϕ_r is again a smooth function, but ϕ_i shows a rapid variation between the wall and the region beyond the critical layer.

Figure 9.6 shows a comparison of calculated and experimental longitudinal velocity fluctuation results for the data of Schubauer and Skramstad. In linearized theory the amplitude of the perturbation is arbitrary; for this reason the computed values of u' were multiplied by a scale factor in order to compare the results with experimental data. The scale factor for the ordinates of the data was found by matching the computed solution with experimental data at the second maximum. The abscissas of the computed solutions were not scaled. The same procedure was followed by Radbill and McCue. The agreement between our results and those given by Radbill and McCue is very good. Similar results were also obtained by Kaplan (1964) and Wazzan et al. (1968). For further comparisons see Cebeci (1977c).

Figure 9.7 shows a stability diagram for Blasius flow obtained by using the temporal amplification theory for values of $c_i = 0$, 0.005, 0.01, and 0.018. The

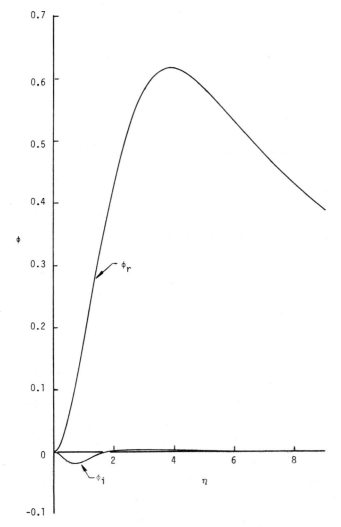

Fig. 9.5 The real and imaginary parts of the eigenfunction, ϕ, corresponding to the lower branch of the neutral curve at $R_{\delta*} = 902$ for Blasius flow.

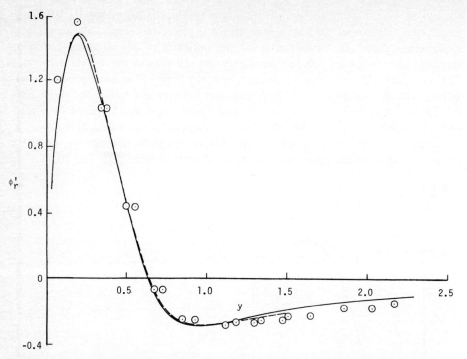

Fig. 9.6 Comparison of calculated and experimental velocity fluctuations for the data of Schubauer and Skramstad (1947). The solid line denotes the numerical solutions of Cebeci (1977c) and the dashed line denotes the numerical solutions of Radbill and McCue (1970). $R_{\delta*} = 902$.

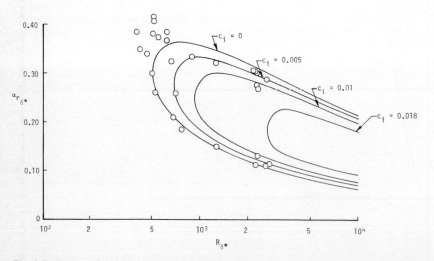

Fig. 9.7 Temporal stability diagram for Blasius flow. The solid lines denote the numerical solutions of Cebeci (1977c), and the symbols denote the experimental data of Schubauer and Skramstad (1947).

figure also shows the experimental data of Schubauer and Skramstad (1947). Similar results were also obtained by Kaplan (1964), Wazzan et al. (1968), Radbill and McCue (1970), and others.

For stability diagrams for other Falkner-Skan velocity profiles, the reader is referred to a very complete report by Wazzan et al. (1968).

9.6 NATURAL TRANSITION

9.6.1 Transition Process

As was mentioned above, the kinetic energy of the small disturbances is supplied by the mean flow, and the latter is therefore distorted; in fact, u^2 changes by an amount of order $\overline{u'^2}$, so u changes by an amount of order $\overline{u'^2}/u$. This second-order effect becomes significant when u' has risen to a few percent of u, and the second-order products of disturbance velocities that were neglected in the small-disturbance equations (9.2.6)–(9.2.8) also become important. For a review of mathematical results for finite-amplitude two-dimensional disturbances, see Stuart (1971). Breakdown to turbulence always involves the onset of three-dimensionality because turbulence is an essentially three-dimensional phenomenon. The finite-amplitude traveling-wave disturbances may acquire significant three-dimensionality either by the prolonged effect of small but finite spanwise variation of streamwise mean velocity or by being themselves unstable to *infinitesimal* three-dimensional disturbances. In either case the initially straight wave front will develop spanwise undulations, which are strengthened by second-order effects. The mechanism is similar to the self-induced distortion of an isolated vortex line with a slight undulation (Fig. 9.9).

In free shear layers (which have points of inflection and are therefore unstable at all but the lowest Reynolds numbers) disturbance growth rates are very high at the Reynolds numbers that occur in practice. Disturbance amplitudes of at least 10% of the mean velocity can be attained before significant three-dimensionality appears, and the dominant nonlinear effect is the generation of subharmonics. At large amplitudes the point on the wave cycle at which $\partial u'/\partial y$ is largest (station AA$'$ in Fig. 9.8) can be regarded as the position of an isolated spanwise vortex representing the disturbance, and the wave train as a whole therefore behaves like a sequence of corotating vortices. Such vortices can capture one another, thus doubling the wavelength of the disturbance and halving its frequency. The process can recur several times before breakdown to turbulence occurs, and it has even been suggested that essentially two-dimensional vortices may survive as a permanent feature of turbulence in free shear layers. For pictures of the vortex-pairing process, see Winant and Browand (1974).

In shear layers that are unstable only by the viscous mechanism, disturbance growth rates are much slower, and three-dimensionality in the mean flow usually distorts the disturbance wave fronts considerably before very high amplitudes are obtained (Klebanoff et al., 1962). Therefore two-dimensional finite-amplitude

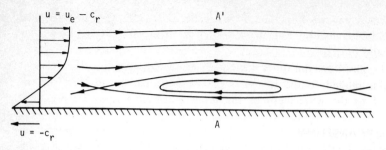

Fig. 9.8 The "cat's eye" diagram: streamlines in an unstable boundary layer seen by an observer moving at the wave velocity c_r. Vertical scale has been magnified about 2 : 1.

theory is not very useful in these cases, except for semiquantitative explanations of behavior at a particular spanwise station. Also, vortex capture has not been observed. The imposition of spanwise undulations on a traveling-wave/traveling-vortex disturbance implies the appearance of streamwise vorticity (Benney and Lin, 1960), which interacts with the main disturbance to concentrate the energy of the latter at "peak" positions (Fig. 9.9), one for every wavelength of the original spanwise undulation (see Klebanoff et al., 1962). Intense internal shear layers (Stuart, 1965) form at the peak spanwise positions and at the streamwise positions of the traveling vortices. In the ideal case of a flow with no disturbances other than exactly periodic traveling waves initially with spanwise wave fronts, perturbed by a sinusoidal spanwise variation of streamwise mean velocity, these shear-layer patches would lie at the mesh points of a rectangular grid in the xz-plane traveling at the wave velocity. In practice the wavelength and amplitude of the traveling wave fluctuate slightly because of background turbulence in the stream, and the spanwise variations of the mean flow are only roughly periodic. Therefore the positions and intensities of the shear-layer patches are slightly irregular. Knapp and Roache (1968) present

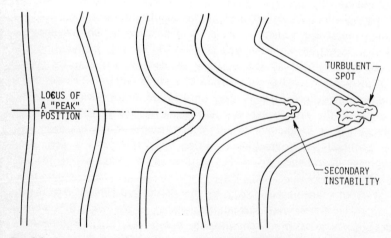

Fig. 9.9 Successive locations of wave fronts in the later stages of transition.

photographs of this stage of transition, and their account of the process as a whole is a very helpful one.

The next stage is that the more intense shear-layer patches become subject to some form of localized instability, which is, of course, strongly three-dimensional. The wavelength is related to the size of the shear-layer patch and therefore still nominally related to the wavelength of the primary traveling wave. However, the localized instability wavelengths are quite small in comparison to the variation in primary wavelength caused by the cumulative effects of background turbulence, so quite large random fluctuations of the phase angle of the localized instability will occur. Interaction between different wavelengths that are not exactly related harmonically leads to further spreading in wavelength. Once the kinetic energy of the three-dimensional disturbance in a shear-layer patch has spread to fill a significant range in wavelength, instead of being concentrated at a small number of discrete, harmonically related wavelengths, the disturbance is effectively turbulent and can be called a "turbulent spot" (Emmons, 1951; Dhawan and Narasimha, 1958; Elder, 1960).

Initially, the turbulent spots are small in comparison to the distance between them. Define the transition intermittency factor, γ_{tr}, as the fraction of the length of a long spanwise line in Fig. 9.9 that is occupied by turbulent spots (strictly speaking, the definition requires the length of the line to tend to infinity). Then γ_{tr} at the right-hand edge of Fig. 9.9 might be 0.1–0.2. As the turbulent spots spread into surrounding nonturbulent fluid, qualitatively in the same way as fully turbulent shear layers spread, γ_{tr} rises. When γ_{tr} is of the order of 0.5, the spots begin to merge with each other, and eventually the whole shear layer is turbulent ($\gamma_{tr} = 1$). The rate of spreading, like the transition process as a whole, is somewhat different for constant-pressure boundary layers and for flows with pressure gradient. The correlation [Eq. (8.3.5)] implies that if $u_e x/\nu$ is 10^6 when turbulent spots first appear, $\gamma_{tr} = 0.99$ is not reached until $u_e x/\nu = 1.8 \times 10^6$.

The details of where and how turbulent spots grow depend on the details of spanwise irregularity and background turbulence in the flow. However, if we adopt the statistical approach and simply seek a formula for γ_{tr} or the position where $\gamma_{tr} = 0.99$, we may expect the formula to be related to the results of small-disturbance theory because the small-disturbance eigenfunction, $\phi(y)$, provides the initial condition for nonlinear development and breakdown to turbulence. A transition correlation based on this argument is presented in Sec. 9.6.2.

Fully turbulent flow is attained only when all memory of the transition process has been lost. In the case of a constant-pressure boundary layer, say, this implies that all properties are determined by the local Reynolds number, $u_e \theta/\nu$, say. However, the latter condition is not quite sufficient, for it is possible that the flow in the final stages of transition, following the stage at which all the turbulent spots have joined up and γ_{tr} is effectively unity, may be in a state of moving equilibrium. This state would depend only on $u_e \theta/\nu$ and the fact that transition had occurred. A boundary layer that had first become fully turbulent some distance upstream and then had its thickness reduced by suction or strong acceleration would have slightly different properties at the same value of $u_e \theta/\nu$. Positive support of this possibility is given by the results of Bushnell et al. (1975), who find markedly different mixing-length distributions in supersonic boundary layers at low Reynolds number according to whether the boundary layer grows on a wind-tunnel wall (transition far upstream followed by favorable pressure gradient) or on a flat plate (transition followed by continuous growth). The latter case is the most common in practice, and a large fraction of the existing data for low-speed flow on flat plates was correlated by Coles (1962), who showed, for instance, that the wake parameter Π (Sec. 6.4) was a unique function of $u_e\theta/\nu$. The departures from that function showed by some experiments could normally be explained by overlarge trip wires or other artificial disturbances in the transition region.

Traveling-wave disturbances with spanwise wave fronts are the most unstable mode in two-dimensional plane shear layers whether the mechanism of instability is viscous or inviscid. Two other types of disturbance modes are often found in unstable shear layers: they are the steady longitudinal contrarotating vortices found in two-dimensional boundary layers on concave surfaces and other unstably curved shear layers and the nearly longitudinal vortices found in three-dimensional boundary layers. Both of these are basically inviscid instabilities.

Elementary analysis shows that a fluid element in a curved stream whose angular momentum decreases with increasing distance from the center of curvature is unstable. If the element is slightly displaced outward from the center, conserving its angular momentum about the center, it will be moving faster than its surroundings (which have smaller angular momentum). Therefore the radial pressure gradient that controls the streamline curvature will be too small to direct the displaced element along a streamline of the main flow, and the element will move even further outward. The converse argument holds for inward displacement. If the angular momentum of the fluid increases with increasing distance from the center of curvature, a displaced element tends to return to its original position. In a real flow with angular momentum decreasing outward, any infinitesimal disturbance will set up both inward and outward movements. When the most unstable mode of disturbance has grown to dominate all others, steady inward and outward movements alternate in the spanwise direction, the spanwise wavelength being roughly equal to the shear-layer thickness. The streamwise wavelength is effectively infinite. This disturbance mode is most easily described as an array of longitudinal vortices. The strength of the vortices grows with distance downstream, assuming that the curvature continues; eventually, strong secondary shear layers form between the vortices, secondary instabilities are excited, and turbulence ensues.

The longitudinal-vortex mode is an unexpected one; spanwise vortices would seem more likely. However, the disturbed-element analysis above implies that a displaced fluid element continues to move in its direction of displacement rather than executing a diverging oscillation like an element passing through an array of spanwise vortices of increasing strength. The instability being basically inviscid, a criterion for its occurrence can be derived as

$$\frac{\text{Typical net ``centrifugal'' force}}{\text{Typical viscous force}} \qquad\qquad (9.6.1)$$

It is called the Taylor number and appears in various forms for different curved-flow problems. For a general derivation see the book by Rosenhead (1963, pp. 497–498). The longitudinal vortices are called Taylor-Görtler vortices.

In a stably curved flow any longitudinal-vortex disturbances would be damped. The growth of spanwise-oriented traveling waves is *unaffected* by curvature (to the order of accuracy of the TSL approximation), so the early stages of transition in the boundary layer on a convex surface are the same as on a flat surface. On a slightly concave surface the critical Reynolds number for traveling-wave instability, $u_e\theta/\nu \simeq 200$ in a constant-pressure flow, will be reached before the critical Taylor number for longitudinal-vortex instability, $(u_e\theta/\nu)\sqrt{(\theta/R)} \simeq 0.3$. This is the case if $u_e R/\nu$ exceeds about 10^8. In general, the mode that becomes unstable first will grow and either suppress the other mode completely or prevent it from reaching a significant amplitude. After three-dimensionality sets in, streamline curvature of either sign will affect further progress toward transition.

In three-dimensional boundary layers it is convenient to resolve the velocity parallel to the surface into a component in the direction of the local free stream and a "cross-flow" component perpendicular to this. The cross-flow velocity is zero at the surface and—by definition—zero in the free stream, reaching the latter value asymptotically. The cross-flow velocity profile therefore has a point of inflection, and if there were no streamwise velocity, the inviscid instability mechanism would generate waves traveling in the cross-flow direction with a velocity close to the cross-flow velocity at the inflection point. These waves could also be regarded as a pattern of nearly longitudinal vortices moving in the cross-flow direction. The instability mode actually observed in a three-dimensional boundary layer is a pattern of fixed nearly longitudinal vortices. The reason is that the purely laminar flow contains temporally steady, spatially varying disturbances arising from nonuniformities in the surface or in the external stream (it is these disturbances that provide the essential three-dimensionality in the latter stages of transition of a nominally two-dimensional flow). Since, as far as the instability modes are concerned, there is nothing special about the cross-flow direction as it is defined above, the velocity component whose inviscid-instability mode dominates is not necessarily the cross-flow component. In fact, the preferred component is the one whose instability mode has zero wave velocity; this is the velocity component whose profile has its point of inflection where the velocity component itself is zero. Obviously, a precondition for this three-dimensional instability is that such a profile shall exist; there is also a critical Reynolds number based on the boundary-layer thickness and, say, the maximum velocity component on the preferred profile. For further details of this and other modes of hydrodynamic stability, see Rosenhead (1963, Chap. 9).

9.6.2 Empirical Correlations for Natural Transition

The prediction of transition in two-dimensional, axisymmetric, or three-dimensional flows is a very difficult problem. In general, transition is affected by free-stream turbulence, pressure gradient, surface curvature, roughness, noise, vibration, surface temperature, compressibility, secondary-flow effects, etc. At the present time there is no exact theoretical method that can account for all—or indeed any—of these effects in predicting transition. There is one empirical method that seems to work reasonably well for two-dimensional and axisymmetric flows. This is the so-called e^9-method, which utilizes the stability theory and some experimental results. It was first used by Smith and Gamberoni (1956) and by Van Ingen (1956). The basic assumption is that transition starts when a small disturbance introduced at the critical Reynolds number has amplified by a factor of e^9 or about 8000. Strictly speaking, the criterion is that amplification predicted by the linear theory should have reached e^9. This method is, of course, wholly empirical. Taken literally, it implies that if transition takes place when the root mean square (rms) disturbance amplitude is 5% of the mean velocity (say), the background amplitude in the range of unstable frequencies is about 0.0006%, probably a lower figure than is ever attained in practice. It is probably better to regard the method as an empirical correlation for the length of the transition region of a boundary layer in a low-turbulence stream, such as the atmosphere or a high-performance tunnel. The method requires a knowledge of the disturbance amplitude, $g(y)$, at each x-station in a boundary layer. Either temporal- or spatial-amplification theory can be used. Often the latter is preferred, since, in a steady mean flow, what can be measured is the amplitude change of disturbance with distance. The amplitude at a fixed point is independent of time. The spatial theory gives this amplitude change in a more direct manner than does the temporal theory.

Recall that in the spatial theory, the typical disturbance function, $g'(x, y, z, t)$, given by Eq. (9.2.12) varies with x as $\exp(-\alpha_i x)$. The logarithmic x-derivative of $|g'|$ yields the spatial amplification rate in the x-direction,

$$\frac{1}{|g'|} \frac{d|g'|}{dx} = -\alpha_i \tag{9.6.2}$$

We note that the disturbances are damped when $\alpha_i > 0$, neutral when $\alpha_i = 0$, and amplified when $\alpha_i < 0$.

The amplitude change with x can be obtained by integrating Eq. (9.6.2). For similar flows we can write (see Prob. 5.8)

$$dx = \frac{2}{m+1} \frac{\delta^*}{(\delta_1^*)^2} dR_{\delta^*} \tag{9.6.3}$$

Substituting Eq. (9.6.3) into Eq. (9.6.2) and integrating the resulting expression, we get the ratio of the amplitudes at two Reynolds numbers, $R_{\delta_1^*}$ and R_{δ^*}, namely,

$$\frac{|g'|}{|g'|_1} \equiv a = \exp\left[-\frac{2}{m+1} \frac{1}{(\delta_1^*)^2} \int_{R_{\delta_1^*}}^{R_{\delta^*}} \alpha_{i\delta^*} \, dR_{\delta^*}\right] \tag{9.6.4}$$

where $\alpha_{i\delta^*} = \alpha_i \delta^*$. The amplification factor, a, varies with x (or R_{δ^*}) because of the dependence on the external-velocity distribution, $u_e(x)$, and the local Reynolds number, $R_\delta{}^*$. Once the external-velocity distribution and the free-stream Reynolds number are given, the velocity profile, $\bar{u}$, can be obtained from the solution of the boundary-layer equations. With this information the solution of the Orr-Sommerfeld equation can then be obtained. For a given $R_\delta{}^*$, the eigenvalue problem needs to be solved for the four remaining scalars, α_r, α_i, c_r, and c_i. Only two of these four scalars are essentially unknown, since in the spatial theory ω is real, that is,

$$\omega_r = \alpha_r c_r - \alpha_i c_i$$
$$\omega_i = \alpha_r c_i + \alpha_i c_r \equiv 0 \qquad\qquad (9.6.5)$$

Thus if we choose α_r and α_i to be the unknown scalars, then for a specified dimensionless frequency, ω_r^*, defined by

$$\omega_r^* = \omega_r \frac{\nu}{u_e^2} \qquad\qquad (9.6.6)$$

from Eq. (9.6.5) we can write

$$c_r = \frac{\omega_r \alpha_r}{\alpha_r^2 + \alpha_i^2} \qquad c_i = -\left(\frac{c_r}{\alpha_r}\right)\alpha_i \qquad\qquad (9.6.7)$$

The unknowns, α_r and α_i, are again obtained by the Newton's method described in Sec. 9.5. For details the reader is referred to Cebeci (1977c).

The procedure for the e^9-method consists of the evaluation of the integral in Eq. (9.6.4) for a set of specified dimensionless frequencies. The stability calculations begin at a Reynolds number, $R_\delta{}^*$, slightly larger than the critical Reynolds number, $R_{\delta^*_{cr}}$ (see Fig. 9.10), on the lower branch of the neutral stability curve. As an example, let us consider the Blasius flow. In this case a choice of $R_\delta{}^*$ equal to 600 is quite satisfactory to initiate the stability calculations, since $R_{\delta^*_{cr}} = 520$. For specified ω_r^* the Orr-Sommerfeld equation is solved for several Reynolds numbers, $R_\delta{}^*$, and the variation of $\ln a$ with x is computed. The calculations stop when the variation of $\ln a$ with x becomes negligible. After that, a new set of calculations is made for a different value of frequency. Figure 9.11 shows the computed amplification factors for values of ω_r^* ranging from 2×10^{-5} to 4×10^{-5}. The envelope of these curves represents the maximum amplification factor, $A(R_\delta{}^*)$. Turbulent spots are assumed to start when $\ln A = 9$ or 10. According to the recent calculations of Cebeci (1977c), at the experimental $(Re_x)_{tr} = 2.84 \times 10^6$, $\ln A = 8.98$. Similar results were also obtained by Smith and Gamberoni (1956) and by Jaffe et al. (1970).

This procedure of predicting transition by the box method has been applied to two-dimensional similar and nonsimilar flows by Cebeci (1977c). The same procedure utilizing a different numerical scheme and a different eigenvalue hunting procedure has been applied to two-dimensional and axisymmetric flows by Jaffe et

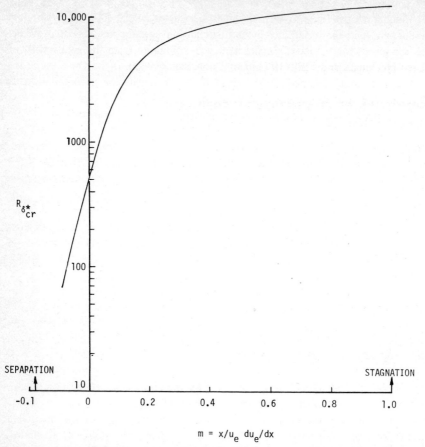

Fig. 9.10 Variation of critical Reynolds number, $R_{\delta^*_{cr}}$, with pressure-gradient parameter, m, after Wazzan et al. (1968).

al. (1970) and by Kaups (1975). In most cases, satisfactory agreement with experiment was observed.

The extension of the e^9-method to three-dimensional flows is not obvious at first because, in addition to disturbances in the form of Tollmien-Schlichting waves, three-dimensional flows also admit vortexlike disturbances that affect transition. However, the transition pattern even in a two-dimensional mean flow becomes three-dimensional well before transition is complete. Preliminary evaluation of the e^9-method for flow over a rotating disk with vortexlike disturbances gives hope that the method may also work for three-dimensional flows if appropriate stability analysis is performed. However, such studies are yet to be conducted.

Several workers have presented prediction methods for the later stages of transition (starting with an assumed value of x_{tr}) using the methods of turbulence modeling. Their assumptions imply that the dimensionless structural parameters of the disturbances (e.g., $\overline{u'v'}/\overline{u'^2}$) can be correlated, for the whole of the transition process following the appearance of finite-amplitude disturbances or turbulent spots, by functions of local Reynolds number such as $u_e\theta/\nu$. This is quite different from the assumption that the transition process is uniquely related to the results of the small-disturbance theory, although the two are not necessarily incompatible. For details, see McDonald and Fish (1973) and Donaldson (1969).

9.7 PROMOTION OF TRANSITION BY FREE-STREAM TURBULENCE OR SURFACE ROUGHNESS

Transition always depends on the presence of velocity fluctuations in the free stream. The e^9-rule implies that the actual level of free-stream disturbance is not very important if it is small, but large free-stream fluctuations can cause transition at a much lower Reynolds number than "natural" transition.

Surface roughness or other steady perturbations cannot cause transition directly but can alter the shape of the mean-velocity profile so that the shear layer becomes unstable at a lower Reynolds number and/or amplifies existing velocity fluctuations more quickly. Since only fluctuations with a wavelength of the order of the shear-layer thickness are amplified, the effect of very long wavelength fluctuations is also to alter the mean-velocity profile into a more unstable form over part of the period. A simple example is the effect of a sinusoidal pulsation in free-stream velocity on a laminar boundary layer (Obremski and Fejer, 1967; Merklin and Thomann, 1975). The pressure gradient that drives the free-stream velocity fluctuations produces larger velocity fluctuations in the boundary layer (Fig. 9.12) so that

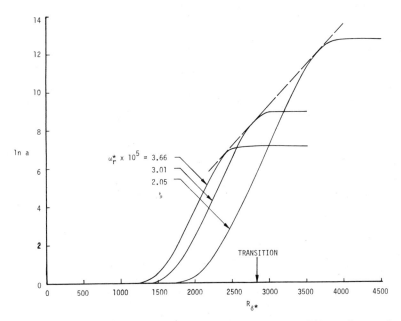

Fig. 9.11 Computed amplification factors for Blasius flow at different frequencies, after Cebeci (1977c).

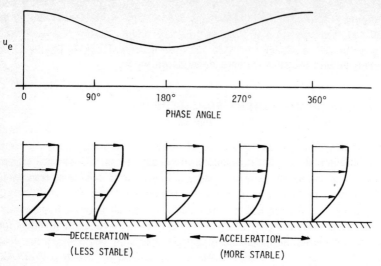

Fig. 9.12 Effect of slow sinusoidal oscillation of free-stream velocity on boundary-layer profiles.

over the decelerating half of the period the profile has a point of inflection and is therefore unstable at a lower Reynolds number than that of the undisturbed profile. Transition will occur only if the sinusoidal fluctuation is slow enough, or large enough, for the small-wavelength disturbances to amplify to breakdown point within half a period of free-stream oscillation; if the Reynolds number is low, the favorable pressure gradient in the next half period may produce "reverse transition," that is, reestablish laminar flow.

The practical interest in promotion of transition takes two forms. First, one may wish to keep the flow laminar to as high a Reynolds number as possible by minimizing free-stream turbulence and, in the case of a wall flow, minimizing surface roughness. Here turbulence includes all forms of free-stream disturbance, including noise, and roughness includes all kinds of surface irregularity, including steps between surface panels. Second, one may wish to turn the flow turbulent, for instance, to forestall laminar separation or to increase heat transfer. Probably the most critical examples of both kinds of transition problem occur in wind-tunnel testing. In small wind tunnels it is usually necessary to force transition at about the same percentage of body length as the position of transition at full scale, so that the boundary-layer development resembles that on the full-scale body as closely as possible. By contrast, in large wind tunnels one may be trying to test a model at or near full-scale Reynolds number, so that free-stream disturbances must be reduced to near the level found at full scale (for most aeronautical purposes the atmosphere can be regarded as nonturbulent). In practice, of course, the designer of a wind tunnel of any size pays at least some attention to the reduction of free-stream disturbances, since the needs of future users cannot be foreseen.

The effect of free-stream turbulence depends on its length scale as well as its rms intensity, as was implied above. A theory of transition due to free-stream turbulence was given by Taylor (1935); see Goldstein (1965) for details. It assumes implicitly that the effect of free-stream turbulence of *any* wavelength is to distort the velocity profile into a highly unstable form. The result is that the transition Reynolds number in zero-pressure gradient depends on

$$\frac{\sqrt{\overline{u'^2}}}{u_e} \left(\frac{\delta}{L}\right)^{1/5} \tag{9.7.1}$$

where L is the length scale of the turbulence, nominally, $(\overline{u'^2})^{3/2}/$(turbulent energy dissipation rate). Most investigators of free-stream turbulence effects have ignored the effect of length scale. Because wind-tunnel geometries are fairly standardized, the ratio of length scale to model size does not vary enough to have a significant effect on transition Reynolds number (Figs. 9.13 and 9.14). Wells (1967) shows the effect of rms intensity on transition in a constant-pressure boundary layer. The definitions of "start" and "end" of transition are somewhat arbitrary. It is not certain that rms intensity of velocity fluctuations is the right variable to correlate the effect of a mixture of true turbulence and sound waves, and the turbulence-reducing devices in modern wind tunnels are so efficient that sound waves may contribute a large part of the total disturbance level when the rms value of the latter is only a few hundredths of 1%. Because the noise level and vibration level rise rapidly with tunnel speed whereas the true turbulence level does not, noise may dominate the disturbance field in high-subsonic tunnels. In supersonic tunnels, fluctuating Mach waves are by far the most important source of disturbance. Therefore Fig. 9.13 is not necessarily a helpful guide to tunnel performance, particularly at high speeds.

Free-stream turbulence is not a convenient means of purposely provoking transition. The usual way of generating free-stream turbulence is to install a grid in the settling chamber of the tunnel or test rig; one grid equals one turbulence level. A unique device used by Favre's group at Marseilles, France, is a vibrating wire ahead of the model; again it is difficult to control the turbulence level. The most popular form of transition stimulator for wall layers is surface roughness, either as a single two-dimensional obstacle, typically a spanwise wire, or a distributed roughness element, typically grit. An intermediate form, elegant but little used, is a spanwise array of air jets blowing normal to the surface (Stone et al., 1970). The effect is much the same as that of an array of solid obstacles, but there may be some effect of turbulence in the jets themselves. The size of disturbance can be controlled very easily, but the engineering difficulties are obvious.

The main requirement of an artificial roughness "trip" is reproducibility so that the right trip for a given case can be chosen on the basis of a set of master experiments. The best-documented geometry is a spanwise wire (Fig. 9.15); if the diameter is d, dimensional analysis shows that in a constant-pressure laminar boundary layer the transition behavior depends on $u_e \delta/\nu$ and d/δ only. An empirical criterion for transition to occur a negligibly small distance behind the wire (Gibbings, 1959) is

$$u \ (y = d) \frac{d}{\nu} \approx 800 \tag{9.7.2}$$

which is a special case of that given above because $u(y = d)/u_e$ is a function of d/δ only. In practice, $u_e\delta/\nu$ at the desired transition position may be not far short of 800 ($u_e x/\nu = 2.6 \times 10^4$ in zero-pressure gradient, more in favorable gradients), so d

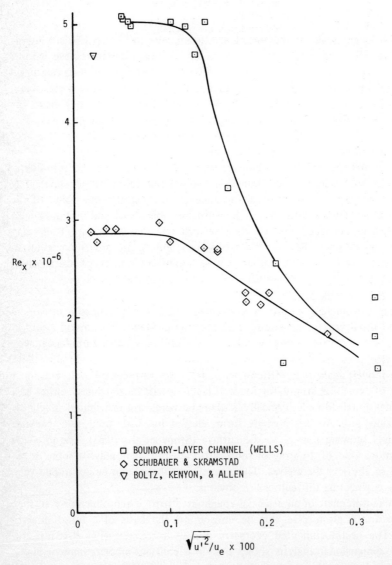

Fig. 9.13 The effect of free-stream turbulence on boundary-layer transition, for low-turbulence intensities. See Van Driest and Blumer (1968) for complete references.

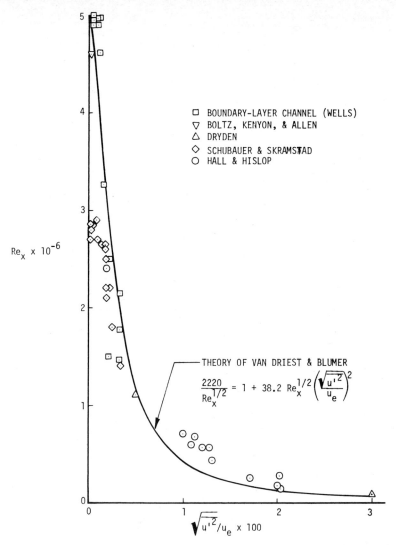

$Re_x \times 10^{-6}$

☐ BOUNDARY-LAYER CHANNEL (WELLS)
▽ BOLTZ, KENYON, & ALLEN
△ DRYDEN
◇ SCHUBAUER & SKRAMSTAD
○ HALL & HISLOP

THEORY OF VAN DRIEST & BLUMER

$$\frac{2220}{Re_x^{1/2}} = 1 + 38.2\, Re_x^{1/2} \left(\frac{\sqrt{\overline{u'^2}}}{u_e}\right)^2$$

$\sqrt{\overline{u'^2}}/u_e \times 100$

Fig. 9.14 The effect of free-stream turbulence on the start of boundary-layer transition compared with the theory of Van Driest and Blumer (1968), who give references in full.

is close to δ, $u(y = d)$ is close to u_e, and the above criterion is not too far from

$$u_e \frac{d}{\nu} = 800 \tag{9.7.3}$$

The purpose of the trip is partly to distort the profile and partly to increase the momentum thickness of the boundary layer (the momentum-thickness Reynolds number at the instability point is about 200). The drag coefficient of a circular wire

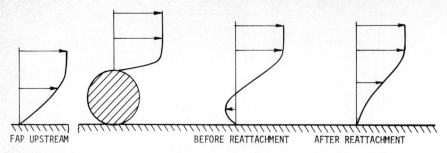

Fig. 9.15 Boundary-layer profiles behind a trip wire (in the absence of transition).

in a uniform shear flow, based on the velocity at the top of the wire, is about 0.75, so the increment in $u_e\theta/\nu$ is about $0.75u_ed/\nu$ if d is as large as δ. Thus a wire chosen in accordance with Eq. (9.7.2) will increase $u_e\theta/\nu$ well above the instability value, and transition to turbulence in the highly unstable mixing layer close behind the wire will be permanent.

It is generally accepted that transition in a boundary layer involves a secondary instability of the Tollmien-Schlichting wave, which is essentially three-dimensional. Therefore a trip that produces three-dimensionality (spanwise periodicity) should precipitate transition more quickly, or with a smaller obstacle size, than a purely two-dimensional trip. Hama (1954) suggested the use of a serrated strip rather like a coarse hacksaw blade placed flat on the surface. However, the extra complication over a circular wire is not often thought worthwhile in low-speed flow because there are no constraints on the trip size other than the need to cause transition. In supersonic flow a transition trip produces a shock wave; even if the velocity at the top of the trip is subsonic, the displacement of the outer flow will produce a shock. Therefore distributed grit is preferred for supersonic flow, especially in tests of transonic airfoils, which depend rather critically on the upper-surface shape near the leading edge. The best-documented proprietary forms of grit are Carborundum (silicon carbide), available in a wide range of nominal sizes, each with about a 2 : 1 range in particle diameter, and Ballotini (glass spheres), with more closely controlled sizes. Different research establishments have different codes of practice for the choice and application of grit roughness (see Versmissen, 1974). For a useful general review with emphasis on supersonic wind tunnels, see Pate (1971, 1974). Reshotko (1975) and the papers that follow outline current transition problems.

Chapter 10

Three-Dimensional and Unsteady Flows

10.1 BOUNDARY SHEETS, BOUNDARY REGIONS, AND UNSTEADY FLOWS

We group three-dimensional and unsteady flows together because the numerical problems are related. Consider an unsteady flow over an oscillating wing. In *xyt*-space the flow is qualitatively like the steady flow over a real wing whose incidence changes periodically in the spanwise (*t*) direction. In some respects, time-dependent flow is easier because there is no doubt about the direction of marching along the *t* axis.

10.1.1 Boundary Sheets

In Sec. 3.2 we distinguished boundary sheets, like the shaded part of the boundary layer and wake in Fig. 10.1, and boundary regions, like the flow near the wing root and wing tip or in the trailing vortex. In the former case, gradients in the *z*-direction are, at most, of the same order of magnitude as those in the *x*-direction. Therefore the TSL approximation is applicable to the *z*-component equation as well as the *x*-component equation and yields Eq. (3.2.3). Equations (3.2.1)-(3.2.3) and (2.2.2), with extra equations for the *x*-component and *z*-component of turbulent shear stress, $-\rho\overline{u'v'}$ and $-\rho\overline{v'w'}$, are a closed set of equations for u, v, and w if p is

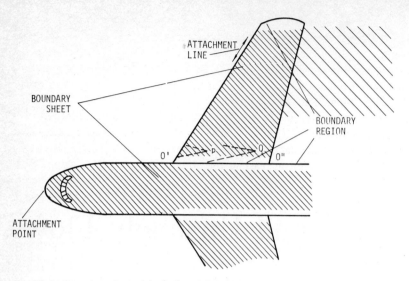

Fig. 10.1 Boundary sheets (shaded) and boundary regions.

given as a function of x and z at the surface or at the outer edge of the shear layer.

A special case with some relevance to aeronautics and to centrifugal turbomachinery is that in which gradients in the z-direction are zero although w is nonzero. The simplest example is a swept wing of constant chord and cross section and of indefinitely high aspect ratio (Fig. 10.2a). If the z-axis is taken parallel to the leading edge, z-wise gradients are zero, but there is, of course, a velocity component in the z-direction. In respect of fluid dynamics this is a fully three-dimensional flow; computationally, it has only two independent variables. The equations for this case are simply Eqs. (3.2.1)–(3.2.3) and (2.2.2) with the z-derivatives neglected, and calculations proceed along any one chordwise line such as the x-axis shown in Fig. 10.2a. The boundary conditions and initial conditions are qualitatively the same as those in two dimensions except that an initial w profile, and in turbulent flow an initial $\overline{v'w'}$ profile, must be supplied. Note that the z-component of the free-stream velocity, w_e, is independent of x as well as of z because the y-component of free-stream vorticity, $\partial u_e/\partial z - \partial w_e/\partial x$, is zero. Two extensions of the infinite-wing concept are its use on straight-tapered wings (Bradshaw et al., 1976), in which radial derivatives in the coordinate system shown in Fig. 10.2b are zero, and in swirling flow in radial diffusers (Wheeler and Johnston, 1972), in which circumferential derivatives are zero (Fig. 10.2c). In both extensions the change of coordinates introduces minor extra terms in the equations of motion. Boundary conditions are essentially the same as those for an infinite swept wing.

Boundary conditions for the general boundary-sheet equations were briefly discussed in Sec. 4.3.3. The domain of dependence of the point P in Fig. 10.1 (outlined by the dotted lines) lies entirely within the boundary sheet, but the domain of dependence of Q overlaps the wing-root region where the boundary-sheet equations are not valid. The only rigorous way of calculating the shear-layer properties at Q is to supply boundary conditions for the boundary-sheet calculations on the line $0'0''$. These boundary conditions can be obtained only from a solution for the boundary region near the root from which the streamlines/characteristics emerge. In turn, the initial conditions for the boundary-region calculations, applied near $0'$, must come from a solution of the boundary-sheet equations for the fuselage ahead of the wing. Flow-visualization pictures, notably by Peake et al. (1972) and Werle (1973), show the large lateral deflection of boundary-layer streamlines that can occur in three-dimensional flow. The largest deflections occur near the surface, generally in a region so thin that it carries little momentum and therefore has little influence on flow development, at least in the absence of separation. Also, deflections are very much less in turbulent flow than in laminar flow (see frontispiece) because, as was explained for two-dimensional flow in Chap. 1, turbulent flows are much better able to resist distortion by pressure gradients. However, it is better to overestimate than to underestimate the importance of domain-of-dependence effects.

Apart from these difficulties with the lateral boundary conditions the problems of three-dimensional boundary sheets are simply the addition of a second momentum equation and, in turbulent flow, the need for a second Reynolds-stress equation. We discuss the first of these problems in Sec. 10.2 and the second in Sec. 10.3.

10.1.2 Boundary Regions

In boundary regions the only terms in the NS equations that can be rigorously neglected are the x-wise stress gradients: $\partial\sigma_{xx}/\partial x$ in the x-component equation,

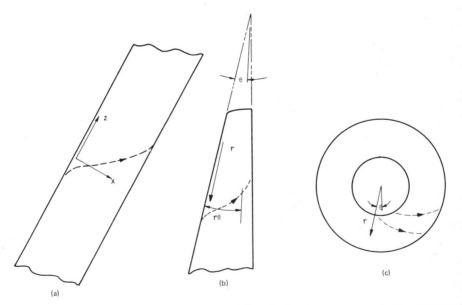

Fig. 10.2 Degenerate cases of three-dimensional flow having three nonzero velocity components but only two independent variables. In each case, y is normal to the page. (a) Infinite swept-wing coordinates (x, y). (b) Tapered swept-wing coordinates $(r\theta, y)$. (c) Radial diffuser coordinates (r, y).

$\partial \sigma_{xy}/\partial x$ in the y-component equation, and $\partial \sigma_{xz}/\partial x$ in the z-component equation. Compared with the case of boundary sheets therefore, the equations for a boundary region contain extra z-wise stress gradients in the x- and z-equations, and the y-component equation resembles the z-equation instead of yielding the trivial result $\partial p/\partial y = 0$. Some simplified cases occur; the case of axisymmetric flow was discussed in Sec. 3.2. Swirling axisymmetric flows, in which the circumferential component of velocity is nonzero but all properties are independent of circumferential position, obey somewhat simplified equations analogous to those for the infinite swept wing, but because of the centripetal acceleration the radial-component equation is nontrivial although its stress terms may be negligible. The equations for continuity and x-component momentum are, respectively,

$$\frac{\partial u}{\partial x} + \frac{1}{r}\frac{\partial rv}{\partial r} = 0 \tag{3.2.7}$$

and

$$\frac{\partial u}{\partial t} + u\frac{\partial u}{\partial x} + v\frac{\partial u}{\partial r} = -\frac{1}{\rho}\frac{\partial p}{\partial x} + \frac{1}{\rho r}\frac{\partial}{\partial r}(r\sigma_{xr}) \tag{10.1.1}$$

where in turbulent Newtonian flow, $\sigma_{xr} = \mu \, \partial u/\partial r - \rho\overline{u'v'}$ as normal, so that Eq. (10.1.1) reduces to Eq. (3.2.6). The equation for r-component momentum is

$$\frac{\partial v}{\partial t} + u\frac{\partial v}{\partial x} + v\frac{\partial v}{\partial r} - \frac{w^2}{r} = -\frac{1}{\rho}\frac{\partial p}{\partial r} + \frac{1}{\rho r}\frac{\partial}{\partial r}(r\sigma_{rr}) \tag{10.1.2}$$

where $\sigma_{rr} = 2\mu \, \partial v/\partial r - \rho\overline{v'^2}$. In most slender swirling flows (changing only slowly in the x-direction) the "centrifugal" approximation

$$\frac{w^2}{r} = \frac{1}{\rho}\frac{\partial p}{\partial r}$$

may be adequate, as in a TSL on the curved surface of an airfoil [see Eq. (3.1.5)]. The θ-component momentum equation is

$$\frac{\partial w}{\partial t} + u\frac{\partial w}{\partial x} + v\frac{\partial w}{\partial r} + \frac{vw}{r} = -\frac{1}{\rho r}\frac{\partial p}{\partial \theta} + \frac{1}{\rho r}\frac{\partial}{\partial r}(r\sigma_{r\theta}) \tag{10.1.3}$$

where $\sigma_{r\theta} = \mu \, \partial w/\partial r - \rho\overline{v'w'}$.

In nonaxisymmetric slender flows such as those in wing-body junctions, v and w will be of the same order and generally much smaller than u. In general, such flows are analyzed in rectangular Cartesian coordinates (but for an elaborate analysis of special coordinates for flow along a streamwise corner see Desai and Mangler, 1974). Streamwise stress gradients will always be negligible, but normal-stress gradients in the y- and z-directions are certainly not negligible in turbulent flow, so in general, the only deletion from Eq. (2.3.4) that is permissible is the term in σ_{xx}. Using the d/dt operator defined in Eq. (2.3.5), the equations for constant-property turbulent flow become

$$\frac{du}{dt} = -\frac{1}{\rho}\frac{\partial p}{\partial x} + \frac{1}{\rho}\frac{\partial}{\partial y}\left(\mu\frac{\partial u}{\partial y} - \rho\overline{u'v'}\right) + \frac{1}{\rho}\frac{\partial}{\partial z}\left(\mu\frac{\partial u}{\partial z} - \rho\overline{u'w'}\right) \tag{10.1.4}$$

$$\frac{dv}{dt} = -\frac{1}{\rho}\frac{\partial p}{\partial y} + \frac{1}{\rho}\frac{\partial}{\partial y}\left(\mu\frac{\partial v}{\partial y} - \rho\overline{v'^2}\right) + \frac{1}{\rho}\frac{\partial}{\partial z}\left(\mu\frac{\partial v}{\partial z} - \rho\overline{v'w'}\right) \tag{10.1.5}$$

$$\frac{dw}{dt} = -\frac{1}{\rho}\frac{\partial p}{\partial z} + \frac{1}{\rho}\frac{\partial}{\partial y}\left(\mu\frac{\partial w}{\partial y} - \rho\overline{v'w'}\right) + \frac{1}{\rho}\frac{\partial}{\partial z}\left(\mu\frac{\partial w}{\partial z} - \rho\overline{w'^2}\right) \tag{10.1.6}$$

where μ and ρ have been left inside the derivatives merely for clarity. The continuity equation is Eq. (2.2.2) as usual. The region of influence of a point P is the half space $x > x_p$, the equations are parabolic. If the method of lines (Sec. 4.3.4) were applied to this flow, elliptic PDEs in the yz-plane would result, and whatever solution technique is used, difficulty is experienced in specifying boundary conditions at large y and z. In the case of flow in a rectangular duct of height h_y and width h_z, symmetrical boundary conditions can be applied, equating z-wise derivatives to zero on the plane $z = \frac{1}{2}h_z$ and, similarly, y-wise derivatives on the plane $y = \frac{1}{2}h_y$. However, in more general problems like the wing-body junction, w at large z depends on the displacement effect of the shear layer at small z, so we are faced with the typical NS difficulty of boundary conditions that depend on the solution.

It will be seen by comparing Eqs. (10.1.4)-(10.1.6) with Eqs. (3.2.1)-(3.2.3) that the boundary-region equations, which, as they are written, do not assume that w is small in comparison with u, include the boundary-sheet equations as a special case. To solve the complete flow over a wing-body combination, we can therefore use the boundary-region equations in a domain extending somewhat beyond the slender shear layers in the wing-body junction and near the wing tip and transfer to the boundary-sheet equations outside that domain.

10.1.3 Secondary Flow

The major new physical features of three-dimensional flow relate to the generation of streamwise vorticity. The reader may like to review Sec. 2.5 at this point.

The equation for the streamwise vorticity, $\xi = \partial w/\partial y - \partial v/\partial z$, in a slender shear layer, valid also in a boundary sheet, is obtained by subtracting the z-derivative of Eq. (10.1.5) from the y-derivative of Eq. (10.1.6). We obtain, after some use of the continuity equation,

$$\frac{d\xi}{dt} = \xi\frac{\partial u}{\partial x} - \frac{\partial u}{\partial y}\frac{\partial w}{\partial x} + \frac{\partial u}{\partial z}\frac{\partial v}{\partial x} + v\left(\frac{\partial^2}{\partial y^2} + \frac{\partial^2}{\partial z^2}\right)\xi + \left(\frac{\partial^2}{\partial y^2} - \frac{\partial^2}{\partial z^2}\right)(-\overline{v'w'})$$
$$+ \frac{\partial^2}{\partial y\partial z}(\overline{v'^2} - \overline{w'^2}) \tag{10.1.7}$$

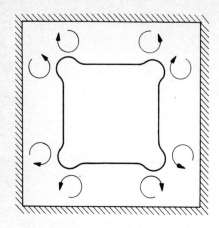

Fig. 10.3 Secondary flow in developing boundary layers in a square duct.

where the products of velocity derivatives are the remains of the "skewing" terms, $\eta \, \partial u/\partial y + \zeta \, \partial u/\partial z$. The equation given by Perkins (1970) retains the x-wise stress gradients.

In laminar flow, Eq. (10.1.7) represents the net effect of viscous diffusion of vorticity and of the stretching and skewing of vorticity by the components of the rate of strain. In nearly parallel laminar flows, such as those in the corner of a straight duct, ξ will be very small in comparison to η and ζ. In accelerating flow (large $\partial u/\partial x$), ξ will be increased by a fraction depending on the total strain, while in curved flows where either $\partial v/\partial x$ or $\partial w/\partial x$ is large, ξ is increased greatly and may approach the order of η or ζ if the flow turns through a large angle.

In turbulent flow the skewing and stretching mechanisms are both present, but the Reynolds-stress terms can actually produce streamwise vorticity instead of merely diffusing it as viscous stresses do. As in all other time-averaged equations for turbulence the integral of the Reynolds-stress terms over the whole yz-plane is zero (since fluctuations vanish at solid boundaries or at infinity), so strictly speaking, these are transport rather than source terms. However, it is usual to pretend that they are source terms that happen to have opposite signs in different parts of the flow. Figure 10.3 shows the secondary flow induced by the Reynolds stresses in a boundary layer in the corner of a straight square duct; the strengths of the two vortices in each corner are equal and opposite by symmetry. There does not seem to be any reliable and simple rule for the direction of the secondary flow, and as usual the relative size of the shear-stress and normal-stress terms depends on the axes. In symmetrical or nonsymmetrical streamwise corners, whether they are concave like a wing-body junction or convex like a wing tip, the secondary flow near the solid surface is away from the corner, while that near the bisector is toward the corner.

The identification of stress-induced secondary flow is due to Prandtl who inferred its presence from the velocity contours measured in noncircular ducts by Nikuradse (see Schlichting, 1968, Chap. 20). Later, Prandtl referred to skew-induced secondary flow as "secondary flow of the first kind" and to stress-induced secondary flow as "secondary flow of the second kind." The names are still in use, but here we prefer the mnemonic terms "skew-induced" and "stress-induced," respectively.

In a boundary sheet the streamwise (strictly speaking, x-component) vorticity is, to a good approximation, equal to $\partial w/\partial y$, and Eq. (10.1.7) reduces, to the same approximation, to

$$\frac{d}{dt}\xi = \frac{\partial w}{\partial y}\frac{\partial u}{\partial x} - \frac{\partial u}{\partial y}\frac{\partial w}{\partial x} + \nu\frac{\partial^2 \xi}{\partial y^2} + \frac{\partial^2}{\partial y^2}(-\overline{v'w'}) \qquad (10.1.8)$$

which adds little to the z-component equation, of which it is simply the y-derivative. However, it is sometimes convenient to discuss the cross flow, $\partial w/\partial y$, in secondary-flow terms, particularly when the effect of streamline curvature in the plane of the boundary sheet, which is related to $\partial w/\partial x$, is considered. From the point of view of the external flow a boundary layer is a vortex sheet whose strength has components $(w, -u)$ in the x- and z-directions.

10.1.4 Separation of Three–Dimensional Boundary Layers

Lateral convergence of streamlines in a shear layer, shown in Fig. 10.4, implies nonzero values of $\partial w/\partial x$ having opposite signs on opposite sides of the centerline. The consequent generation of streamwise vorticity can lead to the formation of a pair of vortices that expel fluid from the plane of the shear layer. In the case of a boundary layer this is a kind of separation, although the streamwise component of skin friction remains positive and the mass-flow deficit in the viscous region remains

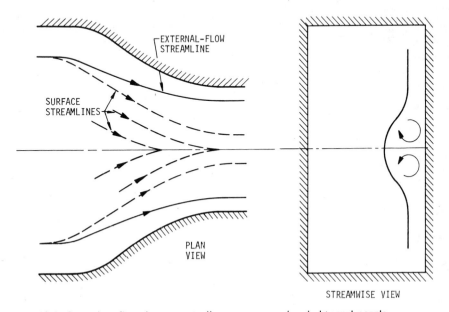

Fig. 10.4 Secondary flow due to streamline convergence in wind-tunnel nozzle.

small. Examples occur in wind-tunnel contractions of rectangular cross section and moderate area ratio, where the streamwise vorticity is augmented by the "stretching" term $\xi \, \partial u/\partial x$ (Fig. 10.4). Note the way in which the surface streamlines converge to the centerline in plan view. Thereafter, they leave the surface along the centerline in the streamwise view. The centerline in plan view is a "ridge line" of separation, so called by analogy with the appearance of map contour lines. A spectacular flow of the same type was observed in an S-bend engine intake by Bansod and Bradshaw (1972), where the rate of growth of boundary-layer thickness in the center plane reached 0.25. A related type of three-dimensional flow separation occurs from bodies like airplane fuselages at incidence. In the cross-sectional plane the flow resembles that in cross flow over an accelerating two-dimensional body (a result that can be made qualitatively useful with the aid of slender-body theory), and an incipient vortex street is fed by secondary vorticity generated upstream. There is not necessarily any point on the body, other than the ordinary stagnation points at the nose and tail, at which the skin friction is zero. Separation, as it was defined above, occurs at points where the cross-flow component of skin friction is zero, as can be seen by the "thought experiment" of increasing the body incidence to 90°.

Lateral divergence of streamlines results in intensification of the lateral (z) component of vorticity. This effect occurs in axisymmetric flow with $dr_0/dx > 0$ but is such a common feature of three-dimensional flow, it merits discussion here. A classical example is the flow around a bluff obstacle in a boundary layer, typified by the frontispiece of Thwaites' book (Thwaites, 1960) and sketched in Fig. 10.5. The flow in the center plane has $w = 0$ by symmetry but cannot usefully be regarded as two-dimensional for any practical purpose. The separation that occurs there is a "saddle point" of skin friction lines (Rosenhead, 1963, Chap. 2). The z-component vorticity, initially equal to the vorticity gradient, $\partial u/\partial y$, in the oncoming two-dimensional flow because $\partial v/\partial x$ is small, is increased by the stretching action of the divergence, $\partial w/\partial z$. On the center plane, differentiation of the x- and y-component momentum equations with respect to y and x, respectively, gives, for

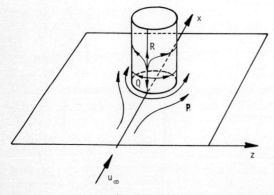

Fig. 10.5 Limiting (surface) streamlines in flow past a flat plate with attached cylinder.

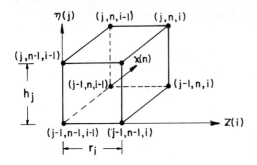

Fig. 10.6 Net cube for the difference equations for three-dimensional flows.

constant-property laminar flow,

$$\frac{d}{dt}\left(\frac{\partial v}{\partial x}\right) = \frac{\partial v}{\partial x}\frac{\partial w}{\partial z} - \frac{1}{\rho}\frac{\partial^2 p}{\partial x \partial y} + \nu\nabla^2\frac{\partial v}{\partial x} \tag{10.1.9}$$

$$\frac{d}{dt}\left(-\frac{\partial u}{\partial y}\right) = -\frac{\partial u}{\partial y}\frac{\partial w}{\partial z} + \frac{1}{\rho}\frac{\partial^2 p}{\partial x \partial y} + \nu\nabla^2\left(-\frac{\partial u}{\partial y}\right) \tag{10.1.10}$$

(Note that these equations apply only to a plane of symmetry.) This shows that a true vortex, with $\partial v/\partial x \sim -\partial u/\partial y$, can arise only by the action of pressure gradients because moderate divergence alone will not increase the order of magnitude of $\partial v/\partial x$. Moderate divergence produces a moderate fractional increase in $\partial u/\partial y$, and then the pressure term transfers part of the increase to $\partial v/\partial x$, which can thus increase to the same order of magnitude as $\partial u/\partial y$. Of course, the z-component vorticity, whose transport equation is the sum of Eqs. (10.1.9) and (10.1.10), is unaffected by pressure gradients. Pressure-driven transfer from one contribution to another does not necessarily occur; in the boundary layer on a swept wing the contribution of $\partial w/\partial y$ to the x-component of vorticity can be quite large while the other contribution, $\partial v/\partial z$, remains small. Even in this case, $\partial v/\partial z$ can rise to the order of $\partial w/\partial y$ in transition from laminar to turbulent flow, but then the transfer is effected mainly by viscous stresses.

The reader should now refer to the analysis of vorticity and separation in Chap. 2 of the book by Rosenhead (1963).

The geometry of attachment points and lines in three-dimensional flow, also discussed by Rosenhead, is similar to that of separation points or lines. Attachment geometries are usually simpler. The commonest cases are the attachment point on the nose of a fuselage and the attachment line on the leading edge of a wing (see Fig. 10.2). The point P in Fig. 10.5 is a simple attachment point with surface streamlines radiating in all directions, while the rest of the line QR is an attachment line from which surface streamlines branch away. In two-dimensional flow the surface streamlines leave the attachment line at right angles, while in three-dimensional flow they are initially tangential to it. Other special cases may occur, but we do not consider them in this book.

10.2 NUMERICAL SOLUTIONS FOR THREE-DIMENSIONAL LAMINAR FLOWS

The governing boundary-layer equations for three-dimensional flows in Cartesian coordinate systems are given by Eqs. (3.2.1)-(3.2.3) and (2.2.2). At the edge of the boundary layer, Eqs. (3.2.1) and (3.2.3) reduce to

$$u_e \frac{\partial u_e}{\partial x} + w_e \frac{\partial u_e}{\partial z} = -\frac{1}{\rho} \frac{\partial p}{\partial x} \tag{10.2.1}$$

$$u_e \frac{\partial w_e}{\partial x} + w_e \frac{\partial w_e}{\partial z} = -\frac{1}{\rho} \frac{\partial p}{\partial z} \tag{10.2.2}$$

The solution of the system given by Eqs. (3.2.1)-(3.2.3), (2.2.2), (10.2.1), and (10.2.2), subject to a set of given boundary conditions, say,

$$y = 0 \quad u, w = 0 \quad v = 0 \tag{10.2.3a}$$

$$y = \delta \quad u = u_e(x, z) \quad w = w_e(x, z) \tag{10.2.3b}$$

requires closure assumptions for the Reynolds stresses appearing in these equations. This solution will be discussed later in Sec. 10.3. They also require initial conditions on two intersecting planes. In some problems, such as the bluff obstacle flow mentioned at the end of Sec. 10.1, we can take advantage of the symmetry conditions and obtain one of the initial conditions by noting that on the line of symmetry the flow is two-dimensional except for the cross-flow derivatives. The flow in that direction is usually referred to as the *attachment-line* flow, because the attachment line is a streamline on the body on which both the cross-flow velocity in the boundary layer, namely, w (see Fig. 10.5) and the cross-flow pressure gradient $(\partial p/\partial z)$ are identically zero, causing the z-momentum equation to be singular on that line. However, differentiation with respect to z yields a nonsingular equation. After performing the necessary differentiation for the z-momentum equation and taking advantage of symmetry conditions, $\partial u/\partial z = \partial v/\partial z = \partial^2 w/\partial z^2 = 0$, we can write the governing attachment-line-flow equations for three-dimensional flows in Cartesian coordinates as

$$u \frac{\partial u}{\partial x} + v \frac{\partial u}{\partial y} = u_e \frac{\partial u_e}{\partial x} + \frac{1}{\rho} \frac{\partial}{\partial y} \left(\mu \frac{\partial u}{\partial y} - \rho \overline{u'v'} \right) \tag{10.2.4}$$

$$u \frac{\partial w_z}{\partial x} + v \frac{\partial w_z}{\partial y} + w_z^2 = u_e \frac{\partial}{\partial x}(w_z)_e + (w_z)_e^2 + \frac{1}{\rho} \frac{\partial}{\partial y} \left[\mu \frac{\partial w_z}{\partial y} - \rho (\overline{v'w'})_z \right] \tag{10.2.5}$$

$$\frac{\partial u}{\partial x} + w_z + \frac{\partial v}{\partial y} = 0, \tag{10.2.6}$$

where $w_z = \partial w/\partial z$ and $(\overline{v'w'})_z = \partial \overline{v'w'}/\partial z$.

For these equations the boundary conditions [Eq. (10.2.3)] become

$$y = 0 \quad u, w_z = 0 \quad v = 0 \tag{10.2.7a}$$

$$y = \delta \quad u = u_e(x,z) \quad w_z = (w_z)_e \equiv w_{ze} \tag{10.2.7b}$$

10.2.1 Transformation of the Equations

As in two-dimensional flows, Eqs. (3.2.1), (3.2.3), and (2.2.2) subject to the boundary conditions [Eq. (10.2.3)] can be solved when they are expressed either in physical coordinates or in transformed coordinates. The transformed coordinates allow large steps to be taken in the x- and z-directions because the profiles expressed in the transformed coordinates do not change rapidly as they do when they are expressed in physical coordinates. The use of transformed variables stretches the coordinate normal to the flow and takes out much of the variation in boundary-layer thickness for laminar flows. In addition, they remove the singularity that the equations in physical coordinates have at $x = 0$ and $z = 0$. For these reasons in three-dimensional flows the use of transformed variables is preferable to the use of physical variables.

There are several transformations that can be used for three-dimensional flows expressed in orthogonal or even nonorthogonal coordinates, see, for example, Cebeci et al. (1973). Here we use a transformation similar to the one used for two-dimensional incompressible flows (see Sec. 5.3). We define the transformed coordinates by

$$x = x \quad z = z \quad \eta = \left(\frac{u_e}{\nu x}\right)^{1/2} y \tag{10.2.8}$$

and introduce a two-component vector potential such that

$$u = \frac{\partial \psi}{\partial y} \quad w = \frac{\partial \phi}{\partial y} \quad v = -\left(\frac{\partial \psi}{\partial x} + \frac{\partial \phi}{\partial z}\right) \tag{10.2.9}$$

In addition, we define dimensionless ψ and ϕ by

$$\psi = (u_e \nu x)^{1/2} f(x,z,\eta) \tag{10.2.10a}$$

$$\phi = \left(\frac{\nu x}{u_e}\right)^{1/2} w_e g(x,z,\eta) \tag{10.2.10b}$$

Then with primes denoting differentiation with respect to η, we get

$$u = \frac{\partial \psi}{\partial y} = \frac{\partial \psi}{\partial \eta}\frac{\partial \eta}{\partial y} = f' u_e$$

$$w = \frac{\partial \phi}{\partial y} = \frac{\partial \phi}{\partial \eta}\frac{\partial \eta}{\partial y} = g' w_e$$

$$\frac{\partial u}{\partial x} = \frac{\partial}{\partial x}(u_e f') + \frac{\partial}{\partial \eta}(u_e f')\frac{\partial \eta}{\partial x} = u_e\left(\frac{\partial f'}{\partial x} + \frac{m_2}{x}f' + f''\frac{\partial \eta}{\partial x}\right)$$

$$\frac{\partial \psi}{\partial x} = \left(\frac{\nu x}{u_e}\right)^{1/2} u_e\left(\frac{\partial f}{\partial x} + m_1\frac{f}{x} + f'\frac{\partial \eta}{\partial x}\right)$$

$$\frac{\partial \phi}{\partial z} = \left(\frac{\nu x}{u_e}\right)^{1/2} w_e\left[\left(m_3 - \frac{m_5}{2}\right)\frac{u_e}{w_e}\frac{g}{x} + g'\frac{\partial \eta}{\partial z} + \frac{\partial g}{\partial z}\right]$$

$$v = -\left(\frac{\nu x}{u_e}\right)^{1/2} u_e\left\{\frac{\partial f}{\partial x} + m_1\frac{f}{x} + f'\frac{\partial \eta}{\partial x} + \frac{w_e}{u_e}\left[\frac{\partial g}{\partial z} + \left(m_3 - \frac{m_5}{2}\right)\frac{u_e}{w_e}\frac{g}{x} + g'\frac{\partial \eta}{\partial z}\right]\right\}$$

$$\frac{\partial u}{\partial y} = \frac{\partial u}{\partial \eta}\frac{\partial \eta}{\partial y} = u_e f''\left(\frac{u_e}{\nu x}\right)^{1/2} \tag{10.2.11a}$$

$$\frac{\partial u}{\partial z} = \frac{\partial u}{\partial z} + \frac{\partial u}{\partial \eta}\frac{\partial \eta}{\partial z} = u_e\left(\frac{m_5}{x}\frac{u_e}{w_e}f' + \frac{\partial f'}{\partial z} + f''\frac{\partial \eta}{\partial z}\right)$$

$$-\frac{1}{\rho}\frac{\partial p}{\partial x} = \frac{x}{u_e}\frac{\partial u_e}{\partial x}\left(\frac{u_e^2}{x}\right) + \frac{x}{u_e}\frac{\partial u_e}{\partial z}\left(\frac{u_e w_e}{x}\right) = m_2\frac{u_e^2}{x} + \frac{m_5}{x}u_e^2$$

Here

$$f' = \frac{u}{u_e} \qquad g' = \frac{w}{w_e}$$

$$m_2 = \frac{x}{u_e}\frac{\partial u_e}{\partial x} \qquad m_1 = \frac{m_2 + 1}{2} \qquad m_3 = \frac{x}{u_e}\frac{\partial w_e}{\partial z} \tag{10.2.11b}$$

$$m_4 = \frac{x}{w_e}\frac{\partial w_e}{\partial x} \qquad m_5 = \frac{w_e}{u_e}\frac{x}{u_e}\frac{\partial u_e}{\partial z}$$

Substituting these relations into Eq. (3.2.1), using Eq. (10.2.1), and assuming laminar flow so that the Reynolds stress, $-\rho\overline{u'v'}$, is zero, we get, after some rearranging,

$$(bf'')' + m_1 ff'' + m_2[1 - (f')^2] + m_5(1 - f'g') + m_6 gf''$$

$$= x\left[f'\frac{\partial f'}{\partial x} - f''\frac{\partial f}{\partial x} + m_7\left(g'\frac{\partial f'}{\partial z} - f''\frac{\partial g}{\partial z}\right)\right] \tag{10.2.12}$$

Here $b = 1$, $m_6 = m_3 - m_5/2$ and $m_7 = w_e/u_e$. By following a similar procedure, Eq. (3.2.3) can also be expressed in transformed coordinates by the following equation:

$$(bg'')' + m_1 fg'' + m_4(1 - f'g') + m_3[1 - (g')^2] + m_6 gg''$$
$$= x\left[f'\frac{\partial g'}{\partial x} - g''\frac{\partial f}{\partial x} + m_7\left(g'\frac{\partial g'}{\partial z} - g''\frac{\partial g}{\partial z}\right)\right] \quad (10.2.13)$$

The boundary conditions [Eq. (10.2.3)] become

$$\eta = 0 \quad f = g = 0 \quad f' = g' = 0 \quad\quad\quad\quad\quad (10.2.14a)$$
$$\eta = \eta_\infty \quad f' = g' = 1 \quad\quad\quad\quad\quad\quad\quad\quad (10.2.14b)$$

The attachment-line equations (10.2.4)–(10.2.7) can also be transformed by a similar procedure. This time, we define the two-component vector potential by

$$u = \frac{\partial \psi}{\partial y} \quad w_z = \frac{\partial \phi}{\partial y} \quad v = -\left(\frac{\partial \psi}{\partial x} + \phi\right) \quad\quad (10.2.15)$$

and again use the expressions given by Eq. (10.2.10) except that now we define ϕ by

$$\phi = \left(\frac{\nu x}{u_e}\right)^{1/2} w_{ze} g(x, z, \eta) \quad\quad\quad\quad\quad (10.2.16)$$

Introducing the expressions (10.2.8), (10.2.10a), (10.2.15), and (10.2.16) into Eqs. (10.2.4) and (10.2.5), and neglecting the Reynolds stresses, we get

$$(bf'')' + m_1 ff'' + m_2[1 - (f')^2] + m_3 gf'' = x\left(f'\frac{\partial f'}{\partial x} - f''\frac{\partial f}{\partial x}\right) \quad (10.2.17)$$

$$(bg'')' + m_1 fg'' + m_4(1 - f'g') + m_3[1 - (g')^2] + m_3 gg'' = x\left(f'\frac{\partial g'}{\partial x} - g''\frac{\partial f}{\partial x}\right) \quad (10.2.18)$$

Here the definitions of the terms are the same as before except for g' and m_4:

$$g' = \frac{w_z}{w_{ze}} \quad m_4 = \frac{x}{w_{ze}}\frac{\partial w_{ze}}{\partial x} \quad\quad\quad\quad (10.2.19)$$

The boundary conditions [Eq. (10.2.7)] become

$$\eta = 0 \quad f = g = 0 \quad f' = g' = 0 \quad\quad\quad\quad\quad (10.2.20a)$$
$$\eta = \eta_\infty \quad f' = g' = 1 \quad\quad\quad\quad\quad\quad\quad\quad (10.2.20b)$$

10.2.2 Numerical Formulation of the Attachment-Line Equations

Numerical formulation of the attachment-line equations is similar to the numerical formulation of the x-momentum equation discussed in Sec. 7.2, except that we now have two equations rather than one. Again we write them in terms of a first-order system of PDEs. We introduce new dependent variables, $u(x, \eta)$, $v(x, \eta)$, $w(x, \eta)$, and $t(x, \eta)$, so that Eqs. (10.2.17) and (10.2.18) can be written as

$$f' = u \tag{10.2.21a}$$

$$u' = v \tag{10.2.21b}$$

$$g' = w \tag{10.2.21c}$$

$$w' = t \tag{10.2.21d}$$

$$(bv)' + m_1 fv + m_2(1 - u^2) + m_3 gv = x\left(u\frac{\partial u}{\partial x} - v\frac{\partial f}{\partial x}\right) \tag{10.2.21e}$$

$$(bt)' + m_1 ft + m_4(1 - uw) + m_3(1 - w^2) + m_3 gt = x\left(u\frac{\partial w}{\partial x} - t\frac{\partial f}{\partial x}\right) \tag{10.2.21f}$$

In terms of the new variables, the boundary conditions [Eq. (10.2.20)] become

$$f(x,0) = 0 \quad g(x,0) = 0 \quad u(x,0) = 0 \quad w(x,0) = 0 \tag{10.2.22a}$$

$$u(x,\eta_\infty) = 1 \quad w(x,\eta_\infty) = 1 \tag{10.2.22b}$$

As before, we write the difference equations by considering one mesh rectangle as shown in Fig. 7.1. We approximate Eqs. (10.2.21a)–(10.2.21d), using centered difference quotients, and average about the midpoint $(x^n, \eta_{j-1/2})$ of the segment $P_1 P_2$:

$$\frac{f_j^n - f_{j-1}^n}{h_j} = u_{j-1/2}^n \tag{10.2.23a}$$

$$\frac{u_j^n - u_{j-1}^n}{h_j} = v_{j-1/2}^n \tag{10.2.23b}$$

$$\frac{g_j^n - g_{j-1}^n}{h_j} = w_{j-1/2}^n \tag{10.2.23c}$$

$$\frac{w_j^n - w_{j-1}^n}{h_j} = t_{j-1/2}^n \tag{10.2.23d}$$

Similarly, Eqs. (10.2.21e) and (10.2.21f) are approximated by centering about the midpoint $(x^{n-1/2}, \eta_{j-1/2})$ of the rectangle $P_1 P_2 P_3 P_4$. This gives

$$h_j^{-1}(b_j^n v_j^n - b_{j-1}^n v_{j-1}^n) + \alpha_1(fv)_{j-1/2}^n - \alpha_2(u^2)_{j-1/2}^n + m_3^n(gv)_{j-1/2}^n$$
$$+ \alpha(v_{j-1/2}^{n-1} f_{j-1/2}^n - f_{j-1/2}^{n-1} v_{j-1/2}^n) = R_{j-1/2}^{n-1} \tag{10.2.23e}$$

$$h_j^{-1}(b_j^n t_j^n - b_{j-1}^n t_{j-1}^n) + \alpha_1(ft)_{j-1/2}^n - \alpha_4(uw)_{j-1/2}^n$$
$$- m_3^n[(w)_{j-1/2}^2 - (gt)_{j-1/2}^n]$$
$$- \alpha(u_{j-1/2}^{n-1} w_{j-1/2}^n - w_{j-1/2}^{n-1} u_{j-1/2}^n - t_{j-1/2}^{n-1} f_j^n{}_{-1/2} + f_{j-1/2}^{n-1} t_{j-1/2}^n)$$
$$= T_{j-1/2}^{n-1} \quad (10.2.23f)$$

where

$$\alpha = \frac{x^{n-1/2}}{k_n} \qquad \alpha_1 = m_1^n + \alpha \qquad \alpha_2 = m_2^n + \alpha \qquad \alpha_4 = m_4^n + \alpha \qquad (10.2.24a)$$

$$R_{j-1/2}^{n-1} = \alpha[-(u^2)_{j-1/2}^{n-1} + (fv)_{j-1/2}^{n-1}] - \{h_j^{-1}(b_j^{n-1} v_j^{n-1} - b_{j-1}^{n-1} v_{j-1}^{n-1})$$
$$+ m_1^{n-1}(fv)_{j-1/2}^{n-1} + m_2^{n-1}[1 - (u^2)_{j-1/2}^{n-1}] + m_3^{n-1}(gv)_{j-1/2}^{n-1}\} - m_2^n \quad (10.2.24b)$$

$$T_{j-1/2}^{n-1} = \alpha[-(uw)_{j-1/2}^{n-1} + (ft)_{j-1/2}^{n-1}] - \{h_j^{-1}(b_j^{n-1} t_j^{n-1} - b_{j-1}^{n-1} t_{j-1}^{n-1})$$
$$+ m_1^{n-1}(ft)_{j-1/2}^{n-1} + m_4^{n-1}[1 - (uw)_{j-1/2}^{n-1}]$$
$$+ m_3^{n-1}[1 - (w^2)_{j-1/2}^{n-1}] + m_3^{n-1}(gt)_{j-1/2}^{n-1}\} - m_3^n - m_4^n \quad (10.2.24c)$$

The boundary conditions [Eq. (10.2.22)] yield, at $x = x^n$,

$$f_0^n = 0 \qquad g_0^n = 0 \qquad u_0^n = 0 \qquad w_0^n = 0 \qquad u_J^n = 1 \qquad w_J^n = 1 \qquad (10.2.25)$$

Equations (10.2.23) are imposed for $j = 1, 2, \ldots, J$. If we assume f_j^{n-1}, u_j^{n-1}, v_j^{n-1}, g_j^{n-1}, w_j^{n-1}, t_j^{n-1} to be known for $0 \leqslant j \leqslant J$, then Eqs. (10.2.23) for $1 \leqslant j \leqslant J$ and the boundary conditions [Eq. (10.2.25)] yield an implicit nonlinear algebraic system of $6J + 6$ equations in as many unknowns ($f_j^n, u_j^n, v_j^n, g_j^n, w_j^n, t_j^n$). This system can be linearized by using Newton's method, discussed in Sec. 7.2.1, and the resulting linear system can be solved by using the block-elimination method discussed in Appendix 7A. Of course, this procedure requires an algorithm similar to the one developed for the third-order system, namely, SOLV3 (see Sec. 7A.2). However, one can save some work at the expense of a less efficient procedure by solving each momentum equation separately by making slight modifications in SOLV3. For example, we can first solve the system Eqs. (10.2.23a), (10.2.23b), and (10.2.23e) for the boundary conditions

$$f_0^n = u_0^n = 0 \qquad u_J^n = 1 \qquad (10.2.26)$$

by defining new coefficients for $(s_1)_j$ to $(s_6)_j$ together with a new definition for $(r_2)_j$ given in Eqs. (7.2.14). Then we can solve the system Eqs. (10.2.23c), (10.2.23d), and (10.2.23f) for the boundary conditions

$$g_0^n = w_0^n = 0 \qquad w_J^n = 1 \qquad (10.2.27)$$

by defining new coefficients for $(s_1)_j$ to $(s_6)_j$ together with a definition for $(r_2)_j$.

We solve each system once and repeat the calculations until the convergence criterion given by Eq. (7A.1.21) is satisfied.

10.2.3 Numerical Formulation of the Three–Dimensional TSL Equations

We write Eqs. (10.2.12) and (10.2.13) as a first-order system. With the new dependent variables defined by Eqs. (10.2.21a)–(10.2.21d) we can write Eqs. (10.2.12) and (10.2.13) as

$$(bv)' + m_1 fv + m_2(1 - u^2) + m_5(1 - uw) + m_6 gv$$

$$= x\left[u\frac{\partial u}{\partial x} - v\frac{\partial f}{\partial x} + m_7\left(w\frac{\partial u}{\partial z} - v\frac{\partial g}{\partial z}\right)\right] \quad (10.2.28a)$$

$$(bt)' + m_1 ft + m_4(1 - uw) + m_3(1 - w^2) + m_6 gt$$

$$= x\left[u\frac{\partial w}{\partial x} - t\frac{\partial f}{\partial x} + m_7\left(w\frac{\partial w}{\partial z} - t\frac{\partial g}{\partial z}\right)\right] \quad (10.2.28b)$$

We now consider the net cube shown in Fig. 10.6 and introduce the new net points by

$$z_0 = 0 \quad z^i = z^{i-1} + r_i \quad i = 1, 2, \dots, I \quad (10.2.29)$$

in addition to the ones introduced in Eq. (7.2.3).

The difference equations that are to approximate Eqs. (10.2.21a)–(10.2.21d) are obtained by averaging about the midpoint $(x^n, z^i, \eta_{j-1/2})$

$$\frac{f_j^{n,i} - f_{j-1}^{n,i}}{h_j} = u_{j-1/2}^{n,i} \quad (10.2.30a)$$

$$\frac{u_j^{n,i} - u_{j-1}^{n,i}}{h_j} = v_{j-1/2}^{n,i} \quad (10.2.30b)$$

$$\frac{g_j^{n,i} - g_{j-1/2}^{n,i}}{h_j} = w_{j-1/2}^{n,i} \quad (10.2.30c)$$

$$\frac{w_j^{n,i} - w_{j-1}^{n,i}}{h_j} = t_{j-1/2}^{n,i} \quad (10.2.30d)$$

where, for example,

$$u_{j-1/2}^{n,i} = \tfrac{1}{2}(u_j^{n,i} + u_{j-1}^{n,i})$$

The difference equations that are to approximate Eqs. (10.2.28a) and (10.2.28b) are rather lengthy. To illustrate the difference equations similar to Eqs.

(10.2.28a) and (10.2.28b), we consider the following model equation:

$$v' + m_1 fv = x\left(u\frac{\partial u}{\partial x} + m_7 w\frac{\partial u}{\partial z}\right) \tag{10.2.31}$$

The difference equations for this equation are

$$h_j^{-1}(\bar{v}_j - \bar{v}_{j-1}) + (m_1)_{i-1/2}^{n-1/2}(\overline{fv})_{j-1/2} = x^{n-1/2}\left[\bar{u}_{j-1/2}\left(\frac{\bar{u}_n - \bar{u}_{n-1}}{k_n}\right)\right.$$

$$\left. + (m_7)_{i-1/2}^{n-1/2}\bar{w}_{j-1/2}\left(\frac{\bar{u}_i - \bar{u}_{i-1}}{r_i}\right)\right] \tag{10.2.32}$$

where, for example,

$$\bar{v}_j = \tfrac{1}{4}(v_j^{n,i} + v_j^{n,i-1} + v_j^{n-1,i-1} + v_j^{n-1,i}) = \tfrac{1}{4}(v_j^{n,i} + v_j^{234}) \tag{10.2.33a}$$

$$\bar{u}_n = \tfrac{1}{4}(u_j^{n,i} + u_j^{n,i-1} + u_{j-1}^{n,i} + u_{j-1}^{n,i-1}) = \tfrac{1}{2}(u_{j-1/2}^{n,i} + u_{j-1/2}^{n,i-1}) \tag{10.2.33b}$$

$$\bar{u}_i = \tfrac{1}{4}(u_j^{n,i} + u_j^{n-1,i} + u_{j-1}^{n,i} + u_{j-1}^{n-1,i}) = \tfrac{1}{2}(u_{j-1/2}^{n,i} + u_{j-1/2}^{n-1,i}) \tag{10.2.33c}$$

$$(m_1)_{i-1/2}^{n-1/2} = \tfrac{1}{4}(m_1^{n,i} + m_1^{n,i-1} + m_1^{n-1,i} + m_1^{n-1,i-1}) \tag{10.2.33d}$$

Here by v_j^{234} we mean $v_j^{234} = v_j^{n,i-1} + v_j^{n-1,i-1} + v_j^{n-1,i}$, the sum of the values of v_j at three of the four corners of the face of the box. The boundary conditions [Eq. (10.2.22)] yield, at $x = x^n$ and at $z = z^i$,

$$f_0^{n,i} = 0 \quad g_j^{n,i} = 0 \quad u_0^{n,i} = 0 \quad w_0^{n,i} = 0 \quad u_J^{n,i} = 1 \quad w_J^{n,i} = 1 \tag{10.2.34}$$

If we assume $(f_j^{n-1,i-1}, u_j^{n-1,i-1}, v_j^{n-1,i-1}, g_j^{n-1,i-1}, w_j^{n-1,i-1}, t_j^{n-1,i-1})$, $(f_j^{n,i-1}, u_j^{n,i-1}, v_j^{n,i-1}, g_j^{n,i-1}, w_j^{n,i-1}, t_j^{n,i-1})$, and $(f_j^{n-1,i}, u_j^{n-1,i}, v_j^{n-1,i}, g_j^{n-1,i}, w_j^{n-1,i}, t_j^{n-1,i})$ to be known for $0 \leqslant j \leqslant J$, then the differenced equations for the general case can be solved by the procedure described for the attachment-line equations.

10.2.4 Sample Calculation

To illustrate the numerical method discussed in the previous sections, we present a sample calculation; we consider a three-dimensional laminar flow past a flat plate with attached cylinder (see Fig. 10.5). For this flow the inviscid-velocity distribution is given by

$$u_e = u_\infty\left(1 + a^2\frac{\Delta_2}{\Delta_1^2}\right) \quad w_e = -2u_\infty a^2\frac{\Delta_3}{\Delta_1^2} \tag{10.2.35}$$

where

$$\Delta_1 = (x - x_0)^2 + z^2 \qquad \Delta_2 = -(x - x_0)^2 + z^2 \qquad \Delta_3 = (x - x_0)z \qquad (10.2.36)$$

Here u_∞ is a reference velocity, a is the cylinder radius, and x_0 denotes the distance of the cylinder axis from the leading edge, $x = 0$. We consider the flow on the plate, upstream of the separation point, P, in Fig. 10.5.

To find the dimensionless pressure gradients, m_1 to m_6, defined in Eq. (10.2.11), we differentiate Eq. (10.2.35) with respect to x and z to get

$$\frac{\partial u_e}{\partial x} = -2(x - x_0)u_\infty a^2 \left(\frac{\Delta_1 + 2\Delta_2}{\Delta_1^3} \right) \qquad \frac{\partial u_e}{\partial z} = 2zu_\infty a^2 \left(\frac{\Delta_1 - 2\Delta_2}{\Delta_1^3} \right)$$

$$\frac{\partial w_e}{\partial x} = -2u_\infty a^2 \left[\frac{-3(x - x_0)^2 z + z^3}{\Delta_1^3} \right] \qquad \frac{\partial w_e}{\partial z} = -2u_\infty a^2 \left[\frac{\Delta_1(x - x_0) - 4z\Delta_3}{\Delta_1^3} \right]$$

$$(10.2.37)$$

These relations can also be used to find the dimensionless pressure gradients m_i ($i = 1$-4) in Eqs. (10.2.17) and (10.2.18) except that now we need to know $\partial w_{ze}/\partial x$ to calculate m_4. It is given by

$$\frac{\partial w_{ze}}{\partial x} = \frac{6}{\Delta_1^2} u_\infty a^2 \qquad (10.2.38)$$

The calculated x- and z-component velocity profiles at $z = 0$ with $u_\infty = 3050$ cm s^{-1}, $a = 6.1$ cm, and $x_0 = 45.7$ cm are shown in Fig. 10.7. These calculations start at $z = 0$ and $x = 0$ and proceed for various values of x up to, say, $x = x_{final}$. Then for another value of z, say $z = 2$, they start at $x = 0$ and continue for the same values of x at $z = 0$ up to $x = x_{final}$. This procedure is repeated for various values of x for a fixed z_i. Note that in Fig. 10.7 at $x = 0$ the initial streamwise and cross-flow profiles are those given by Blasius flow. As the calculations proceed in the x-direction for a given z-station, the velocity profiles change considerably under the influence of the pressure gradient imposed by the cylinder.

10.3 TURBULENCE MODELS FOR THREE-DIMENSIONAL FLOWS

10.3.1 Boundary Sheets

The x- and z-component momentum equations for a turbulent boundary sheet with $d\delta/dx$ and $d\delta/dz$ both small in comparison with unity, Eqs. (3.2.1) and (3.2.3), contain only two Reynolds stresses. They are the usual shear stress acting in the x-direction on a plane parallel to the xz-plane, $-\rho\overline{u'v'}$, and the shear stress acting in the z-direction on the same plane, $-\rho\overline{w'v'}$, usually written $-\rho\overline{v'w'}$. Almost all workers have inferred, from the fact that the choice of direction of the axes in the xz-plane is arbitrary, that the assumptions made for $\rho\overline{v'w'}$ should be closely

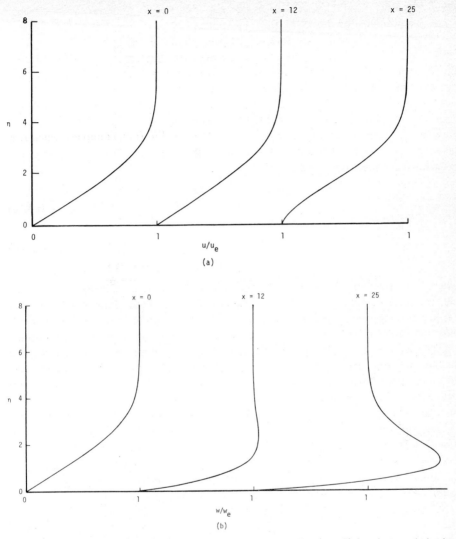

Fig. 10.7 The computed velocity profiles on the attachment line ($z = 0$) for the sample laminar flow problem of Sec. 10.2.4. (a) Streamwise velocity profiles. (b) z-component velocity profiles.

analogous to those made for $\rho\overline{u'v'}$. Mathematically, they assume that the turbulence-model equation for $\rho\overline{v'w'}$ should be obtainable from that for $\rho\overline{u'v'}$ by cyclic interchange of symbols. However, it is not so obvious that the equations for $\rho\overline{u'v'}$ can be simply derived from the models used for two-dimensional flow (Chap. 6). The argument commonly used, explicitly or implicitly, is that turbulence, being instantaneously three-dimensional, should not be seriously affected by moderate three-dimensionality of the mean flow. There is, of course, a loss of symmetry—for instance, $\overline{u'w'}$ is zero in two-dimensional flow but not in three-dimensional flow—and Rotta (1977) has recently shown that the asymmetry can noticeably affect the modeling of the shear-stress equations.

The law of the wall [Eq. (6.2.1)] and the mixing-length formula [Eq. (6.2.4)] with $f = 1$ are the foundations of most methods for two-dimensional flow. Clearly, Eq. (6.2.1) requires modification, since the velocity now has an extra component, but the "local equilibrium" arguments leading to Eq. (6.2.4) suggest that it should still be valid in three-dimensional flow if the x-axis is taken to coincide with the direction of the shear stress at height y. The assumption of local equilibrium between the magnitudes of τ and $\partial u/\partial y$ that leads to Eq. (6.2.4) implies, when it is taken at face value, that there should be local equilibrium, that is, coincidence, between their directions. This leads to

$$\frac{\partial u}{\partial y} = \frac{-\overline{u'w'}}{(\tau/\rho)^{1/2}\kappa y} \tag{10.3.1a}$$

$$\frac{\partial w}{\partial y} = \frac{-\overline{v'w'}}{(\tau/\rho)^{1/2}\kappa y} \tag{10.3.1b}$$

where $\tau = [(\rho\overline{u'v'})^2 + (\rho\overline{v'w'})^2]^{1/2}$. The argument is not, of course, very convincing—the local equilibrium concept is an approximation whose limits of validity need further investigation by experiment. Experiments in three-dimensional flow, particularly measurements of $\overline{v'w'}$, are difficult, and there is evidence both for and against Eq. (10.3.1). A safe position to take is that local-equilibrium concepts are not likely to fail catastrophically as soon as the mean flow becomes slightly three-dimensional, and indeed the calculation methods that use Eq. (10.3.1) seem to agree acceptably with most of the experimental data not too near separation.

An undeniable difficulty in treating three-dimensional wall layers is that the viscous sublayer is *not* a local-equilibrium region; there is a transfer of turbulent energy toward the wall by the turbulent fluctuations themselves to compensate for viscous dissipation. Therefore conditions at one value of y depend on conditions at the other values of y, and although the directions of velocity gradient and of shear stress coincide at the surface (Reynolds stresses negligible) and—according to Eq. (10.3.1)—again coincide outside the sublayer, they may differ within the sublayer. As a result the direction of the *velocity* outside the sublayer may differ from that of the shear stress or velocity gradient. In practical terms the constant of integration in any velocity profile derived from Eq. (10.3.1), or the damping constant, A^+, of Sec. 6.4, will have two components. The effect will be significant only if the direction of the shear stress changes significantly across the sublayer. Since at the surface, $\partial\tau_x/\partial y = \partial p/\partial x$ and $\partial\tau_z/\partial y = \partial p/\partial z$, this will occur only if there is a significant pressure gradient normal to the wall-stress vector, as for instance in a boundary layer flowing into a lateral bend ($\partial w_e/\partial x \neq 0$). Van den Berg (1975) has proposed a dimensionally correct empirical correlation: taking the x-axis in the direction of the wall shear stress, the z-component velocity at the outer edge of the sublayer is $12u_\tau(\nu/\rho u_\tau{}^3)\,\partial p/\partial z$. If a boundary layer with $u_e\theta/\nu = 10^4$ enters a sharp lateral bend of radius 20δ, the predicted value of w/u_e at the edge of the sublayer is only 0.005, so the effect of the correction is small, except in flow around obstacles like the circular cylinder discussed in Sec. 10.1.

The outer layer, like the sublayer, is not a local-equilibrium region, and the direction of the shear stress will lag behind the direction of the velocity gradient if the latter changes with x. Several experiments have shown angles between the shear stress and velocity-gradient vectors of the same order as that of the cross-flow angle (i.e., the angle between the external velocity and the surface shear stress). However, the accuracy of prediction of the boundary layer thickness and the surface shear-stress vector does not depend critically on the shear-stress direction in the outer layer, and good agreement has been found between the available data and an extension of the eddy-viscosity formulation of Sec. 8.3, in which the velocity defect used in Eq. (8.3.7) is just taken as the magnitude of the vector $u_e - u$ at given y. The same eddy viscosity is used in

$$-\rho\overline{u'v'} = \rho\epsilon_m \frac{\partial u}{\partial y} \tag{10.3.2a}$$

and in

$$-\rho\overline{v'w'} = \rho\epsilon_m \frac{\partial w}{\partial y} \tag{10.3.2b}$$

so that the directions of shear stress and velocity gradient are equated. According to Cebeci (1974), a generalization of the eddy-viscosity formulation of Sec. 8.3 for three-dimensional boundary layers is

$$(\epsilon_m)_i = L^2 \left[\left(\frac{\partial u}{\partial y}\right)^2 + \left(\frac{\partial w}{\partial y}\right)^2 \right]^{1/2} \qquad 0 \leqslant y \leqslant y_c \tag{10.3.3a}$$

$$(\epsilon_m)_o = 0.0168 \left| \int_0^\infty [u_t - (u^2 + w^2)^{1/2}]\, dy \right| \qquad y_c \leqslant y \leqslant \delta \tag{10.3.3b}$$

Here L is given by Eq. (8.3.2a) with A defined by Eq. (8.3.2b) and N by Eq. (8.3.2c), except now

$$u_\tau = \left(\frac{\tau_t}{\rho}\right)^{1/2} \qquad \frac{\tau_t}{\rho} = \nu \left[\left(\frac{\partial u}{\partial y}\right)_w^2 + \left(\frac{\partial w}{\partial y}\right)_w^2 \right]^{1/2} \tag{10.3.4}$$

In Eq. (10.3.3b), u_t is the total edge velocity defined by

$$u_t = (u_e^2 + w_e^2)^{1/2} \tag{10.3.5}$$

Little attention, theoretical or experimental, has been paid to three-dimensional turbulent free shear layers, with the exception of axisymmetric swirling flows. Hackett and Cox (1970) performed a unique experiment on the mixing layer between two nonparallel streams of equal speed. The object was to see if the shear stress was everywhere in the plane of maximum velocity difference, as it would be in a laminar flow. Within the limits of experimental accuracy, it was, despite the presence of significant Reynolds normal-stress gradients normal to that plane.

10.3.2 Boundary Regions

Axisymmetric turbulent swirling flows, notably jets and trailing vortices, have received much attention recently. The increased mixing rate of swirling jets is valuable in many industrial processes, while the rate of decay of trailing vortices fixes the minimum safe distance between aircraft in the approach pattern. The increase of mixing rate by swirl results from centrifugal instability. The angular momentum of a swirling jet in still air decreases outward, except near the axis, and therefore fluid elements displaced in a radial direction tend to increase their displacement (Bradshaw, 1973). As a result, all components of the Reynolds stress are increased. In a trailing vortex, on the other hand, the angular momentum *increases* outward, except for a small region of decrease near the perimeter, because the angular momentum outside the vortex is nonzero and equal to the circulation around the vortex. Therefore turbulence is reduced; even aircraft trailing vortices have virtually nonturbulent cores, clearly visible in smoke photographs, surrounded by a turbulent region.

In turbulent swirling flows therefore the problems of predicting the shear-stress components, $\sigma_{xr} \equiv -\rho \overline{u'v'}$ and $\sigma_{r\theta} \equiv -\rho \overline{v'w'}$, are increased by poorly understood centrifugal effects. It is generally agreed that the apparent eddy viscosity is different for the two stress components, but at present the difference—or the equivalent difference in other kinds of turbulence models—has to be represented by an empirical function of a local swirl parameter. For a more detailed review see Bradshaw (1973).

The prediction of secondary flow in nonaxisymmetric turbulent slender shear layers is one of the most difficult problems in fluid dynamics. The x-component mean vorticity is independent of the pressure gradient, although it depends on a delicate balance of Reynolds stresses, and the appearance of second derivatives increases the demands on the turbulent model. However, the mean-velocity components in the yz-plane do depend on the pressure field, which, as was explained in Sec. 10.1.4, controls the partition of the vorticity between the two velocity gradients that contribute to it, in this case, $\partial w/\partial y$ and $\partial v/\partial z$. (In turbulent flow, second derivatives of Reynolds stress appear as well but do not explicitly transfer contributions to vorticity from one velocity gradient to the other.) Therefore a proper prediction of the secondary flow involves calculating the mean pressure field. An equation for the Laplacian of the pressure, $\nabla^2 p$, is obtainable as the sum of the x-, y-, and z-derivatives of Eqs. (10.1.4), (10.1.5), and (10.1.6), respectively; the x-derivatives of Reynolds stress resulting from Eq. (10.1.4) are probably negligible, but all the Reynolds stresses that appear in the x-component vorticity equation reappear in the pressure equation, again in second derivatives.

These various ways of looking at the secondary flow all emphasize the need for an accurate turbulence model. Unfortunately, none exists; an isotropic eddy-viscosity model cannot predict secondary flow, and an empirical extension of the mixing-length concept by Launder and Ying (1972) was too closely linked to its prototype problem, the square duct, to be of general use. We may expect that transport equation models, nominally capable of predicting all the Reynolds stresses, will soon be applied to secondary flow calculations. For a pioneering attempt see Naot et al. (1974), who solved transport equations for Reynolds stresses but used an arbitrary distribution of length scale.

10.4 NUMERICAL SOLUTIONS FOR THREE-DIMENSIONAL TURBULENT FLOWS (BOUNDARY SHEETS)

For laminar layers the accuracy of the method depends on the accuracy of the numerical scheme. According to the study conducted by Cebeci (1975), accurate solutions can be obtained by taking 25–30 points across the boundary layer. Although a higher degree of accuracy can be obtained with less points by using Richardson extrapolation, the accuracy obtained with 25–30 points is sufficient for most laminar layers.

For turbulent flows the accuracy of any calculation method depends also on the accuracy of the turbulence model for the Reynolds stresses. With the use of eddy-viscosity or mixing-length concepts, the governing boundary-layer equations for turbulent flows can be put in a form very similar to laminar flows, as we demonstrated in Chap. 8 for two-dimensional flows; the same is true for three-dimensional flows. By defining $b = 1 + \epsilon_m^+$, and by using suitable formulas for ϵ_m^+, we can solve the governing turbulent boundary-layer equations by the procedure described for laminar flows. This has been done extensively by Cebeci and his associates; see, for example, Cebeci (1974), Cebeci (1975), and Cebeci et al. (1976, 1977). His eddy-viscosity formulation was given in Sec. 10.3. We may recall that that formulation worked well for two-dimensional and axisymmetric turbulent flows, as was demonstrated in Chap. 8; it has also worked quite well for three-dimensional flows, as will be demonstrated in this section.

To check the accuracy of the method for turbulent flows, it is necessary to compare the solutions with experiment. While the experimental data for full three-dimensional turbulent boundary layers are limited, there are a number of experimental data for restricted quasi-three-dimensional flows, such as flows over infinite swept wings. For this reason the method described in Sec. 10.2 was first applied to such quasi-three-dimensional turbulent flows before it was applied to full three-dimensional flows.

10.4.1 Infinite Swept-Wing Calculations

Equations (3.2.1), (3.2.3), and (2.2.2) are applicable to three-dimensional flows in Cartesian coordinates. For flows with yawed infinite cylinder or swept-wing approximations in Cartesian coordinates, where the spanwise flow is independent of the z-coordinate, these equations simplify considerably. The continuity and the

x-momentum (chordwise) equations reduce to the two-dimensional forms, respectively; the z-momentum (spanwise) equation becomes

$$u \frac{\partial w}{\partial x} + v \frac{\partial w}{\partial y} = v \frac{\partial}{\partial y}\left(\frac{\partial w}{\partial y} - \overline{w'v'}\right) \tag{10.4.1}$$

By using the Falkner-Skan transformation discussed in Sec. 5.3, and by using the relations defined by Eq. (10.3.2), the governing equations can be put in a form identical to those given by Eqs. (10.2.12) and (10.2.13) provided that we set the coefficients m_i ($i = 3$-7) equal to zero. The boundary conditions [Eq. (10.2.14)] remain unchanged, and $m_2 = m$ and $m_1 = (m + 1)/2$, as in Sec. 5.3.

In a study reported by Cebeci (1974) a number of flows that fall in this class were considered. Here we present the results for the 45° infinite swept-wing data of Bradshaw and Terrell (1969). This experiment was set up especially to test the outer-layer assumptions made in extending the boundary-layer calculation method of Bradshaw et al. (1967) from two dimensions to three (see Bradshaw, 1971b). Measurements were made only on the flat rear of the wing in a region of nominally zero-pressure gradient and decaying cross flow. See the sketch in Fig. 10.8a. Spanwise and chordwise components of mean velocity and shear stress, and all three components of turbulence intensity, were measured at streamwise distances of $x' = 0, 4, 10, 16,$ and 20 in. from the start of the flat portion of the wing (see Fig. 10.8). The surface shear stress, measured with a Preston tube, was constant along a generator at the start of the flat part of the wing, except for a few inches at each end and except for small undulations of small spanwise wave length caused by residual nonuniformities in the tunnel flow.

For a three-dimensional flow the velocity vector at any y-location in the boundary layer differs in direction from the full-stream vector when both are projected on the surface. In that case the cross-flow velocity, w, within the boundary layer differs from zero (except at the wall). The departure of the velocity vector within the boundary layer is conveniently represented by the cross-flow angle, β. It is defined by the following expression:

$$\beta \equiv \tan^{-1}\left(\frac{w}{u}\right) \tag{10.4.2}$$

The above formula becomes indeterminate at $y = 0$; however, with the use of L'Hôpital's rule it can be written as

$$\beta = \tan^{-1}\left(\frac{\partial w}{\partial y}\right)_w \left(\frac{\partial u}{\partial y}\right)_w^{-1} \tag{10.4.3}$$

Figure 10.8 shows the results calculated by the yawed infinite cylinder approximation compared with experimental results and those obtained by Bradshaw's method (1971b). These calculations were started at $x' = 0$ by inputting the initial velocity profiles (see Cebeci et al., 1976).

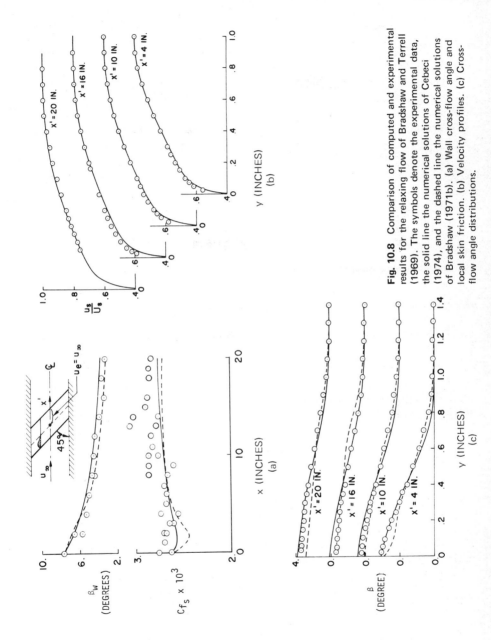

Fig. 10.8 Comparison of computed and experimental results for the relaxing flow of Bradshaw and Terrell (1969). The symbols denote the experimental data, the solid line the numerical solutions of Cebeci (1974), and the dashed line the numerical solutions of Bradshaw (1971b). (a) Wall cross-flow angle and local skin friction. (b) Velocity profiles. (c) Cross-flow angle distributions.

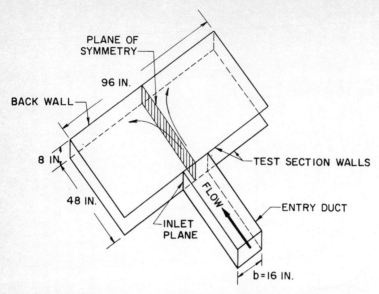

Fig. 10.9 Schematic drawing of Johnston's test geometry.

10.4.2 Three-Dimensional Flows in Cartesian Coordinates

To check the accuracy of the method discussed in the previous sections for fully three-dimensional turbulent boundary layers, we now present a comparison of calculated and experimental results reported by Cebeci (1975) for the data of Johnston (1957).

Johnston's experimental apparatus consisted of a rectangular inlet duct from which an issuing jet impinged on an end wall 48 in. from the outlet of the channel.

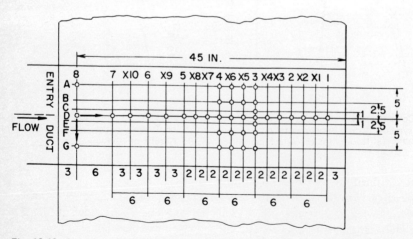

Fig. 10.10 Sketch showing the measured stations in Johnston's experiment.

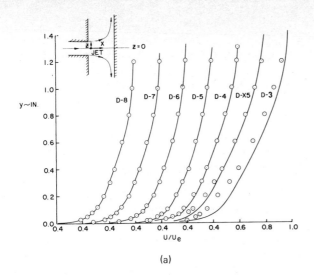

(a)

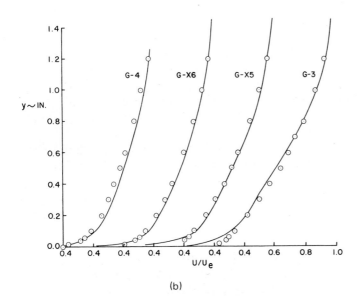

(b)

Fig. 10.11 Comparison of calculated and experimental velocity profiles for Johnston's data (1957). The symbols denote experimental data, and the solid line denotes numerical solutions of Cebeci (1975). (a) Results for the attachment-line flow. (b) Results off the line of symmetry, $z = 5$ in.

The jet was confined to the top and bottom of flat surfaces, and the boundary layer that developed on the floor of the test section was probed. Figure 10.9 shows the schematic drawing of Johnston's test geometry, and Fig. 10.10 shows the locations where the measurements were conducted.

In this case the calculations start on the attachment line at $x = 0$, $z = 0$, and proceed downstream. At station $x = 0$, $z = 0$, the flow is laminar, and it becomes turbulent at any specified station where $x > 0$, $z > 0$. The calculations can be started at any x-location by inputting the initial-velocity profiles (see Cebeci, 1974). However, the calculations reported by Cebeci (1975) were started at $x = 0$, $z = 0$, and the initial experimental velocity profile was matched as closely as possible. The inviscid external-velocity distribution was obtained by the formulas given by Milne-Thompson (1960) (see Cebeci, 1974).

Figure 10.11 shows a comparison of calculated and experimental velocity profiles on the line of symmetry ($z = 0$) and off the line of symmetry ($z = 5$ in.). The calculations were made by taking $z = 1$, 2.5, and 5 in.

10.5 NUMERICAL SOLUTIONS FOR UNSTEADY FLOWS

The continuity and momentum equations for two-dimensional, incompressible unsteady laminar and turbulent boundary layers are given by Eqs. (2.2.3) and (3.2.12). The boundary conditions for the wall (no mass transfer) and for the outer edge of the boundary layer are

$$y = 0 \qquad u, v = 0 \tag{10.5.1a}$$

$$y \to \infty \qquad u \to u_e(x, t) \tag{10.5.1b}$$

To complete the formulation of the problem, initial conditions must be specified on some upstream surface, which is most conveniently taken as $t = T(x)$, say, normal to the xt-plane. Here we consider the simple case where this surface is made up of the two surfaces $t = t_a = $ const and $x = x_a = $ const. The required boundary conditions are

$$\text{for} \quad t = t_a \quad \text{and} \quad x \geqslant x_a \quad u = u_a(x, y) \tag{10.5.2a}$$

$$\text{for} \quad x = x_a \quad \text{and} \quad t \geqslant t_a \quad u = u_a(t, y) \tag{10.5.2b}$$

Thus at some initial time, $t = t_a$ conditions are specified everywhere, and for all subsequent times, conditions are specified on some upstream station, $x = x_a$, for all y.

The solution of unsteady flows for laminar and turbulent boundary layers requires the solution of the system given by Eqs. (2.2.3), (3.2.12), (10.5.1), and (10.5.2) with closure assumptions for the Reynolds shear stress term, if any. The unsteady flows of most practical interest fall into two separate categories; these consist of flows with (a) oscillatory motion and (b) impulsive motion starting from rest. We shall discuss the calculation of such flows in the next two sections. Any general nonimpulsive unsteady flow can be regarded as part of the first cycle of an oscillatory but nonsinusoidal motion. For a detailed discussion of unsteady laminar flows the reader is referred to Stewartson (1960) and Rosenhead (1963).

10.5.1 Oscillatory Boundary Layers for Laminar and Turbulent Flows

As in steady boundary layers, it is also desirable to solve the governing equations in transformed coordinates and transformed variables. One convenient transformation (see Cebeci, 1977a) defines the transformed coordinate, η, by

$$\eta = \sqrt{\frac{u_0}{\nu x}}\, y \tag{10.5.3a}$$

and the dimensionless stream function, $f(x, t, \eta)$, by

$$\psi = \sqrt{\nu x u_0}\, f(x, t, \eta) \tag{10.5.3b}$$

Here $u_0(x)$ is some reference velocity, and ψ is the stream function that satisfies the continuity equation.

With the relations defined by Eq. (10.5.3), and with the definition of eddy viscosity, ϵ_m, it can be shown that the momentum equation (3.2.12) for laminar and turbulent flows can be written as

$$(bf'')' + \frac{m+1}{2} ff'' - m(f')^2 + m_3 = x\left(f' \frac{\partial f'}{\partial x} - f'' \frac{\partial f}{\partial x} + \frac{1}{u_0}\frac{\partial f'}{\partial t}\right) \tag{10.5.4}$$

Here primes denote differentiation with respect to η and

$$f' = \frac{u}{u_0} \qquad m = \frac{x}{u_0}\frac{du_0}{dx} \qquad m_3 = \frac{x}{u_0{}^2}\left(u_e \frac{\partial u_e}{\partial x} + \frac{\partial u_e}{\partial t}\right) \qquad b = 1 + \epsilon_m^+ \tag{10.5.5}$$

The boundary conditions given by Eq. (10.5.1) become

$$\eta = 0 \qquad f = f' = 0 \tag{10.5.6a}$$

$$\eta \to \eta_\infty \qquad f' = \frac{u_e}{u_0} \tag{10.5.6b}$$

Similarly, the initial conditions given by Eq. (10.5.2) can be transformed. For some problems where the flow starts as laminar, they can be obtained from Eq. (10.5.4). Along the t-axis, Eq. (10.5.4) reduces to

$$f''' + \frac{m+1}{2} ff'' - m(f')^2 + m_3 = \frac{x}{u_0}\frac{\partial f'}{\partial t} \tag{10.5.7}$$

Note that, in general, the right-hand side of Eq. (10.5.7) is not equal to zero. For example, although for a flat-plate flow it is equal to zero, for stagnation-point flow ($u_0 = Ax$) it is not. For generality we will keep it in the form shown.

At time $t = 0$, Eq. (10.5.4) reduces to

$$(bf'')' + \frac{m+1}{2} ff'' - m(f')^2 + m_3 = x\left(f'\frac{\partial f'}{\partial x} - f''\frac{\partial f}{\partial x}\right) \tag{10.5.8}$$

where, for example,

$$m_3 = \frac{x}{u_0^2} u_e \frac{du_e}{dx} \tag{10.5.9}$$

Numerical Formulation of the Momentum Equation

The numerical formulation of the two momentum equations, Eqs. (10.5.7) and (10.5.8), for the initial conditions is very similar to the numerical formulation of the momentum equation for two-dimensional flows (see Sec. 7.2). For this reason we shall only discuss the numerical formulation of the momentum equation for the general case, namely, Eq. (10.5.4). Its formulation is very similar to the formulation of either of the two momentum equations for three-dimensional flows (see Sec. 10.2). Therefore only a brief description of it will be presented.

To make the description easier, let us denote x and t coordinates by z and x, respectively. Then using the new dependent variables $u(x, z, \eta)$ and $v(x, z, \eta)$ defined by Eqs. (10.2.21a) and (10.2.21b) we can write Eq. (10.5.4) as

$$(bv)' + \frac{m+1}{2} fv - mu^2 + m_3 = z\left(u\frac{\partial u}{\partial z} - v\frac{\partial f}{\partial z} + \frac{1}{u_0}\frac{\partial u}{\partial x}\right) \tag{10.5.10}$$

Next we consider the net cube shown in Fig. 10.6 and write the difference equations for Eqs. (10.2.21a), (10.2.21b), and (10.5.10). For Eqs. (10.2.21a) and (10.2.21b) they are given by Eqs. (10.2.30a) and (10.2.30b). For Eq. (10.5.10), by using the procedure followed for three-dimensional flows, it can be shown that the difference equations are

$$(bv)_j - (bv)_{j-1} + h_j \Big\{(m_1)_{n-1/2}^{i-1/2}(fv)_{j-1/2} - (m_2)_{n-1/2}^{i-1/2}(u^2)_{j-1/2}$$

$$- \frac{\alpha_i}{2}[u_{j-1/2}(u_{j-1/2} + u_{j-1/2}^{234} + u_{j-1/2}^{i,n-1} - 2\bar{u}_{i-1}) - v_{j-1/2}f_{j-1/2}$$

$$- (f_{j-1/2}^{i,n-1} - 2\bar{f}_{i-1})v_{j-1/2} - v_{j-1/2}^{234}f_{j-1/2}] - 2\beta_n u_{j-1/2}\Big\} = T_{j-1/2}^{i-1,n-1} \tag{10.5.11}$$

Here

$$m_1 = \frac{m+1}{2} \qquad m_2 = m \qquad \alpha_i = \frac{x_{i-1/2}}{r_i} \qquad \beta_n = \frac{x_{i-1/2}}{k_n u_0^{i-1/2}}$$

$$T_{j-1/2}^{i-1,n-1} = (bv)_{j-1}^{234} - (bv)_j^{234} + h_j\Big\{-(m_1)_{n-1/2}^{i-1/2}(fv)_{j-1/2}^{234}$$

$$+ (m_2)_{n-1/2}^{i-1/2}(u^2)_{j-1/2}^{234} - 4(m_3)_{n-1/2}^{i-1/2}$$

$$+ \frac{\alpha_i}{2}[u_{j-1/2}^{234}(u_{j-1/2}^{i,n-1} - 2\bar{u}_{i-1}) - v_{j-1/2}^{234}(f_{j-1/2}^{i,n-1} - 2\bar{f}_{i-1})] + 2\beta_n(u_{j-1/2}^{i-1,n} - 2\bar{u}_{n-1})\Big\}$$

Note that, for simplicity, we have dropped the superscripts i and n in the above equations.

The boundary conditions given by Eq. (10.5.6) become

$$f_0 = 0 \quad u_0 = 0 \quad u_J = \frac{u_e}{(u_0)_n} \tag{10.5.12}$$

If we assume $(f_j^{n-1,i-1}, u_j^{n-1,i-1}, v_j^{n-1,i-1})$, $(f_j^{n,i-1}, u_j^{n,i-1}, v_j^{n,i-1})$ and $(f_j^{n-1,i}, u_j^{n-1,i}, v_j^{n-1,i})$ to be known for $0 \leqslant j \leqslant J$, then the differenced equations for the general case can be solved by the procedure described for the attachment-line equations in Sec. 10.2.2. For details see Cebeci (1977a).

Laminar Flows

There are several problems in which it is necessary to account for the fluctuations in the external flow. These fluctuations may change both in direction and in magnitude. A simpler case is one in which the external flow fluctuates only in magnitude and not in direction. To illustrate the numerical scheme, we consider this problem, which has been previously studied analytically by Lighthill (1954) for a flat-plate flow. According to his analysis, for an external flow in the form

$$u_e(t) = u_0(1 + B \cos \omega t) \tag{10.5.13}$$

the reduced skin-friction coefficient $(c_f/2) \sqrt{\mathrm{Re}_x}$ is given by two separate formulas depending on whether the local Strouhal number $\tilde{\omega}$ ($\equiv \omega x/u_0$) is much smaller or much greater than one, $(\mathrm{Re}_x = u_0 x/\nu)$.

$$\frac{c_f}{2} \sqrt{\mathrm{Re}_x} = \begin{cases} 0.332 + B(0.498 \cos \omega t - 0.849 \tilde{\omega} \sin \omega t) & \tilde{\omega} \ll 1 \qquad (10.5.14a) \\ 0.332 + B \tilde{\omega}^{1/2} \cos\left(\omega t + \frac{\pi}{4}\right) & \tilde{\omega} \gg 1 \qquad (10.5.14b) \end{cases}$$

In many problems it is often desirable to find the phase angle, ϕ, between the external flow and, say, the reduced skin-friction coefficient, which in terms of the transformed variables defined by Eq. (10.5.3) is f_w''. To determine this numerically for a fixed $x = x_0$, let us consider the general case in which the external flow is given by

$$u_e(x, t) = u_0(x)(1 + B \cos \omega t) \tag{10.5.15}$$

and let $f_w''(x, t) = g(x, t)$. We use the following procedure to determine the phase angle between $u_e(x_0, t)$ and $g(x_0, t)$. We first compute $u_0(x_0)$ and $\bar{g}(x_0)$ from

$$u_0(x_0) = \frac{1}{P} \int_{t_0}^{t_0+P} u_e(x_0, t) \, dt \tag{10.5.16a}$$

$$\bar{g}(x_0) = \frac{1}{P} \int_{t_0}^{t_0+P} g(x_0, t)\, dt \tag{10.5.16b}$$

Here P denotes the period defined by $P = 2\pi/\omega$, and t_0 must be taken sufficiently large so that all transient effects have decayed. Usually $t_0 \approx P$ is adequate.

From Eq. (10.5.15) we can write $[A \equiv u_0(x_0)B]$

$$u_e(x_0, t) - u_0(x_0) = A \cos \omega t \tag{10.5.17}$$

Similarly, we can write

$$g(x_0, t) - \bar{g}(x_0) = C \cos[\omega t + \phi(x_0)]$$
$$= C[\cos \omega t \cos \phi(x_0) - \sin \omega t \sin \phi(x_0)] \tag{10.5.18}$$

with $\phi(x_0)$ denoting the phase angle between u_e and g at $x = x_0$. If we take the product of Eqs. (10.5.17) and (10.5.18) and integrate the resulting expression, we find $\cos \phi(x_0)$ to be given by

$$\cos \phi(x_0) = \frac{\displaystyle\int_{t_0}^{t_0+P} \{[u_e(x_0, t) - u_0(x_0)][g(x_0, t) - \bar{g}(x_0)]\}\, dt}{AC(\pi/\omega)} \tag{10.5.19}$$

Here

$$A^2 = \frac{\omega}{\pi} \int_{t_0}^{t_0+P} [u_e(x_0, t) - u_0(x_0)]^2\, dt \tag{10.5.20a}$$

$$C^2 = \frac{\omega}{\pi} \int_{t_0}^{t_0+P} [g(x_0, t) - \bar{g}(x_0)]^2\, dt \tag{10.5.20b}$$

Using this procedure we find that the phase angle [between the external velocity and reduced skin friction, Eqs. (10.5.14) and (10.5.15), respectively] according to Lighthill's analysis is

$$\cos \phi = \begin{cases} \dfrac{1}{(1 + 2.9064\tilde{\omega})^{1/2}} & \tilde{\omega} \ll 1 \\[2mm] 0.707 & \tilde{\omega} \gg 1 \end{cases} \tag{10.5.21}$$

Figure 10.12 shows a comparison of results obtained by the numerical solution of the boundary layer equations and those obtained from Eq. (10.5.21). The

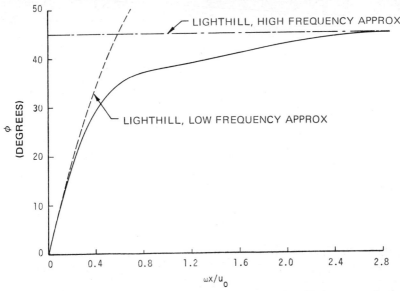

Fig. 10.12 Variation of phase angle between wall shear and oscillating external velocity for a laminar flow over a flat plate. The solid line denotes the numerical solutions of Cebeci (1977a).

numerical calculations were made by Cebeci (1977a) for an external flow given by Eq. (10.5.13) with $B = 0.150$ and $u_0 = 17.5$. A total of 41 t-stations and 26 x-stations with $\Delta t = 0.20$ and $\Delta x = 0.1$ were taken. Initially, η_∞ was taken as 10 with $\Delta \eta = 0.28$. These numerical calculations are in good agreement with those computed by Ackerberg and Phillips (1972); they show that if $B \leqslant 0.150$ Lighthill's low-frequency approximations to the phase angle are accurate for $\omega \leqslant 0.20 u_0/x$, and his high-frequency approximation is accurate for $\omega \geqslant 2.6 u_0/x$.

Oscillating laminar flow in a pipe, which is of great importance in the study of viscous attenuation of sound waves, is discussed by Rosenhead (1963, Chap. 7). An analytic solution of the NS equation can be obtained in the form of a Bessel function. For high frequencies of oscillation the viscous layer is confined close to the pipe wall, corresponding to the reduction of boundary-layer thickness with increasing frequency in oscillatory external flows.

Turbulent Flows

The experimental data on unsteady turbulent boundary layers are very limited. Karlsson's flow (1959) consists of an oscillating free stream in a zero pressure gradient turbulent flow. In order to compare the numerical calculations with these data, we take the external flow to be the same as that of Eq. (10.5.15) and use almost the same eddy-viscosity formulation of Sec. 8.5; the only change occurs in the damping length constant where the pressure gradient, $\partial p/\partial x$, is now $-\rho(\partial u_e/\partial t + u_e\, \partial u_e/\partial x)$. For a flow with no mass transfer, Eq. (8.3.2e) is

$$N = [1 - 11.8(p^+ + p_t^+)]^{1/2} \qquad (10.5.22)$$

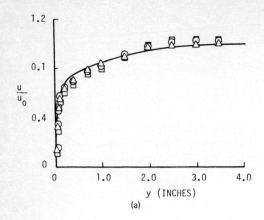

(a)

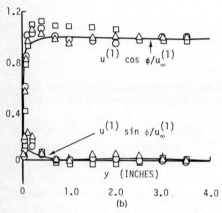

(b)

Fig. 10.13 (a) Comparison of calculated and experimental velocity profiles at $R_\delta^* \equiv 3.6 \times 10^3$. Symbols denote u/u_0 for values of $\tilde{B} = 0.292, 0.202,$ and 0.147 from Karlsson (1959). The solid line denotes the numerical solutions of Cebeci (1977a). (b) Experimental data from Karlsson (1959) are for $\omega/2\pi = 4.0$ cycles s^{-1}; $u_\infty^{(1)}/u_0 = 26.4\%$ (circles), 13.6% (triangles), and 6.2% (squares). The solid lines denote the numerical solutions of Cebeci (1977a) for $u_\infty^{(1)}/u_0 = 26.4\%$.

Here

$$p_t^+ = \frac{\nu}{u_e^3} \frac{\partial u_e}{\partial t} \qquad (10.5.23)$$

Figure 10.13 shows the results for Karlsson's data. These calculations were started as laminar at $x = 0$ for the external flow given by Eq. (10.5.15) with $u_0 = 17.5$ ft/s. The Δx-spacing for turbulent flows was 1 ft. The turbulent-flow calculations were started at $x = 0.1$ ft and at $x \approx 12$ ft, the experimental data [corresponding to $R_\delta^* (\equiv u_0 \delta^*/\nu)$ of 3.6×10^3] were matched as shown in Fig. 10.13a. Figure 10.13b shows a comparison between calculated and experimental values of the in-phase and out-of-phase components, that is, $u^{(1)} \cos \phi/u_\infty^{(1)}$ and $u^{(1)} \sin \phi/u_\infty^{(1)}$, respectively, according to Karlsson's notation.

10.5.2 Boundary–Layer Motion Started Impulsively from Rest

In problems involving boundary-layer motion from rest we are interested in the development of the unsteady boundary layer when the body is given, impulsively, a free-stream velocity. In general, the external velocity can be expressed as

$$u_e(x, t) = \begin{cases} 0 & t \leqslant 0 \\ u_e(x) & t > 0 \end{cases} \tag{10.5.24}$$

The governing boundary-layer equations are given by Eqs. (2.2.3) and (3.2.12). They are subject to the conditions given by Eqs. (10.5.1a) and (10.5.24).

At the *beginning* of the motion the boundary layer is of zero thickness. For this reason we can set the convective terms in Eq. (3.2.12) equal to zero to get

$$\frac{\partial u}{\partial t} = \nu \frac{\partial^2 u}{\partial y^2} \tag{10.5.25}$$

The solution of Eq. (10.2.25) subject to Eqs. (10.5.1a) and (10.5.24) is

$$u = u_e \, \mathrm{erf}\,(\zeta) \tag{10.5.26}$$

Here the variable ζ and the error function are given by

$$\zeta = \frac{y}{2\sqrt{\nu t}} \qquad \mathrm{erf}\,(\zeta) = \frac{2}{\sqrt{\pi}} \int_0^{\zeta} e^{-\zeta^2} \, d\zeta \tag{10.5.27}$$

This problem was first considered by Lord Rayleigh (Lamb, 1932) for a plate that moved impulsively from rest; it was extended to the case of a semi-infinite plate by Howarth (1950).

To find the development of the flow for later times ($t \to \infty$ and steady-state conditions), we need to solve the system given by Eqs. (2.2.3), (3.2.12), (10.5.1a), and (10.5.24). To illustrate the calculation of such flows, we shall now consider two simple cases, the impulsive motion of a flat plate and that of a circular cylinder. Analogous axisymmetric cases could be treated by using the Mangler transformation (Sec. 3.3).

Impulsively Started Flat Plate

The development of a boundary layer over an impulsively started semi-infinite flat plate has two features. Initially, the flow is identical to that given by Eq. (10.5.26), and the flow tends ultimately to that given by Blasius (see Sec. 5.3). The problem in the calculation of this flow and flows of this type is to describe the evolution of the flow from the initial to the steady state.

The problem was first studied by Stewartson (1951). He found that the structure of the boundary layer at a fixed distance x from the leading edge is different, according to $\tau \lessgtr 1$, where $\tau = u_e t / x$ or, more generally, $t / \int_0^x dx' / u_e$. If $\tau < 1$, it is the same as for an infinite plate and independent of x. If $\tau > 1$, x enters the structure, which approaches the steady Blasius solution as $\tau \to \infty$. This means that the disturbance caused by the presence of the leading edge travels down the boundary layer with the maximum local velocity at any station of x, u_e. Hence the effect of the leading edge is felt when $\tau > x/u_e$ or, in general, $\tau > \int dx/u_e$. Note that there is a fundamental difference between sharp-edged plates and bodies with a front stagnation point near which $u_e \sim x$, so that the integral becomes infinite (Stewartson, 1973).

The first numerical solutions to the problem were obtained by Hall (1969). He solved the systems Eqs. (2.2.3) and (3.2.12) subject to the sharp-edged plate problem. At $x = x_0$ and $t \geqslant t_a$ he assumed initial-velocity profiles $u = u_a(t, y)$ and used the Rayleigh solution [Eq. (10.5.26)] at $t = t_a$ and $x \geqslant x_a$. Then making use of a similarity condition that the solutions must satisfy at some x-station, he used an iteration procedure to improve (or update) his initial-velocity profiles, $u = u_a(t, y)$. He observed that his iteration procedure converged rapidly.

Figure 10.14 shows the variation of wall shear parameter $f_w'' [\equiv \mathrm{Re}_x^{-1/2} x/u_e (\partial u/\partial y)_w]$ with dimensionless time $t^* (\equiv u_e^2 t/\nu)$. Note the rapidity of the transition from the Rayleigh to the Blasius state. Mathematically, an infinite time is required for the attainment of the steady state. Numerically, for all practical purposes the steady state is reached by $t^* = 4$, as shown in Fig. 10.14. The figure also indicates that there is no appreciable departure from the Rayleigh state until $t^* = 2$. These solutions were later confirmed by Dennis (1972), who solved the governing equations in the similarity form.

Impulsively Started Circular Cylinder

To illustrate the numerical solution of boundary layers impulsively started from rest, we shall consider the calculation of an impulsively started circular cylinder

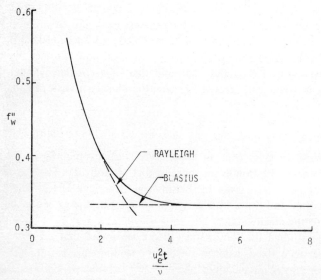

Fig. 10.14 Variation of wall-shear parameter, f_w'', with dimensionless time for an impulsively started flat-plate flow. Calculations were made by Hall (1969).

problem. However, we shall see that our solution procedure is a general one in that it can be applied to other problems, for example, impulsively started flows over a flat plate or airfoil. By considering the governing equations for axisymmetric flows, it can also be applied to axisymmetric bodies that start from rest.

As in previous examples, we use transformed variables and coordinates and define

$$\eta = \frac{y^*}{\sqrt{\nu t^*}} \qquad \psi = \nu \sqrt{t^*}\, f(\xi, \eta, t^*)$$

$$t^* = \frac{u_\infty^2 t}{\nu} \qquad y^* = \frac{y u_\infty}{\nu} \qquad \xi = \frac{x u_\infty}{\nu} \tag{10.5.28}$$

Then it can be shown that for a laminar flow, the system Eqs. (2.2.3) and (3.2.12) and their boundary conditions can be written as

$$f''' + \frac{\eta}{2} f'' + t\left(u_e^* \frac{\partial u_e^*}{\partial x} + \frac{\partial u_e^*}{\partial t}\right) = t\left(\frac{\partial f'}{\partial t} + f' \frac{\partial f'}{\partial x} - f'' \frac{\partial f}{\partial x}\right) \tag{10.5.29}$$

$$\eta = 0 \quad f = f' = 0; \quad \eta = \eta_\infty \quad f' = \frac{u_e}{u_\infty} \tag{10.5.30}$$

Here the primes denote differentiation with respect to η, $u_e^* = u_e/u_\infty$, and we have set $\xi = x$, $t^* = t$ for convenience.

The numerical formulation of Eqs. (10.5.29) and (10.5.30) is very similar to the general case of Eqs. (10.5.4) and (10.5.6). Along the t-axis or x-axis, we do not need to solve special equations, because for impulsively started flows the initial conditions ($t = 0$) along the x-axis are either known or assumed. In the case of a circular cylinder, $u \equiv 0$ at the stagnation point, so $f(0, \eta, t)$ and its derivatives are all zero on the (t, η)-plane.

For a circular cylinder the inviscid velocity distribution in the steady state is (see Ex. 5.5)

$$u_e(x) = 2u_\infty \sin\left(\frac{x}{r_0}\right) \tag{10.5.31}$$

This is valid also in unsteady flow (although the *pressure* distribution is different). The initial conditions along the x-axis are given by the Rayleigh solution. In our case

$$f'(x, \eta, 0) = u_e^* \operatorname{erf}\left(\frac{\eta}{2}\right) \tag{10.5.32}$$

so that $f(x, \eta, 0)$ is obtained by integrating Eq. (10.5.32), and $f''(x, \eta, 0)$ is obtained by differentiating Eq. (10.5.32) with respect to η.

Figure 10.15 shows the (x/r_0)-stations as functions of dimensionless time $u_\infty t/d$ where the wall shear [$\equiv \mu(\partial u/\partial y)_w$] becomes equal to zero. The solid lines

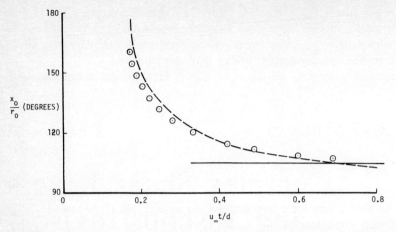

Fig. 10.15 Variation of zero wall-shear location, (x_0/r_0), as a function of dimensionless time, $u_\infty t/d$, on a circular cylinder started impulsively from rest. The symbols denote the numerical solutions of Cebeci (1977b), and the dashed lines denote the results according to Blasius' formula, Eq. (10.5.34). The solid line denotes the steady-state separation, 105° (see Fig. 8.2).

denote the numerical solutions of Cebeci (1977b) and the dashed lines denote the simple formula given by Blasius (see Schlichting, 1968). According to this formula for flows impulsively started from rest a rough estimate of the time at which the wall shear becomes zero can be obtained from

$$t = \frac{-1}{(1 + 4/3\pi)(du_e/dx)} = -\frac{0.702}{du_e/dx} \qquad (10.5.33)$$

Here t denotes the time at which the wall shear is zero. For the external-velocity distribution given by Eq. (10.5.31), we can write Eq. (10.5.33) as

$$\frac{x}{r_0} = \cos^{-1}\left(\frac{-0.1755}{tu_\infty/d}\right) \qquad (10.5.34)$$

The figure also shows the steady-state separation point, 105° (see Fig. 8.2).

We must distinguish between points of zero skin friction ($\tau_w = 0$) and points of separation. Clearly, zero skin friction in unsteady flow on an oscillating flat plate is not associated with large rate of growth of boundary-layer thickness ($d\delta/dx$ of order one, say). Therefore it does not necessarily follow that zero skin friction at a given point implies $d\delta/dx$ of order one at the same point in the case of a circular cylinder in unsteady flow. For two reviews with different viewpoints, see Sears and Telionis (1975) and Riley (1975).

Chapter 11

Complex Shear Layers and Viscous/Inviscid Interactions

There have been several hints in earlier chapters that shear layers can be more complicated than the idealized cases shown in Fig. 1.1 and that shear layers can interact with the external stream. One obvious complication, discussed in Sec. 3.5, is that a shear layer may grow too rapidly for the TSL equations, based on $d\delta/dx \ll 1$, to apply. In duct flows (Fig. 1.1d), boundary layers grow on the walls and eventually meet near the center of the duct. The resulting interaction between the two shear layers can be quite complicated in the turbulent case. The flow in a wing-body junction (Sec. 10.1, Fig. 10.1) is also an example of interaction between two shear layers (the boundary layers on the body and wing) and is again very complicated.

The presence of a shear layer affects the external flow by displacing its stream-lines outward. In the case of a TSL the displacement near the edge of the shear layer equals the displacement thickness defined in Chap. 1, but the effect far from the edge is more complicated. The displacement thickness of the boundary layer and wake near the trailing edge of an airfoil considerably reduces the lift at given incidence, thus affecting the pressure distribution and the whole development of the boundary layer. If the shear layer is not thin, it is likely to have a large effect on the free stream, which again cannot usefully be represented as a local displacement effect.

In Sec. 11.1 we discuss the main examples of complex shear layers, those that disobey the TSL equations because of rapid changes and those that interact with other shear layers. The reader must bear in mind that there are two related but distinct problems: the effect on the equations for the mean motion (in laminar or turbulent flow) and the effect on the Reynolds stresses (in turbulent flow only). In some cases the effect of mean-flow distortion on the turbulence is significant well before the TSL equations break down.

In Sec. 11.2 we deal with interactions between a shear layer and the neighboring, usually inviscid, flow. Strong interactions are often associated with breakdown of the TSL approximation, but again the reader must distinguish between the shear-layer equations as such and the problem of combining them with the external-flow equations so as to solve a complete problem. The TSL equations can still be applied to quite complicated shear layers, even including some separating flows in which the flow leaves the surface smoothly (Fig. 1.6). However, a separation point is inevitably a point of singularity of the TSL equations, and great care is needed to obtain a solution that is valid through separation. This difficulty can be avoided by solving the full NS equations, of course, but the time needed to compute solutions is much longer.

11.1 COMPLEX SHEAR LAYERS

11.1.1 Interactions between Shear Layers

It is usual to think of jets, wakes, duct flows, and wall jets (Figs. 1.1 and 11.1) as single shear layers even though the sign of the mean shear ($\partial U/\partial y$) changes in midlayer. Strictly speaking, as can be seen by considering the inlet regions of a duct or the initially uniform flow from a jet nozzle (Fig. 1.1), these are double shear layers formed by two interacting shear layers. In a duct two boundary layers merge at their high-velocity edges, while in the wake of a streamlined airfoil two boundary layers merge at their low-velocity edges. A plane jet is the result of two mixing layers merging at their high-velocity sides. (The wake of a bluff body is formed by two mixing layers meeting at their low-velocity sides but only after violent distortion in a vortex street, so this is not a helpful concept.) Finally, a wall jet (Fig. 11.1a) is composed of a mixing layer and a boundary layer; this flow is essentially asymmetrical, and the previous examples could be asymmetrical at least near the source.

More complicated examples occur when larger numbers of shear layers meet or when the flow is three dimensional. The wake of an airfoil with a slotted flap (Fig. 11.1b) consists of the remains of two or more pairs of boundary layers, while multiple jet flows (Fig. 11.1c) occur in many different forms. The most important of the intrinsically three-dimensional interactions is the flow in a streamwise corner where two boundary layers meet (Fig. 10.1), but mildly three-dimensional versions of the previous examples are common.

Even in laminar flow, interacting shear layers are not, in general, simple superpositions of their constituents because the equations of motion are nonlinear. There

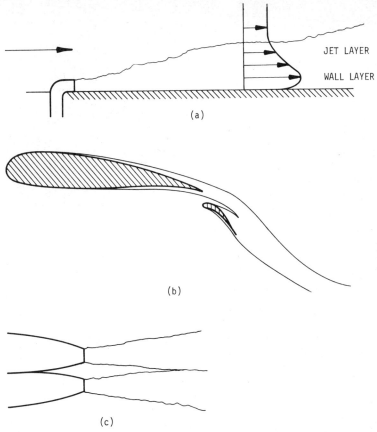

JET LAYER

WALL LAYER

(a)

(b)

(c)

Fig. 11.1 Interacting shear layers. (a) Wall jet. (b) Flapped airfoil. (c) Multiple jets. (See also Fig. 1.1.)

may not appear to be much difference in shape between the velocity profile in two-dimensional duct flow and in two-dimensional stagnation flow (which like a duct flow has a shear-layer thickness independent of streamwise distance), and indeed the shape factors are not too different, at $H = 2.5$ and $H = 2.2$, respectively. However, the shear-stress gradient, $\mu\, \partial^2 u/\partial y^2$, goes to zero at the edge of the stagnation-flow boundary layer, while the shear-stress gradient in the duct flow is everywhere equal to the pressure gradient. Better than regarding a duct flow as two abutting boundary layers is to think of it as two overlapping boundary layers, but this is not immediately helpful. An approximate superposition analysis for merging free-shear layers was proposed by Reichardt (1943) and has been used for interacting jet studies by Knystautas (1964) and others, but there appear to be no approximate theoretical treatments of interacting laminar shear layers. Numerical solutions of Eqs. (5.1.1) and (5.1.2), or of the more complicated set for three-dimensional flow, are not significantly more difficult in interacting flows than in isolated shear layers. The only likely complication is the need to change the boundary condition at the point of merging. Of course, merging shear layers may take part in a viscous/ inviscid interaction as well, as in ducts or airfoil wakes, but this is a separate problem (see Sec. 11.2).

Interacting turbulent shear layers pose a problem concerning turbulence modeling: how should the assumptions made about Reynolds stress in an isolated shear layer be extended to a pair of interacting shear layers? Is it necessary to insert extra empirical information, or can the turbulence model be generalized purely algebraically? A simple example of a turbulence model that fails is the application of the eddy-viscosity formula [Eq. (6.3.1)] to an asymmetrical jet or a wall jet. Because the shear stress and velocity gradient in general go to zero at different points on the profile, the eddy viscosity deduced from experimental results is negative in between those points and goes to infinity where the velocity gradient is zero. In symmetrical interactions the apparent eddy viscosity is well behaved, since both shear stress and velocity gradient vary linearly near the axis of symmetry; however, the apparent mixing length, $(\tau/\rho)^{1/2}/(\partial u/\partial y)$, goes to infinity as (distance from axis)$^{-1/2}$.

More refined turbulence models such as that developed by Hanjalic and Launder (1972; see also Launder et al., 1975) will handle interacting flows without any special modification. Dean (1974) and Sabot and Comte-Bellot (1974) have shown that the turbulence in the interaction region has some of the "intermittent" properties of turbulence near the edge of an isolated shear layer. Large eddies erupt into the interaction region from opposite sides more or less alternately; the dividing line between fluid originating in the upper shear layer (Fig. 11.2) and fluid originating in

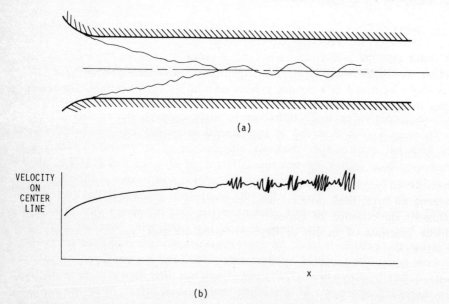

(a)

VELOCITY
ON
CENTER
LINE

x

(b)

Fig. 11.2 Interaction near centerline of duct (vertical scale enlarged). Note that large-amplitude fluctuations occur when eruptions from *either* side penetrate across the centerline. (a) Geometry. (b) Typical appearance of instantaneous velocity on centerline.

the lower shear layer is therefore highly corrugated like the instantaneous edge of an isolated shear layer. In Fig. 11.2, which for simplicity shows an interaction that is symmetrical in the mean, the turbulent velocity fluctuations on the centerline are large when an eruption has penetrated far beyond the centerline but small near the dividing line. This "time-sharing" means that the changes in the turbulence structure caused by interaction are not very large. (Indeed, a "superposition" analysis of the turbulence field works quite well, the superposition and time-sharing being indistinguishable in the statistical average equations.)

If Figs. 1.1c and 1.1d represent axisymmetric flows, the interaction takes place in a cylindrical region near the axis and cannot helpfully be regarded as the confluence of a finite number of simple shear layers.

Strongly three-dimensional interacting shear layers are found near streamwise corners (e.g., wing roots) or edges (e.g., wing tips). In most practical cases, changes in flow direction in, or upstream of, the corner will cause transfer of vorticity from the other two directions into the streamwise direction (i.e., the generation of skew-induced secondary flow of the first kind as discussed in Sec. 10.1.3). In turbulent flow, stress-induced secondary flow of the second kind always appears. Although it may be qualitatively useful to regard, say, the flow in a wing–body junction or an interaction between the nearly perpendicular boundary layers on the body and on the wing, the presence of one or both kinds of secondary flow dominates the flow development.

11.1.2 Distorted Shear Layers

The TSL equations of Chap. 3 were based on the assumption that streamwise gradients were small in comparison to cross-stream gradients. This led, paradoxically, to the result that the cross-stream gradients of pressure could be neglected. In this subsection we discuss shear layers with significant cross-stream pressure gradients, such as might be found in a boundary layer on a highly curved surface. The reader may like to review Sec. 3.1 before continuing, noting the careful distinction drawn between plane and curved shear layers. In boundary layers on plane or nearly plane walls, the streamline curvature is necessarily very small close to the wall, and the TSL equations are applicable to quite strong perturbations.

Except at very low Reynolds numbers or in certain regions of turbulent flow the cross-stream pressure gradient will be balanced by acceleration, stress-gradient terms being no larger than those in undistorted shear layers and therefore negligible. The cross-stream equation of motion therefore reduces to Eq. (3.1.5). Similarly, streamwise gradients of viscous or Reynolds stress are small and can therefore be neglected, so Eq. (3.1.16) is an acceptable approximation for the streamwise motion if the pressure, p, is understood to be a function of y as well as of x. In most numerical calculation methods based on these equations the neglected stress-gradient terms could be grafted onto the main solution of Eqs. (3.1.5) and (3.1.16) in an informal or approximate fashion (Sec. 3.5) without disturbing the properties of the main scheme.

When a shear layer is significantly curved, like the wake in Fig. 11.1b, the

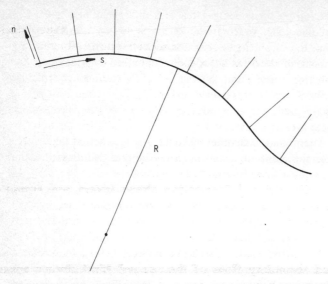

Fig. 11.3 (s, n)-coordinates. Note that n is measured on straight lines normal to the s-axis and that R is a function of s.

streamwise and cross-stream directions mentioned above do not coincide in general with a linear coordinate system. For algebraic, if not numerical, analysis, it is simplest to use semicurvilinear coordinates (s, n), the curved line $n = 0$ (Fig. 11.3) being an arbitrarily defined shear-layer axis and the $s = $ const lines being straight and normal to the $n = 0$ line. The system becomes nonunique at values of n larger than the local radius of curvature $R(s)$, but except near stagnation points, δ/R will be considerably less than unity and the multivalued region is well outside the shear layer. This, of course, is the system used without thought in boundary-layer calculations on curved surfaces. In the latter case, δ/R is very small in comparison to unity, and the terms in n/R that appear in the (s, n) system are negligible. If δ/R is no longer negligible in comparison to unity, these terms cannot be neglected. The equations for a two-dimensional TSL in constant-property flow, using the notation of Chap. 2 for the stress tensor, σ, are

$$u\frac{\partial u}{\partial s} + \left(1 + \frac{n}{R}\right)v\frac{\partial u}{\partial n} + \frac{uv}{R} = -\frac{1}{\rho}\frac{\partial p}{\partial s} + \left(1 + \frac{n}{R}\right)\frac{\partial \sigma_{xy}}{\partial n} + \frac{2\sigma_{xy}}{R} \tag{11.1.1}$$

$$u\frac{\partial v}{\partial s} = -\left(1 + \frac{n}{R}\right)\frac{1}{\rho}\frac{\partial p}{\partial n} \tag{11.1.2}$$

$$\frac{\partial u}{\partial s} + \left(1 + \frac{n}{R}\right)\frac{\partial v}{\partial n} + \frac{v}{R} = 0 \tag{11.1.3}$$

Some of the approximations that can be made in the above equations when δ/R is small but not very small are discussed by Van Dyke (1975); the study of Eqs.

(11.1.1)–(11.1.3) and the other equations for perturbed shear layers constitutes "higher-order boundary-layer theory." To some extent this study has been superseded by modern numerical methods that can treat the full NS equations; however, a more economical treatment of distorted shear layers imbedded in an inviscid flow can be achieved by matching some form of Eqs. (11.1.1)–(11.1.3) to an outer inviscid-flow calculation. We return to this subject in Sec. 11.2.

Shear-layer curvature is only one of the possible modes of distortion. In (x, y)-coordinates it implies an extra rate of strain, $\partial v/\partial x$, in addition to the main velocity gradient $\partial u/\partial y$. There are seven other possible components of the velocity gradient, but most of them present no problem in choosing a coordinate system, and Eq. (11.1.3) connects the extensional rates of strain. Providing that the extra rates of strain are small in comparison with $\partial u/\partial y$, that is, providing that the shear layer is still distinguishable as such, distorted shear layers do not in general present severe problems in laminar flow.

11.2 VISCOUS/INVISCID INTERACTIONS

11.2.1 Interaction of Thin Shear Layers with an External Inviscid Flow

The classical problem is the interaction of the boundary layer and wake of an unstalled airfoil with the external flow (Fig. 11.4). With the exception of a small region near the trailing edge, the TSL approximation can be used. This implies the neglect of $\partial p/\partial y$ in the shear layer, which in turn implies that if the shear layer were absent, the inviscid-flow velocity would change negligibly between $y = 0$ and $y = \delta$. Therefore the simple definition of displacement thickness, δ^*, in Eq. (1.2.17) can be used, and the flow in the region $y > \delta$ is closely the same as the inviscid flow around a body whose thickness is increased, on each surface, by δ^*. Near the trailing edge, and especially in the early part of the wake where the wall constraint $u = 0$ is suddenly released, large streamwise rates of change occur. Strictly speaking, the full NS equations should be used in this region. Most engineering calculation methods refer to turbulent flow, where the offending region is confined

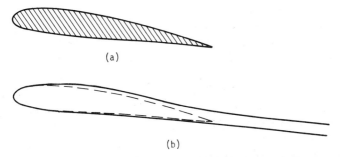

(a)

(b)

Fig. 11.4 The displacement-surface concept. (a) Airfoil alone. (b) Displacement surface (airfoil plus δ^*).

to the viscous sublayer and its continuation in the wake and is therefore very thin so that a crude treatment suffices. In laminar flow, using the TSL equations, the singularity at separation is more influential; the singularity as such can be avoided by prescribing δ^* rather than p in the viscous/inviscid matching procedure.

Since δ^* depends on the inviscid flow, which determines the external-stream velocity, $u_e(x)$, any solution technique that uses separate equations for the inviscid flow and the shear layer must be iterative. The usual sequence is

1. Calculate the inviscid flow around the solid body, neglecting or crudely approximating the displacement effect of the shear layer.
2. Using the external-stream velocity distribution obtained from step 1, calculate the shear layers (the upper- and lower-surface boundary layer and the wake).
3. Add δ^*, obtained from step 2, to the body shape to form a new displacement surface and recalculate the inviscid flow. Repeat steps 2 and 3 until the results converge.

As was stated in Chap. 4, the inviscid-flow equations are elliptic if the flow is subsonic and hyperbolic if the flow is supersonic. Unless "direct" (Fourier-Galerkin) methods are used (Sec. 4.3.2), the calculation of a subsonic inviscid flow is itself iterative. In most cases the inviscid calculation will take longer than the shear-layer calculation, and computer time can be saved by using a lenient convergence criterion for the inviscid calculation in the first few passes through the sequence above. The reason for inserting a crude approximation of the displacement thickness in step 1 is that complete neglect of δ^* leads to a stagnation point at the trailing edge if the trailing-edge angle is finite. The first shear-layer calculation would inevitably predict separation just ahead of the trailing edge, leading to breakdown of the computation. An acceptable way of allowing for displacement thickness in step 1 is to do the inviscid-flow calculation with $\delta^* = 0$ and overwrite the sharp pressure rise near the trailing edge with a smoother variation. The final converged answer is unaffected.

Since the displacement thickness of the wake affects the pressure distribution over the airfoil, the pressure distribution in the wake region must be extracted from the inviscid calculation. This presents difficulties because the wake centerline is not known (the nominal centerline is the inviscid-flow streamline that leaves the trailing edge) and because conventional inviscid-flow programs often yield only the pressure distribution on the solid body. Only the early part of the wake, up to say, one-third of the chord downstream of the trailing edge, need be calculated accurately. As $x \to \infty$, $H \to 1$ in the wake, and because θ tends to a constant proportional to the drag, δ^* also tends to a constant.

The main effect of the shear layer on the pressure distribution on a lifting airfoil arises from the difference between the displacement thicknesses of the upper- and lower-surface boundary layers near the trailing edge. The result is that the centerline of the displacement surface has less camber than the centerline of the solid airfoil and therefore less lift at given incidence. Typically, the lift-curve slope, $dC_L/d\alpha$, is reduced by 10% or so, but the effects are frequently larger. Figure 11.5

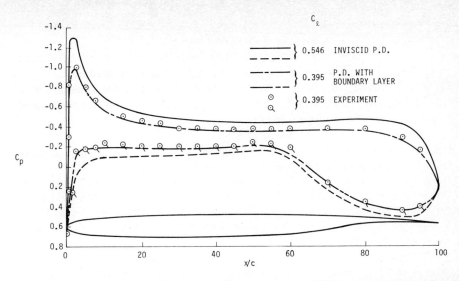

Fig. 11.5   Effect of boundary layer on the lift of a Douglas airfoil at a constant angle of attack.

shows the very large reduction in lift experienced by a "rear-loaded" airfoil, that is, one with a low upper-surface pressure up to a high percentage of the chord, followed by a strong adverse pressure gradient, which causes rapid boundary-layer growth. Pressure distributions of this sort are desirable in transonic flow to yield as much lift as possible without developing a large supersonic region with the danger of shock-wave formation. The real-flow pressure distribution looks closely the same as the inviscid-flow distribution at the same *lift*, but that lift is obtained at a higher *incidence* in the real flow.

Of course, the maximum attainable lift is limited by separation of the upper-surface boundary layer (Fig. 11.6) either at the trailing edge or in the strong adverse pressure gradient that develops near the leading edge of airfoils at high incidence. At the stall, small changes in incidence cause very large changes in δ^*, and it follows that the iterative process described above fails to converge at an incidence slightly below the stall because δ^* changes excessively from one iteration to the next.

Calculation of the viscous/inviscid interaction on nonlifting bodies is nominally easier, but in practice, nonlifting bodies often have large areas of separation, in which case they are called "bluff bodies" (as opposed to streamlined bodies). Not only does the TSL approximation break down in a separated-flow region, but in many cases the flow is oscillatory, that is, a vortex street is formed behind the body. An iterative calculation method would fail to surmount the second difficulty even if the first could be overcome. A method like that described above would, at best, fail to converge but continue to oscillate in a rather poor imitation of the real flow. Clearly, a time-dependent method is needed for oscillating flows. (Note that a time-dependent method is an acceptable, though probably not optimal, iterative method for predicting a steady flow: each time step corresponds to one iteration.)

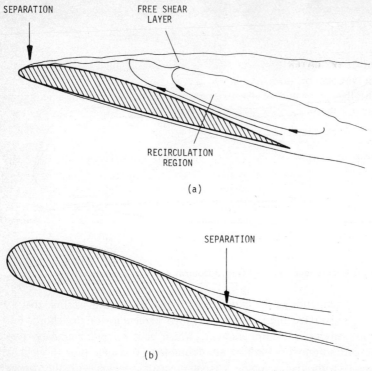

Fig. 11.6 (a) Separation near leading edge (thin airfoil). (b) Separation near trailing edge (thick airfoil).

A vortex street, or a less strongly periodic form of "buffeting," is most likely to occur when boundary layers separate from smoothly curved surfaces (Fig. 11.7a) and a perturbation in pressure gradient causes a perturbation in separation point. If separation occurs only at sharp edges, as in the case of a square-section prism, the effect of perturbations in pressure gradient on the separation region is weaker and the feedback mechanism that leads to a vortex street is less likely to occur. Also, weaker vortex streets are found behind asymmetrical bodies with one separation point at a sharp corner and the other on a smoothly curved surface (Fig. 11.5a). For instance, the vortex street behind a circular cylinder can be suppressed by a "splitter plate" extending downstream on the centerline. At the present time, calculation methods for separated flows with vortex streets are mostly either advanced time-dependent numerical schemes or simple quasi-steady models in which the vortex street is replaced by fixed rows of point vortices or by "free streamlines," constant-pressure surfaces separating the outer potential flow from a supposedly stationary separated-flow region. In the latter case, shear-layer concepts do not enter, and in the former case the time dependence of shear-layer position makes a shear-layer/inviscid-flow "matching" procedure almost as time-consuming as a full NS calculation [see Mehta and Lavan (1975) for an advanced matching method for separated flows over airfoils].

11.2.2 The Triple-Deck Model for Strongly Perturbed Boundary Layers

When a boundary layer enters a region of strong pressure gradient or experiences some other form of strong perturbation such as a sudden change in surface roughness, changes in viscous or turbulent stress are at first important only near the surface. In qualitative or analytical discussion, and possibly in numerical computation, it is helpful to divide the flow into three layers (Fig. 11.8) or "decks" as they are called by Stewartson and Williams (1969):

1. The lower deck is next to the surface, in which the boundary conditions at the surface enforce a rapid response to perturbations to maintain $u = 0$ and $\partial\tau/\partial y = \partial p/\partial x$ at the surface. The TSL equations apply and some further simplifications may be permissible; even in flows with high free-stream Mach number, the Mach number in the lower deck will be low, at least if the flow is laminar.
2. The main deck comprises the rest of the boundary-layer thickness. If the pressure gradient is much larger than the stress gradients, the latter can be neglected. A refinement that is useful in the case of turbulent flow is to assume that the shear stress remains constant along a given streamline (so that $\partial\tau/\partial y$ varies as the streamlines converge or diverge in the xy-plane). In turbulent flow either $\partial v/\partial x$ or longitudinal strain, $\partial u/\partial x$, may have large effects on shear stress, and in

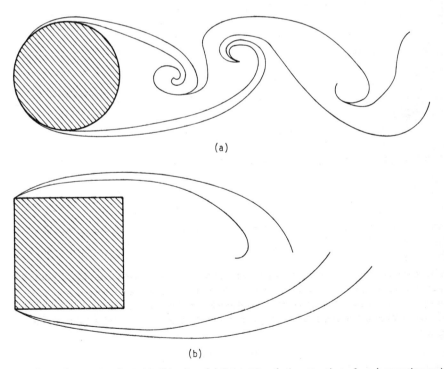

(a)

(b)

Fig. 11.7 Separation from bluff bodies. (a) Separation from smooth surface (separation points free to move). (b) Separation from sharp corner (separation points fixed).

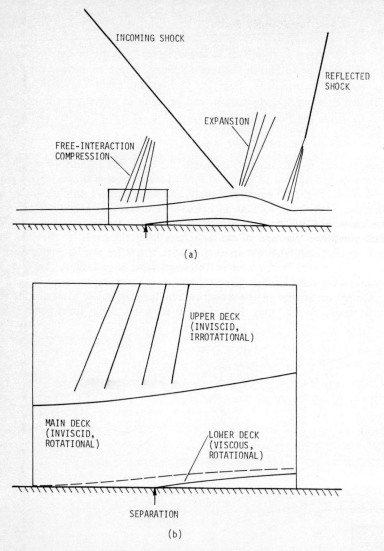

Fig. 11.8 The triple-deck model. (a) Interaction of boundary layer and incoming shock wave. (b) Triple-deck structure of free-interaction region.

compressible flow, variation of μ and ρ along a streamline in a pressure gradient causes further complications. However, the frozen-stress assumption will almost always be an improvement on neglecting stress gradients in the main deck.
3. The upper deck, outside the boundary layer, is the deck in which the inviscid-flow equations apply.

There are several versions of the triple-deck model. The concepts were introduced by Lighthill (1953) and used for low-speed flows by Stratford (1954) and

Townsend (1962) and for high-speed flows by Gadd (1953). The qualitative concept as described above is useful in a wide range of flows. The most rigorous treatment is for laminar flows at large Reynolds numbers; if the surface is nearly flat (so that the TSL equations would apply in the absence of a sudden perturbation), the normal-pressure gradient may be negligible in both the lower deck and the main deck, greatly simplifying the theory. For a review, see Stewartson (1974). In this case the thickness of the three decks can be shown to be of order $Re^{-5/8}$, $Re^{-1/2}$, and $Re^{-3/8}$, respectively, where Re is the Reynolds number based on distance from the boundary-layer origin. The thickness of the upper deck (i.e., the perturbed part of the inviscid flow) is of roughly the same order as the streamwise length of the strongly perturbed region. The triple-deck concept is relevant only if this streamwise length is fairly small, as in a shock-wave/boundary-layer interaction. In principle, the concept is applicable in subsonic flows also, although the absence of the simple supersonic-flow relation between pressure and flow angle makes the inviscid-flow solution more complicated. This is not quite the same as saying that the inviscid-flow equations are elliptic instead of hyperbolic as in the supersonic case; it is still possible to have a fairly localized region of viscous/inviscid interaction in subsonic flow, although care is needed when treating airfoils, whose whole pressure distribution can be changed by perturbations near the trailing edge.

In turbulent flow there are no rigorous order-of-magnitude arguments for the relative thickness of the three decks. In the classical case of boundary-layer perturbation by an incident shock (of strength Δp, say) the interaction region extends upstream for a distance (l, say) that is much shorter in turbulent flow than in laminar flow. This is true because a much larger pressure gradient is needed to produce separation, or even significant thickening, in the former case, and the pressure gradient is necessarily of order $\Delta p/l$. The nominal thickness of the upper deck is, as in laminar flow, of order l. Perturbations of Reynolds stress propagate normal to the streamlines at a speed of order u_τ, which is always much less than u, so that the lower deck is still thin in comparison to the main deck and the qualitative concepts of the triple deck are valid, although they cannot be used to obtain analytical solutions as in the laminar case.

The triple-deck model, as such, is confined to boundary layers because solid-surface boundary conditions are required to generate the lower deck. However, the initial region of an airfoil wake or a separated flow can also be regarded as a triple deck, with a lower deck resulting from the sudden withdrawal of solid-surface boundary conditions. The reversed-flow region that appears between the solid surface and the lower deck in separation bubbles (Fig. 11.8) forms a fourth deck in which a downstream-marching solution is not possible. Alternative ways of avoiding this problem are to march upstream in this deck, matching to a downstream-marching solution for the other decks, or to neglect the transport term, $u\,\partial u/\partial x$, when it is negative (Williams, 1975). The wake of an airfoil (see the right-hand enlargement in Fig. 1.1b) has a well-behaved lower deck, otherwise known as the inner wake. On either side is a main deck with an upper deck outside. In the symmetrical case the configuration is simply two triple decks back to back, and at least some of the results for single laminar boundary layers could be recovered.

However, the practical case is turbulent and usually asymmetrical, and so the triple-deck concept merely licenses the neglect of changes in viscous or turbulent stress outside the lower deck in the initial portion of an airfoil wake. A different but related approach to strong interactions and separated regions in supersonic flow is that developed by Lees and collaborators (Reeves and Lees, 1965; Alber and Lees, 1968). In this case, upstream influence is introduced by trial-and-error adjustment of the upstream conditions; the equations solved are actually the (parabolic) TSL equations and the (hyperbolic) supersonic inviscid-flow equations. The novel features are more mathematical than physical, and the reader is referred to the original papers for details.

11.2.3 Physics of Separation Bubbles

Separation of a boundary layer usually leads to the formation of a broad wake, often carrying a vortex street, which it is not helpful to discuss in terms of shear layers. However, there are some cases in which the separated shear layer reattaches to the surface. As well as in obvious cases like a forward-facing step on a long wall, reattachment can often occur if the pressure gradient decreases rapidly soon after separation, so that a strong reversed flow is not established. Thus the relatively small entrainment flow into the underside of the separated shear layer is larger than the available reversed flow coming from far downstream, and the shear layer literally sucks itself back onto the surface; the result is the separation bubble shown in Fig. 11.9. Note that the entrainment flow is necessarily supplied—in two-dimensional flow—by an upstream-going stream of fluid originating at the reattachment point. One streamline joins the separation and reattachment points. In three-dimensional configurations (and even in attempts to simulate two-dimensional conditions in a wind tunnel) fluid can enter or leave the bubble laterally. Because the rate of exchange of fluid between bubble and shear layer is quite low, lateral ventilation can have large effects (as can suction or injection through the surface).

The key to bubble behavior is transition in the free shear layer. It is rare for a shear layer that is laminar at separation to reattach as a laminar layer except on a

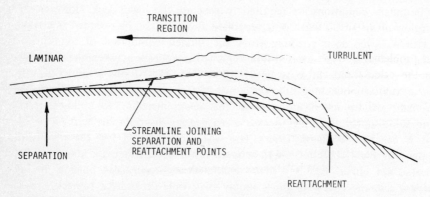

Fig. 11.9 Transitional separation bubble.

highly concave surface; the entrainment rate is not strong enough to overcome the effect of the continuing adverse pressure gradient in forcing fluid upstream and so widening the separated-flow region. The same applies to an initially turbulent layer; although the entrainment rate is larger than in a laminar flow, so is the pressure gradient required to produce separation. However, if an initially laminar shear layer (caused to separate by a weak pressure gradient) goes turbulent soon after separation, the entrainment rate rises rapidly and reattachment is more likely. The entrainment rate into the low-velocity side of a turbulent mixing layer in still air is only about 0.03 of the velocity difference across the layer, so that reattachment will occur only if the shear layer would remain close to the surface even in the absence of mixing.

Although free shear layers have a point of inflection in their velocity profiles and therefore become unstable at very low Reynolds numbers, it takes a considerable streamwise distance before the small naturally occurring disturbances have amplified to the point at which the entrainment rate approaches the fully turbulent value. Separation bubbles therefore occur only if the Reynolds number is high enough for transition to be nearly complete before the shear layer has moved too far from the surface. Of course, an upper limit for the appearance of transitional separation bubbles is the Reynolds number at which transition occurs upstream of the point at which laminar separation would occur. The range of Reynolds number in which bubbles occur depends on geometry, but a large number of typical airfoil sections exhibit separation bubbles at typical wind-tunnel Reynolds numbers and even in flight.

In this elementary account we do not discuss the details of separation bubbles and their effect on the characteristics of airfoils or other bodies. For a review see Tani (1964).

11.3 INVERSE BOUNDARY-LAYER CALCULATIONS

As shown in previous sections and chapters, boundary-layer calculations are performed for given boundary conditions, which, for two-dimensional incompressible flows, are usually provided in the form

$$u(x,0) = 0 \quad v(x,0) = v_w(x) \quad u(x,\delta) = u_e(x) \tag{11.3.1}$$

In some problems, however, the external-velocity distribution may not be known prior to the calculations and is to be determined in order to satisfy an additional boundary condition such as the displacement thickness, $\delta^*(x)$, or the wall shear, $\tau_w(x)$. These problems, sometimes called inverse boundary-layer methods, find applications in many practical problems. For example, the airfoils discussed in an article by Liebeck (1976) are designed on the principle that, in certain regions of an airfoil, the flow deceleration is such that the wall shear is nearly equal to zero. Other important applications of inverse boundary-layer procedures occur in the calculation of duct flows, such as those discussed in Secs. 5.7 and 7.3, where the displacement effect determines the pressure distribution.

A particularly important application of inverse boundary-layer procedures is their potential ability to compute separating and reattaching flows. It is well known that, for a prescribed external velocity distribution, the boundary-layer equations are singular at separation. For an excellent review, see Brown and Stewartson (1969). To compute flows with regions of negative wall shear (i.e., flows including separation points) by using the boundary-layer equations, it is necessary to specify either the displacement thickness or the wall shear and compute the external velocity distribution. For example, Catherall and Mangler (1966) solved the laminar boundary-layer equations in the usual way until the separation point was approached. Then by assuming that the displacement thickness behaved in a prescribed and continuous manner in the region of the separation point, they calculated the external velocity distribution in that region for a prescribed displacement-thickness distribution. Their numerical solutions did not reveal singular behavior at separation.

In recent years, a number of studies of inverse boundary-layer calculations have been conducted; see, for example, Cebeci and Keller (1973), Klineberg and Steger (1974), Carter (1974, 1975), Williams (1975), and Cebeci (1976a, b). In this section we describe a new inverse boundary-layer procedure called the Mechul function approach, which differs from the procedure discussed in Sec. 7.3 in that the pressure is treated as an unknown. While the procedure of Sec. 7.3 works only for flows with positive wall shear, the new procedure works well for flows with positive and negative wall shear.

11.3.1 Mechul Function Approach

To illustrate the Mechul function approach, we consider an incompressible laminar flow with specified displacement-thickness distribution $\delta^*(x)$. Using the definition of stream function, for two-dimensional flows the boundary-layer equations (5.1.1) and (5.1.2) can be written as a third-order equation in dimensionless form:

$$\psi''' - p_x^* = \psi' \frac{\partial \psi'}{\partial x} - \psi'' \frac{\partial \psi}{\partial x} \qquad (11.3.2)$$

where $\psi = F$. Equation (11.3.2), regarded as an eigenvalue problem for p^*, can be solved subject to the boundary conditions

$$Y = 0 \quad \psi = \psi' = 0 \qquad (11.3.3a)$$

$$Y = Y_e \quad \psi' = u_e^* \quad \psi = u_e^* [Y_e - \tilde{\delta}^*(x)] \qquad (11.3.3b)$$

Here the primes denote differentiation with respect to Y, $u_e^* = u_e/u_0$, $\tilde{\delta}^*(x) = (\delta^*/L)\sqrt{R_L}$, and the subscript refers to the boundary-layer edge [see Eqs. (5.7.4), (5.7.7), and (5.7.8)]. The external velocity, u_e^*, is related to the static pressure, p^*, through Bernoulli's equation, namely,

$$p^* + \frac{(u_e^*)^2}{2} = \text{const} \qquad\qquad (11.3.4)$$

In order to use the box method (Sec. 7.2), we write Eqs. (11.3.2) and (11.3.3) as a first-order system. With the function p^* treated as an unknown and with the help of the y-momentum equation (3.1.15), we can write

$$\psi' = u \qquad\qquad (11.3.5a)$$

$$u' = v \qquad\qquad (11.3.5b)$$

$$p' = 0 \qquad\qquad (11.3.5c)$$

$$v' - p_x = u\frac{\partial u}{\partial x} - v\frac{\partial \psi}{\partial x} \qquad\qquad (11.3.5d)$$

Similarly, Eqs. (11.3.3) become

$$y = 0 \quad \psi = u = 0 \qquad\qquad (11.3.6a)$$

$$y = y_e \quad u = u_e \quad \psi = u_e\left[y_e - \delta^*(x)\right] \qquad\qquad (11.3.6b)$$

Here for convenience we have dropped the asterisks on u_e and p and have denoted Y by y.

The finite-difference equations for the above system are very similar to those given in Sec. 7.2. Again, the finite-difference approximations of Eqs. (11.3.5a)–(11.3.5c) are written for the midpoint $(x^n, y_{j-1/2})$ of the segment P_1P_2 shown in Fig. 11.10. These approximations are

$$\frac{\psi_j^n - \psi_{j-1}^n}{h_j} = u_{j-1/2}^n \qquad\qquad (11.3.7a)$$

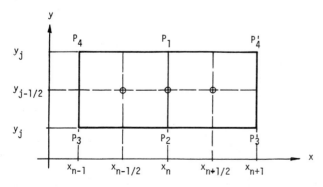

Fig. 11.10   Net rectangle for difference approximations.

$$\frac{u_j^n - u_{j-1}^n}{h_j} = v_{j-1/2}^n \tag{11.3.7b}$$

$$\frac{p_j^n - p_{j-1}^n}{h_j} = 0 \tag{11.3.7c}$$

where $h_j = \Delta y_j$. Similarly, Eq. (11.3.5d) is approximated by centering about the midpoint $(x^{n-1/2}, y_{j-1/2})$ of the rectangle $P_1 P_2 P_3 P_4$ to get

$$\frac{v_j^n - v_{j-1}^n}{h_j} + \alpha[(\psi v)_{j-1/2}^n + v_{j-1/2}^{n-1} \psi_{j-1/2}^n - (u^2)_{j-1/2}^n - \psi_{j-1/2}^{n-1} v_{j-1/2}^n]$$
$$- 2\alpha(p_{j-1/2}^n - p_{j-1/2}^{n-1}) = R_{j-1/2}^{n-1} \tag{11.3.7d}$$

Here

$$R_{j-1/2}^{n-1} = -\frac{v_j^{n-1} - v_{j-1}^{n-1}}{h_j} + \alpha[(\psi v)_{j-1/2}^{n-1} - (u^2)_{j-1/2}^{n-1}]$$

$$\alpha = \frac{1}{x_n - x_{n-1}}$$

Similarly, the boundary conditions Eq. (11.3.6) and the relation Eq. (11.3.4) become

$$\psi_0^n = 0 \qquad u_0^n = 0 \qquad u_J^n = u_e^n \qquad \psi_J^n = u_e^n[y_e - \delta^*(x)] \tag{11.3.8}$$

$$p^n + \frac{(u_e^2)^n}{2} = p^{n-1} + \frac{(u_e^2)^{n-1}}{2} \tag{11.3.9}$$

Equations (11.3.7)–(11.3.9) form a system of $4J + 4$ equations for the solution of $4J + 4$ unknowns, $(\psi_j^n, u_j^n, v_j^n, p_j^n)$, $j = 0, 1, \ldots, J$. Before the system is solved by the block elimination method discussed in Appendix 7A, the nonlinear difference equations are linearized by Newton's method (see Sec. 7.2.1). For further details, see Cebeci (1976b).

Note that the solution of the governing equations (11.3.2) and (11.3.3) requires initial conditions that can more readily be provided by solving the equations in transformed variables or by using the procedure described in Ex. 8B.3.

For flows with negative wall shear, it is necessary to make approximations to the governing equations to continue the calculations through separation in the downstream direction (see Sec. 4.3). Here we use the approximation first suggested by Reyhner and Flugge-Lotz (1968). This approximation, referred to as FLARE by Williams (1975), neglects the $u \, \partial u/\partial x$ term in the region of negative u-velocity. All inverse boundary-layer procedures, including the Mechul function procedure, use this approximation for regions of separated flow. Once a solution is obtained with this approximation, however, more accurate calculations can be made by reinstating the $u \, \partial u/\partial x$ term. This can be done by using a forward and backward difference scheme. In the region of reverse flow ($u_j \leqslant 0$), we write the difference approximations of Eq. (11.3.5d) for midpoint ($x^{n+1/2}, y_{j-1/2}$) of the rectangle $P_1 P_2 P_3' P_4'$ to get

$$\frac{v_j^n - v_{j-1}^n}{h_j} - \beta[(\psi v)_{j-1/2}^n + v_{j-1/2}^{n+1}\psi_{j-1/2}^n - (u^2)_{j-1/2}^n - \psi_{j-1/2}^{n+1}v_{j-1/2}^n]$$
$$- 2\beta(p_{j-1/2}^{n+1} - p_{j-1/2}^n) = R_{j-1/2}^{n+1} \quad (11.3.10)$$

Here

$$R_{j-1/2}^{n+1} = -\frac{v_j^{n+1} - v_{j-1}^{n+1}}{h_j} + \beta[(\psi v)_{j-1/2}^{n+1} + (u^2)_{j-1/2}^{n+1}]$$

$$\beta = \frac{1}{x_{n+1} - x_n}$$

With this procedure the calculations are now repeated, starting at the x-station where the calculations first indicated reverse flow and proceeding downstream all the way to the final x-station, to obtain new solutions. Then a second sweep is started with these new solutions to obtain another new solution; this process is repeated until the solutions do not indicate any changes. This procedure, referred to as DUIT (downstream, upstream iteration) by Williams, usually gives converged solutions with two or three DUITs.

11.3.2 Calculated Results for Flows with Prescribed Displacement Thickness

Several sample calculations have been made, using the above procedure, for both laminar and turbulent flows with prescribed displacement thickness (see Cebeci, 1976b). The turbulent flow calculations were made with the eddy-viscosity formulation described in Sec. 8.3. In Figs. 11.11 and 11.12 we present two of the calculations reported by Cebeci (1976b) for laminar flows. The first one, Fig. 11.11, corresponds to a flow previously computed by Carter (1975) who used a different numerical procedure and a slightly different FLARE assumption. The calculations of Cebeci (1976b) were made with the FLARE approximation discussed here. Carter's calculations used both FLARE and DUIT. The results shown in Fig. 11.12 were obtained with a prescribed δ^*, which was deduced by Briley (1971) from the solution of the NS equations for a flow involving separation and reattachment. As is seen from these comparisons, the agreement between the local skin-friction coefficient obtained by the inverse boundary-layer procedure and the coefficients obtained from the solution of the NS equations is excellent.

11.4 EXPLANATION OF THE FRONTISPIECE

The frontispiece photographs show the flow in an idealized wing–body junction, a very common type of streamwise corner flow (Sec. 10.1.3). The body is a

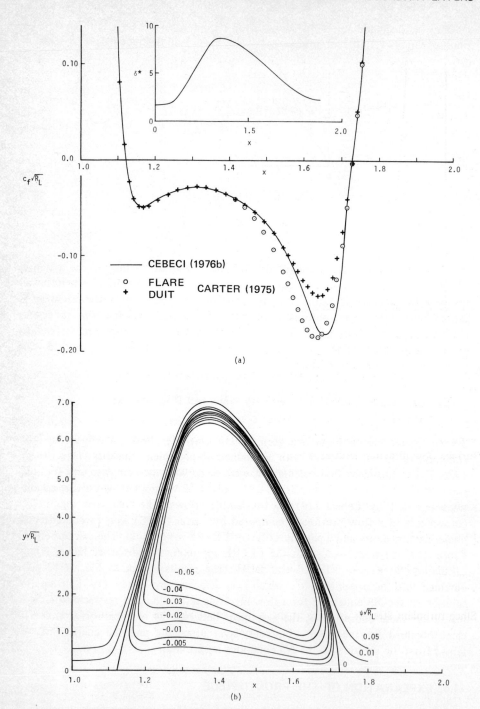

Fig. 11.11 (a) Local skin-friction distribution for a laminar flow with prescribed displacement thickness. (b) Streamline pattern in separation bubble.

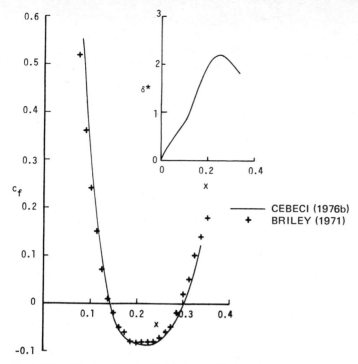

Fig. 11.12 Calculated local skin-friction coefficient distributions for separating and reattaching flow.

wind-tunnel wall, taken as the xz-plane with x in the streamwise direction, and the wing is a two-dimensional constant-chord airfoil operating at small positive lift. The surface flow direction is at an angle

$$\lim_{y \to 0} \tan^{-1} \frac{w}{u}$$

to the x-axis. The limit is of type 0/0 and can be replaced by

$$\lim_{y \to 0} \tan^{-1} \frac{\partial w / \partial y}{\partial u / \partial y}$$

Since turbulent stresses are zero at the surface (Sec. 6.2) this is the same as

$$\lim_{y \to 0} \tan^{-1} \frac{\tau_z}{\tau_x}$$

in laminar or turbulent flow. In the photographs the tunnel wall was covered with a mixture of oil and dye, which forms streaks in the direction of the surface shear-stress vector whose components are (τ_x, τ_z). The above analysis identifies this direction as that of the limiting streamline at the surface.

Comparing the upper parts of the two photographs, we see that the pressure field created by the airfoil produces cross-flow angles (Sec. 10.1.3) much larger in a laminar boundary layer than in a turbulent one. The external streamlines are very nearly the same in the two cases and are deflected even less than the limiting streamlines in the turbulent case. The cross flow is toward the airfoil because the pressure on the upper surface of a lifting airfoil is lower than ambient. This is a spectacular example of the greater ability of a turbulent boundary layer to withstand pressure gradients.

At about midchord in the upper picture, transition occurs over most of the tunnel wall. The mechanism is the three-dimensional (longitudinal-vortex) instability described in Sec. 9.6.1. The pronounced streaks in the transition region probably show the location of the vortices but may be partly attributable to agglomerations of dye. In some parts of the field the laminar-flow streamlines cross the vortex streaks. This is because the vortex direction is that of the stream at the profile inflection point (Sec. 9.6.1), which is closer to the free-stream direction than to the limiting streamline direction. The fact that *both* directions appear in the oil-flow picture emphasizes that transition is an intermittent process, the vortices producing turbulent spots for only part of the time. The oil flow, of course, shows a long-time average. Near the airfoil, transition occurs upstream of the leading edge and the dark, wavelike streaks seen just above the leading edge are probably caused by transitional vortices.

The bright "bow-wave" streak ahead of the airfoil is a ridge line of separation (Sec. 10.1.4) with surface streamlines approaching from either side. Near the bottom of the top picture, where the streamlines from the front and rear approach the ridge line at right angles, is a saddle point of separation at which the skin friction is zero. Naturally, the oil-flow pattern is rather confused in this region. The ridge line is much further from the airfoil in the laminar case than in the turbulent case, another example of the sensitivity of laminar flows to pressure gradients.

Very close to the leading edge the skin friction is quite high, as is seen by the well-formed dye streaks in the top photograph. Skewing of the boundary-layer flow (through an angle greater than $90°$ in some places) causes the generation of streamwise vorticity (secondary flow of the first kind, Sec. 10.1.3) and a "horseshoe" vortex is wrapped around the airfoil. The direction of the vortex on the top surface is clockwise viewed from upstream. However, the ridge line waves toward the airfoil (instead of away from it, as would be expected from the vortex-induced motion alone) because of the strong cross flow from the tunnel wall boundary layer. Particularly in the lower (turbulent) picture, a very strong cross flow induced by the vortex can be seen to start at the trailing edge. In the laminar case there appears to be a region of separation (in the two-dimensional sense, $\tau_w = 0$) just above the trailing edge. The convergence of streamlines near the trailing edge suggests considerable boundary-layer thickening, and evidently the slow-moving fluid cannot surmount the pressure rise needed to reach the trailing edge.

Photographs similar to these, taken at higher incidences, show very strong interactions between the wing and body boundary layers and the trailing vorticity shed by the wing. In the present case, however, viscous/inviscid interaction is not very important, but most of the other phenomena discussed in this book appear.

Oil-flow photographs are frequently used to infer *free-stream* flow directions. This must be done with great caution; the frontispiece photographs show that the behavior of boundary layers is much more complicated, and much more interesting, than the behavior of inviscid flows. We hope that the reader of this book has maintained an interest in shear layers while becoming more aware of their complications.

References

Ackerberg, R. C., and J. H. Phillips: The Unsteady Laminar Boundary Layer on a Semi-Infinite Flat Plate Due to Small Fluctuations in the Magnitude of the Free-Stream Velocity, *J. Fluid Mech.*, vol. 51, p. 137, 1972.

Alber, I. E., and L. Lees: Integral Theory for Supersonic Turbulent Base Flows. *AIAA J.*, vol. 6, p. 1343, 1968.

Bansod, P., and P. Bradshaw: The Flow in S-Shaped Ducts, *Aero. Quart.*, vol. 23, p. 131, 1972.

Barbin, A. R., and J. B. Jones: Turbulent Flow in the Inlet Region of a Smooth Pipe, *J. Basic Eng.*, vol. 85, p. 29, 1963.

Benney, D. J., and C. C. Lin: On the Secondary Motion Induced by Oscillation in a Shear Flow, *Phys. Fluids*, vol. 3, p. 656, 1960.

Birch, S. F., D. H. Rudy, and D. M. Bushnell (eds.): Free Turbulent Shear Flows, *NASA* SP-321, 1972.

Bird, R. B., W. E. Stewart, and E. N. Lightfoot: "Transport Phenomena," chap. 2, John Wiley, New York, 1960.

Bradshaw, P.: "Experimental Fluid Mechanics," Pergamon Press, New York, 1970.

Bradshaw, P.: "An Introduction to Turbulence and Its Measurement," Pergamon Press, New York, 1971a.

Bradshaw, P.: Calculation of Three-Dimensional Turbulent Boundary Layers, *J. Fluid Mech.*, vol. 46, p. 417, 1971b.

Bradshaw, P.: The Understanding and Prediction of Turbulent Flow, *Aero. J.*, vol. 76, p. 403, 1972.

Bradshaw, P.: Effect of Streamline Curvature on Turbulent Flow, *AGARDograph* 169, 1973.

Bradshaw, P.: Suggested Origin of Prandtl's Mixing-Length Theory, *Nature*, vol. 249, p. 135, 1974.

Bradshaw, P.: Complex Turbulent Flows, *J. Fluids Eng.*, vol. 97, p. 146, 1975.

Bradshaw, P. (ed.): "Topics in Applied Physics, Turbulence," Volume 12; Springer, Heidelberg, 1976.

Bradshaw, P., and D. H. Ferriss: The Response of a Retarded Equilibrium Boundary Layer to the Sudden Removal of Pressure Gradient, *NPL Aero Rept.* 1145, 1965.

Bradshaw, P., and D. H. Ferriss: Applications of a General Method of Calculating Turbulent Shear Layers, *J. Fluids Eng.*, vol. 94, p. 345, 1972.

Bradshaw, P., and M. G. Terrell: The Response of a Turbulent Boundary Layer on an "Infinite" Swept Wing to the Sudden Removal of Pressure Gradient, *NPL Aero. Rept.* 1305, ARC 31514, 1969.

Bradshaw, P., D. H. Ferriss, and N. P. Atwell: Calculation of Boundary-Layer Development Using the Turbulent Energy Equation, *J. Fluid Mech.*, vol. 28, p. 593, 1967.

Bradshaw, P., G. Mizner, and K. Unsworth: Calculation of Compressible Turbulent Boundary Layers on Straight-Tapered Swept Wings, *AIAA J.*, vol. 14, p. 399, 1976.

Briley, W. R.: A Numerical Study of Laminar Separation Bubbles Using the Navier-Stokes Equations, *J. Fluid Mech.*, vol. 47, no. 4, pp. 713–736, 1971.

Brown, S. N., and K. Stewartson: Laminar Separation, *Ann. Rev. Fluid Mech.*, vol. 1, p. 45, 1969.

Bushnell, D. M., and I. E. Beckwith: Calculation of Nonequilibrium Hypersonic Turbulent Boundary Layers and Comparisons With Experimental Data, *AIAA J.*, vol. 8, p. 1462, 1970.

Bushnell, D. M., A. M. Cary, and B. B. Halley: Mixing Length in Low-Reynolds-Number Compressible Turbulent Boundary Layers, *AIAA J.*, vol. 13, p. 1119, 1975.

Carter, J. E.: Solutions for Laminar Boundary Layers with Separation and Reattachment, *AIAA Pap.* 74-583, 1974.

Carter, J. E.: Inverse Solutions for Laminar Boundary-Layer Flows with Separation and Reattachment, *NASA Rept.* TR R-447, 1975.

Catherall, D. and K. W. Mangler: The Integration of the Two-Dimensional Laminar Boundary-Layer Equations Past the Point of Vanishing Skin Friction, *J. Fluid Mech.*, vol. 26, p. 163, 1966.

Cebeci, T.: Behavior of Turbulent Flow Near a Porous Wall With Pressure Gradient, *AIAA J.*, vol. 8, p. 2152, 1970a.

Cebeci, T.: Laminar and Turbulent Incompressible Boundary Layers on Slender Bodies of Revolution in Axial Flow, *J. Basic Eng.*, vol. 92, p. 545, 1970b.

Cebeci, T.: Variation of the Van Driest Damping Parameter With Mass Transfer, *AIAA J.*, vol. 11, p. 237, 1973a.

Cebeci, T.: Eddy-Viscosity Distribution in Thick Axisymmetric Turbulent Boundary Layers, *J. Fluids Eng.*, vol. 95, p. 319, 1973b.

Cebeci, T.: Kinematic Eddy Viscosity at Low Reynolds Numbers, *AIAA J.*, vol. 11, p. 102, 1973c.

Cebeci, T.: Calculation of Three-Dimensional Boundary Layers, pt. 1, Swept Infinite Cylinders and Small Cross Flow, *AIAA J.*, vol. 12, p. 779, 1974.

Cebeci, T.: Calculation of Three-Dimensional Boundary Layers, pt. 2, Three-Dimensional Flows in Cartesian Coordinates, *AIAA J.*, vol. 13, p. 1056, 1975.

Cebeci, T.: An Inverse Boundary-Layer Method for Compressible Laminar and Turbulent Boundary Layers, *J. Aircraft*, vol. 13, p. 709, 1976a.

Cebeci, T.: Separated Flows and Their Representation by Boundary-Layer Equations, *Mech. Eng. Rept.* ONR-CR215-234-2, California State University at Long Beach, 1976b.

Cebeci, T.: Calculation of Unsteady Two-Dimensional Laminar and Turbulent Boundary Layers With Fluctuations in External Velocity, *Proc. Roy. Soc. Ser. A*, in press, 1977a.

Cebeci, T.: On the Solution of Laminar Flow Over a Circular Cylinder Started Impulsively from Rest, in "Professor A. Walz' 70th Birthday Anniversary Volume," to be published, 1977b.

Cebeci, T.: Paper in Preparation, 1977c.

Cebeci, T., and P. Bradshaw: "Convective Heat Transfer in Boundary Layers," Volume in Preparation.

Cebeci, I., and K. C. Chang: An Efficient Method for Laminar and Turbulent Boundary Layers in Duct Flows, Paper in Preparation, 1977.

Cebeci, T., and H. B. Keller: Shooting and Parallel Shooting Methods for Solving the Falkner-Skan Boundary-Layer Equation, *J. Comp. Phys.*, vol. 7, p. 289, 1971.

Cebeci, T., and H. B. Keller: Laminar Boundary Layers With Assigned Wall Shear, in H. Cabannes and R. Temam (ed.), "Lecture Notes in Physics," vol. 18, pt. II, "Proceedings of the Third International Conference on Numerical Methods in Fluid Dynamics," pp. 79–85, Springer-Verlag, Berlin, 1973.

Cebeci, T., and H. B. Keller: Paper in Preparation, 1977.

Cebeci, T., and A. M. O. Smith: "Analysis of Turbulent Boundary Layers," Academic Press, New York, 1974.

Cebeci, T., K. Kaups, G. J. Mosinskis, and J. A. Rehn: Some Problems of the Calculation of Three-Dimensional Boundary-Layer Flows on General Configurations, *NASA Rept.* CR-2285, 1973.

Cebeci, T., K. Kaups, and A. Moser: Calculation of Three-Dimensional Boundary Layers, pt. 3, Three-Dimensional Incompressible Flows in Curvilinear Orthogonal Coordinates, *AIAA J.*, vol. 14, p. 1090, 1976.

Cebeci, T., K. Kaups, and J. A. Ramsey: A General Method for Calculating Three-Dimensional Compressible Laminar and Turbulent Boundary Layers on Arbitrary Wings, *NASA Rept.* CR-2777, 1977.

Chapman, S., and T. G. Cowling: "The Mathematical Theory of Non-Uniform Gases," University Press, Cambridge, 1970.

Clauser, F. H.: Turbulent Boundary Layers in Adverse Pressure Gradient, *J. Aero. Sci.*, vol. 21, no. 91, 1954.

Clauser, F. H.: The Turbulent Boundary Layer, *Advan. Appl. Mech.*, vol. 4, p. 1, 1956.

Coles, D.: The Law of the Wake in the Turbulent Boundary Layer, *J. Fluid Mech.*, vol. 1, p. 191, 1956.

Coles, D.: The Turbulent Boundary Layer in a Compressible Fluid, *Rand Corp., Santa Monica, Rept.* R-403-PR, appendix A, 1962.

Coles, D.: A Survey of Data for Turbulent Boundary Layers With Mass Transfer, *AGARD Conf. Proc.*, *93*, *Turbulent Shear Flows* (also AD 738-102, N72-20273), 1971.

Coles, D., and E. A. Hirst: "Computation of Turbulent Boundary Layers—1968, AFOSR-IFP-Stanford Conference," vol. 2, Thermosciences Division, Stanford University, Stanford, Calif., 1969.

Cornish, J. J., III, and D. W. Boatwright: Application of Full-Scale Boundary-Layer Measurements to Drag Reduction of Airships, *Aerophys. Dept., Mississippi State Univ., Jackson, Rept. 28,* 1960.

Courant, R. E., and D. Hilbert: "Methods of Mathematical Physics," Interscience Publishers, Inc., New York, 1961.

Dean, R. B.: The Application of a Conditional Sampling Technique to the Understanding of Turbulent Interacting Shear Layers in Duct Flow, *Proc. 5th Australian Conf. on Hydraul. and Fluid Mech., Christchurch, New Zealand,* 1974 (see also *J. Fluid Mech.*, vol. 76, p. 641, 1976).

Dean, R. B.: Turbulent flow in Ducts and Pipes: A Literature Survey, *J. Fluids Eng.*, in press, 1976.

Dennis, S. C. R.: The Motion of a Viscous Fluid Past an Impulsively Started Semi-infinite Flat Plate. *J. Inst. Math. Applicat.*, vol. 10, pp. 105–117, 1972.

Desai, S. S., and K. W. Mangler: Incompressible Laminar Boundary-Layer Flow Along a Corner Formed by Two Intersecting Planes, RAE TR 74063, ARC 35505, 1974.

Dhawan, S., and R. Narasimha: Some Properties of Boundary-Layer Flow During the Transition From Laminar to Turbulent Motion, *J. Fluid Mech.*, vol. 3, p. 418, 1958.

Donaldson, C. duP.: A Computer Study of an Analytical Model of Boundary-Layer Transition, *AIAA J.*, vol. 7, p. 271, 1969.

Drazin, P. G. and L. N. Howard: Hydrodynamic Stability of Parallel Flow of Inviscid Fluid, *Advan. Appl. Mech.*, vol. 9, pp. 1–89, 1966.

Dunn, D. W., and C. C. Lin: On the Stability of the Laminar Boundary Layer in a Compressible Fluid, *J. Aero. Sci.*, vol. 22, p. 455, 1955.

Durst, F., A. Melling, and J. H. Whitelaw: "Principles and Practice of Laser-Doppler Anemometry," Academic Press, New York, 1976.

Elder, J. W.: An Experimental Investigation of Turbulent Spots and Breakdown to Turbulence, *J. Fluid Mech.*, vol. 9, p. 235, 1960.

Emmons, H. W.: The Laminar-Turbulent Transition in a Boundary Layer, *J. Aero. Sci.*, vol. 18, p. 490, 1951.

Falkner, V. G., and S. W. Skan: Some Approximate Solutions of the Boundary-Layer Equations, ARC R&M 1314, 1930 (see also *Phil. Mag.*, vol. 12, p. 865, 1931).

Finley, P. J., K. C. Phoe, and C. J. Poh: Velocity Measurements in a Thin Turbulent Water Layer, *Houille Blanche*, vol. 21, p. 713, 1966.

Gadd, G. E.: Interactions Between Wholly Laminar or Wholly Turbulent Boundary Layer and Shock Waves Strong Enough to Cause Separation, *J. Aero. Sci.*, vol. 20, p. 11, 1953.

Garabedian, P.: "Partial Differential Equations," Interscience Publishers, Inc., New York, 1964.

Gibbings, J. C.: *Aero. Res. Council Current Pap. 462,* 1959.

Goldstein, S.: "Modern Developments in Fluid Dynamics," Dover, New York, 1965.

Görtler, H.: A New Series for the Calculation of Steady Laminar Boundary-Layer Flows, *J. Math. Mech.*, vol. 6, p. 1, 1957 (see also Zahlentafeln universeller Funktionen zur neuen Reihe für die Berechnung laminarer Grenzschichten, Bericht 34 of the Deutsche Versuchsanstalt für Luftfahrt).

Granville, P. S.: The Calculation of the Viscous Drag of Bodies of Revolution, *David W. Taylor Model Basin Rept.* 849, 1953.

Granville, P. S.: The Determination of the Local Skin Friction and the Thickness of Turbulent Boundary Layers From the Velocity Similarity Laws, *David W. Taylor Model Basin Rept.* 1340, 1959.

Granville, P. S.: The Prediction of Transition From Laminar to Turbulent Flow in Boundary Layers on Bodies of Revolution, in R. D. Cooper and S. W. Doroff, (eds.), "Tenth Symposium Naval Hydrodynamics," U.S. Government Printing Office, Washington, D.C., pp. 705–729, 1974.

Granville, P. S.: Similarity-Law Entrainment Method for Turbulent Boundary Layers in Pressure Gradients, *David Taylor Naval Ship Res. and Devel. Center Rept.* 4657, 1976a.

Granville, P. S.: Similarity-Law Entrainment Method for Thick Axisymmetric Turbulent Boundary Layers in Pressure Gradients, *David Taylor Naval Ship Res. and Devel. Center Rept.* 4525, 1976b.

Hackett, J. E., and D. K. Cox: The Three-Dimensional Mixing Layer Between Two Grazing Perpendicular Streams, *J. Fluid Mech.*, vol. 43, p. 77, 1970.

Hall, M. G.: A Numerical Method for Calculating Unsteady Two-Dimensional Laminar Boundary Layers, *Ing. Arch.*, vol. 38, p. 97, 1969.

Hama, F. R.: Boundary-Layer Characteristics for Smooth and Rough Surfaces, *Trans. Soc. Naval Arch. Marine Eng.*, vol. 62, p. 333, 1954.

Hanjalic, K., and B. E. Launder: A Reynolds-Stress Model of Turbulence, and Its Application to Thin Shear Flows, *J. Fluid Mech.*, vol. 52, p. 609, 1972.

Hansen, A. G.: "Similarity Analysis of Boundary-Value Problems in Engineering," Prentice-Hall, Englewood Cliffs, N.J., 1964.

Hartree, D. R.: On an Equation Occurring in Falkner and Skan's Approximate Treatment of the Equations of the Boundary Layer, *Proc. Cambridge Phil. Soc.*, vol. 33, p. 223, 1937.

Head, M. R.: Entrainment in the Turbulent Boundary Layers, ARC R&M 3152, 1958.

Head, M. R., and V. C. Patel: Improved Entrainment Method for Calculating Turbulent Boundary-Layer Development, ARC R&M 3643, 1969.

Hoerner, S. F.: "Fluid Dynamic Drag," Hoerner Fluid Dynamics, Brick Town, N.J., 1958.

Howarth, L.: Rayleigh's Problem for a Semi-Infinite Plate, *Proc. Cambridge Phil. Soc.*, vol. 46, p. 127, 1950.

Iglisch, R.: Exact Calculation of Laminar Boundary Layer in Longitudinal Flow Over a Flat Plate With Homogeneous Suction, *NACA Tech. Memo.* 1205, 1949.

Ioselevich, V. A., and V. I. Pilipenko: Logarithmic Velocity Profile for Flow of a Weak Polymer Solution Near a Rough Surface, *Sov. Phys. Dokl.*, vol. 18, p. 790, 1974.

Isaacson, E., and H. B. Keller: "Analysis of Numerical Methods," John Wiley, New York, 1966.

Jaffe, N. A., and T. T. Okamura: The Transverse Curvature Effect of the Incompressible Laminar Boundary Layer for Longitudinal Flow over a Cylinder, *Z. Angew. Math. Phys.*, vol. 19, p. 564, 1968.

Jaffe, N. A., T. T. Okamura, and A. M. O. Smith: Determination of Spatial Amplification Factors and Their Application to Predicting Transition, *AIAA J.*, vol. 8, p. 301, 1970.

Johnston, J. P.: Three-Dimensional Turbulent Boundary Layer, *M.I.T. Gas Turbine Lab. Rept.* 39, 1957.

Kaplan, R. E.: The Stability of Laminar Incompressible Boundary Layers in the Presence of Compliant Boundaries, M.I.T. *Aero-Elastic and Structures Res. Lab. Rept.* ASRL-TR-116-1, 1964.

Karlsson, S. K. F.: An Unsteady Turbulent Boundary Layer, *J. Fluid Mech.*, vol. 5, p. 622, 1959.

Kaups, K.: Transition Prediction on Bodies of Revolution, *Douglas Aircraft Co., Long Beach, Calif., Rept.* MDC J6530, 1975.

Keller, H. B.: "Numerical Methods for Two-Point Boundary Value Problems," Ginn-Blaisdell, Waltham, Mass., 1968.

Keller, H. B.: A New Difference Scheme for Parabolic Problems, in J. Bramble (ed.), "Numerical Solutions of Partial Differential Equations," vol. II, Academic Press, New York, 1970.

Keller, H. B.: Accurate Difference Methods for Nonlinear Two-Point Boundary-Value Problems, *SIAM J. Numer. Anal.*, vol. 11, pp. 305–320, 1974.

Keller, H. B.: "Numerical Solution of Two-Point Boundary Value Problems," Regional Conference Series in Applied Mathematics, Society for Industrial and Applied Mathematics, Philadelphia, Pa., 1976.

Keller, H. B. and T. Cebeci: Accurate Numerical Methods for Boundary-Layer Flows, pt. 1, Two-Dimensional Laminar Flows, in "Lecture Notes in Physics, 8, Proceedings of the Second International Conference on Numerical Methods in Fluid Dynamics," p. 92, Springer-Verlag, New York, 1971.

Keller, H. B., and T. Cebeci: Accurate Numerical Methods for Boundary-Layer Flows, pt. 2, Two-Dimensional Turbulent Flows, *AIAA J.*, vol. 10, p. 1193, 1972.

Keller, H. B., and T. Cebeci: Numerical Methods for the Orr-Sommerfeld Equation, Paper in Preparation, 1977.

Kelly, H. R.: A Note on the Laminar Boundary Layer on a Circular Cylinder in Axial Incompressible Flow, *J. Aeronaut. Sci.*, vol. 21, p. 634, 1954.

Klebanoff, P. S.: Characteristics of Turbulence in a Boundary Layer With Zero Pressure Gradient, *NACA Rept.* 1247, 1955.

Klebanoff, P. S., K. D. Tidstrom, and L. M. Sargent: The Three-Dimensional Nature of Boundary-Layer Instability, *J. Fluid Mech.*, vol. 12, p. 1, 1962.

Kline, S. J., M. V. Morkovin, G. Sovran, and D. J. Cockrell: "Computation of Turbulent Boundary Layers—1968 AFOSR-IFP-Stanford Conference," vol. 1, Thermosciences Division, Stanford University, Stanford, Calif., 1969.

Klineberg, J. M. and J. L. Steger: On Laminar Boundary-Layer Separation, *AIAA Pap.* 74-94, 1974.

Knapp, C. F., and P. J. Roache: A Combined Visual and Hot-Wire Anemometer Investigation of Boundary-Layer Transition, *AIAA J.*, vol. 6, p. 29, 1968.

Knystautas, R.: The Turbulent Jet From a Series of Holes in Line, *Aero. Quart.*, vol. 15, p. 1, 1964.

Lamb, H.: "Hydrodynamics," University Press, Cambridge, 1932.

Laufer, J.: The Structure of Turbulence in Fully-Developed Pipe Flow, *NACA Rept.* 1174, 1954.

Launder, B. E., and W. M. Ying: Secondary Flows in Ducts of Square Cross-Section, *J. Fluid Mech.*, vol. 54, p. 289, 1972.

Launder, B. E., G. J. Reese, and W. Rodi: Progress in the Development of a Reynolds-Stress Turbulence Closure, *J. Fluid Mech.*, vol. 68, p. 537, 1975.

Liebeck, R. H.: On the Design of Subsonic Airfoils for High Lift, *AIAA Pap.* 76-406, 1976.

Lighthill, M. J.: On Boundary Layers and Upstream Influence, pt. 1, A Comparison Between Subsonic and Supersonic Flows, *Proc. Roy. Soc.*, vol. 217, pp. 478–507, 1953.

Lighthill, M. J.: The Response of Laminar Skin Friction and Heat Transfer to Fluctuations in the Stream Velocity, *Proc. Roy. Soc. Ser. A*, vol. 224, p. 1, 1954.

Lock, R. C.: Hydrodynamic Stability of the Flow in the Laminar Boundary Layer Between Parallel Streams, *Proc. Cambridge Phil. Soc.*, vol. 50, p. 105, 1954.

Ludwieg, H., and W. Tillmann: Investigations of the Wall Shearing Stress in Turbulent Boundary Layers, *NACA Rept.* TM 1285, 1949 (English translation).

Mack, L. M.: A Numerical Study of the Temporal Eigenvalue Spectrum of the Blasius Boundary Layer, *J. Fluid Mech.*, vol. 73, p. 497, 1976.

McCroskey, W. J., and J. J. Philippe: Unsteady Viscous Flow on Oscillating Airfoils, *AIAA J.*, vol. 13, p. 71, 1975.

McDonald, H., and R. W. Fish: Practical Calculations of Transitional Boundary Layers, *Int. J. Heat Mass Transfer*, vol. 16, no. 9, pp. 1729–1744, 1973.

Mehta, U. B., and Z. Lavan: Starting Vortex, Separation Bubbles and Stall—A Numerical Study of Laminar Unsteady Flow Around an Airfoil, *J. Fluid Mech.*, vol. 67, pp. 227–256, 1975.

Mellor, G. L., and H. J. Herring: A Method of Calculating Compressible Turbulent Boundary Layers, *NASA Rept.* CR-1144, 1968.

Merkli, P., and H. Thomann: Transition to Turbulence in Oscillating Pipe Flow, *J. Fluid Mech.*, vol. 68, pp. 567–576, 1975.

Merzkirch, W.: "Flow Visualization," Academic Press, New York, 1974.

Michel, R.: Etude de la Transition sur les Profiles d'Aile; Establissement d'un Critére de Determination de Point de Transition et Calcul de la Trainée de Profile Incompressible, *ONERA Rept.* 1/1578A, 1951.

Milne-Thompson, L. M.: "Theoretical Hydrodynamics," MacMillan, New York, 1960.

Mojola, O. O., and A. D. Young: An Experimental Investigation of the Turbulent Boundary Layer Along a Streamwise Corner, *AGARD Conf. Proc.* 93, 1971.

Moss, G. F.: Some Examples of R.A.E. Investigations of Transition Fixing Effects, *R.A.E. Tech. Memo. Aero* 1165, 1969.

Naot, D., A. Shavit, and M. Wolfstein: Numerical Calculation of Reynolds Stresses in a Square Duct With Secondary Flow, *Wärme- Stoffübertrag.*, vol. 7, p. 155, 1974.

Newman, B. G.: Some Contributions to the Study of the Turbulent Boundary Layer Near Separation, *Austr. Dept. Supply Rept.* ACA-53, 1951 (see Coles and Hirst, 1969).

Obremski, H. J., and A. A. Fejer: Transition in Oscillating Boundary Layer Flows, *J. Fluid Mech.*, vol. 29, p. 93, 1967.

Patankar, S. V., and D. B. Spalding: "Heat and Mass Transfer in Boundary Layers," Intertext Books, London, 1970.

Pate, S. R.: Measurements and Correlations of Transition Reynolds Numbers on Sharp Slender Cones at High Speeds, *AIAA J.*, vol. 9, p. 1082, 1971.

Pate, S. R.: Comparison of NASA Helium Tunnel Transition Data With Noise-Transition Correlation, *AIAA J.*, vol. 12, p. 1615, 1974.

Peake, D. J., W. J. Rainbird, and E. G. Atraghji: Three-Dimensional Flow Separation on Aircraft and Missiles, *AIAA J.*, vol. 10, p. 567, 1972.

Perkins, H. J.: The Formation of Streamwise Vorticity in Turbulent Flow, *J. Fluid Mech.*, vol. 44, p. 721, 1970.

Perry, A. E., W. H. Schofield, and P. N. Joubert: Rough Wall Turbulent Boundary Layers, *J. Fluid Mech.*, vol. 37, p. 383, 1969.

Pfenninger, W.: Further Laminar Flow Experiments in a 40-Foot Long Two-Inch Diameter Tube, *Northrop Aircraft, Hawthorne, Calif., Rept.* AM-133, 1951.

Prandtl, L.: Turbulent Flow, *NACA Tech. Memo.* 435, 1926 (originally delivered to 2nd Internat. Congr. Appl. Mech., Zurich, 1926).

Prandtl, L., and O. G. Tietjens: "Applied Hydro- and Aero-Mechanics," McGraw-Hill Book Company, New York, 1954.

Pratap, V. S., and D. B. Spalding: Numerical Computation of the Flow in Curved Ducts, *Aero. Quart.*, vol. 26, p. 219, 1975.

Radbill, J. R., and G. A. McCue: "Quasilinearization and Nonlinear Problems in Fluid and Orbital Mechanics," American Elsevier, New York, 1970.

Reichardt, H.: On a New Theory of Free Turbulence, *J. Roy. Aero. Soc.*, vol. 47, p. 167, 1943.

Reichert, J. K., and R. S. Azad: Nonasymptotic Behavior of Developing Turbulent Pipe Flow, *Can. J. Phys.*, vol. 54, p. 268, 1976.

Reeves, B. L., and L. Lees: Theory of Laminar Near Wake of Blunt Bodies in Hypersonic Flow, *AIAA J.*, vol. 3, pp. 2061, 1965.

Reshotko, E.: A Program for Transition Research, *AIAA J.*, vol. 13, p. 261, 1975.

Reyhner, T. A. and I. Flugge-Lotz: The Interaction of a Shock Wave With a Laminar Boundary Layer, *Int. J. Non-Linear Mech.*, vol. 3, no. 2, pp. 173–199, 1968.

Richman, J. W., and R. S. Azad: Developing Turbulent Flow in Smooth Pipes, *Appl. Sci. Res.*, vol. 28, p. 302, 1974.

Richmond, R.: "Experimental Investigation of Thick Axially-Symmetric Boundary Layers on Cylinders at Subsonic and Hypersonic Speeds," Ph.D. thesis, California Institute of Technology, Pasadena, California, 1957.

Riley, N.: Unsteady Laminar Boundary Layers, *SIAM Rev.*, vol. 17, p. 274, 1975.

Roache, P. J.: "Computational Fluid Dynamics," Hermosa Publishers, Albuquerque, New Mexico, 1972.

Rosenhead, L.: "Laminar Boundary Layers," Clarendon Press, Oxford, 1963.

Rotta, J. C.: A Family of Turbulence Models for Three-Dimensional Thin Shear Layers, DFVLR-AVA Göttingen IB251-76A25, 1977 (to appear in *J. Fluid Mech.*).

Sabot, J., and G. Comte-Bellot: Internal Intermittency in the Core Region of Pipe Flow, Paper Presented at Colloquium on Coherent Structures in Turbulence, Southampton, England, 1974 (see also *J. Fluid Mech.*, vol. 74, p. 767, 1976).

Schlichting, H.: "Boundary-Layer Theory," McGraw-Hill Book Company, New York, 1968.

Schoenherr, K. E.: Resistance of Flat Surfaces Moving Through a Fluid, *Trans. Soc. Nav. Architects Mar. Eng.*, vol. 40, p. 279, 1932.

Schubauer, G. B.: Air Flow in the Boundary Layer of an Elliptic Cylinder, *NACA Rept.* 652, 1939.

Schubauer, G. B., and H. K. Skramstad: Laminar Boundary-Layer Oscillations and Transition on a Flat Plate, *NACA Tech. Rept.* 909, 1947.

Sears, W. R., and D. P. Telionis: Boundary-Layer Separation in Unsteady Flow, *SIAM J. Appl. Math.*, vol. 28, p. 215, 1975.

Seban, R. A., and R. Bond: Skin Friction and Heat Transfer Characteristics of a Laminar Boundary Layer on a Circular Cylinder in Axial Incompressible Flow, *J. Aero. Sci.*, vol. 18, p. 671, 1951.

Shapiro, A. H., R. Siegel, and S. J. Kline: Friction Factor in the Laminar Entry Region of a Round Tube, *Proc. Second U.S. Nat. Congr. Appl. Mech., ASME*, p. 733, 1954.

Simpson, R. L., W. M. Kays, and R. J. Moffat: The Turbulent Boundary Layer on a Porous Plate: An Experimental Study of the Fluid Dynamics With Injection and Suction, *Stanford Univ. Mech. Eng. Dept. Rept.* HMT-2, 1967.

Smith, A. M. O., and N. Gamberoni: Transition, Pressure Gradient, and Stability Theory, *Proc. Internat. Congr. Appl. Mech., 9, Brussels, Belgium*, vol. 4, p. 234 (also AD 125 559), 1956.

Sparrow, E. M., H. Quack, and C. J. Boerner: Local Nonsimilarity Boundary-Layer Solutions, *AIAA J.*, vol. 8, p. 1936, 1970.

Stevenson, T. N.: A Modified Velocity Deflect Law for Turbulent Boundary Layers With Injection, *College of Aero. Cranfield Rep. Aero.* 170 (also N64-19324), 1963.

Stewartson, K.: On the Impulsive Motion of a Flat Plate in a Viscous Fluid, *Quart. J. Mech. Appl. Math.*, vol. 4, p. 182, 1951.

Stewartson, K.: Further Solutions of the Falkner-Skan Equation, *Phil. Mag.*, vol. 12, p. 865, 1954.

Stewartson, K.: The Asymptotic Boundary Layer on a Circular Cylinder, *Quart. Appl. Math.*, vol. 13, p. 13, 1955.

Stewartson, K.: The Theory of Unsteady Laminar Boundary Layers, *Advan. Appl. Mech.*, vol. 6, p. 1, 1960.

Stewartson, K.: On the Impulsive Motion of a Flat Plate in a Viscous Fluid, pt. 2, *Quart. J. Mech. Appl. Math.*, vol. 26, p. 143, 1973.

Stewartson, K.: Multistructured Boundary Layers on Flat Plates and Related Bodies, *Advan. Appl. Mech.*, vol. 14, p. 145, 1974 (C. S. Yih, (ed.) Academic Press, New York).

Stewartson, K., and P. G. Williams: Self-Induced Separation, pt. 2, *Mathematika*, vol. 20, p. 98, 1969.

Stone, D. R., A. M. Cory, Jr., and E. L. Morrisette: Use of Discrete Sonic Jets as Boundary-Layer Trips in Hypersonic Flow, *AIAA J.*, vol. 8, p. 945, 1970.

Stratford, B. S.: Flow in Laminar Boundary Layer Near Separation, *Rep. Memo. Aero. Res. Counc., London*, 3002, 1954.

Stuart, J. T.: The Production of Intense Shear Layers by Vortex Stretching and Convection, *Nat. Physical Lab. Aero. Rept.* 1147, 1965.

Stuart, J. T.: Nonlinear Stability Theory, *Annu. Rev. Fluid Mech.*, vol. 3, p. 347, 1971.

Tani, I.: Low-Speed Flows Involving Bubble Separations, *Progr. Aero. Sci.*, vol. 5, pp. 70–103, 1964.

Taylor, E. S.: "Dimensional Analysis for Engineers," Clarendon Press, Oxford, 1973.

Thwaites, B.: Approximate Calculation of the Laminar Boundary Layer, *Aero. Quart.*, vol. 1, p. 245, 1949.

Thwaites, B. (ed): "Incompressible Aerodynamics," Clarendon Press, Oxford, 1960.

Townsend, A. A.: The Behavior of a Turbulent Boundary Layer Near Separation, *J. Fluid Mech.*, vol. 12, p. 536, 1962.

Townsend, A. A.: "The Structure of Turbulent Shear Flow," Cambridge University Press, New York, 1976.

Van den Berg, B.: A Three-Dimensional Law of the Wall for Turbulent Shear Flows, *J. Fluid Mech.*, vol. 70, p. 149, 1975.

Van Driest, E. R.: On Turbulent Flow Near a Wall, *J. Aero. Sci.*, vol. 23, p. 1007, 1956.

Van Driest, E. R., and C. B. Blumer: Boundary-Layer Transition at Supersonic Speeds: Roughness Effects With Heat Transfer, *AIAA J.*, vol. 6, p. 603, 1968.

Van Dyke, M. D.: "Perturbation Methods in Fluid Mechanics," Parabolic Press, Stanford, 1975.

Van Ingen, J. L.: A Suggested Semi-Empirical Method for the Calculation of the Boundary-Layer Transition Region, *Rept.* V.T.H. 71, V.T.H. 74, Delft, Holland, 1956.

Versmissen, J. J.: Some Experiments on the Performance and the Drag Penalty of Various Distributed Roughness Bands Used for Boundary Layer Transition Fixing, *Netherlands NLR* TR 74009 U, 1974.

Wang, K. C.: On the Determination of the Zones of Influence and Dependence for Three-Dimensional Boundary-Layer Equations, *J. Fluid Mech.*, vol. 48, p. 397, 1971.

Wazzan, A. R., T. T. Okamura, and A. M. O. Smith: Spatial and Temporal Stability Charts for the Falkner-Skan Boundary-Layer Profiles, *Douglas Aircraft Long Beach, Calif. Rept.* DAC-67086, 1968.

Wells, C. S., Jr.: Effects of Freestream Turbulence on Boundary-Layer Transition, *AIAA J.*, vol. 5, p. 172, 1967.

Werle, H.: Hydrodynamic Flow Visualization, *Annu. Rev. Fluid Mech.*, vol. 5, p. 361, 1973.

Wheeler, A. J., and J. P. Johnston: Three-Dimensional Turbulent Boundary Layers—Data Sets for Two-Space-Coordinate Flows, *Stanford Univ. Rept.* MD-32, 1972.

White, F. M.: An Analysis of Axisymmetric Turbulent Flow Past a Long Cylinder, *J. Basic Eng.*, vol. 94, p. 200, 1972.

Wieghardt, K.: Zum Reibungswiderstand rauher Platten, *Kaiser-Wilhelm-Institut für Stromungsforschung, Göttingen,* UM 6612, 1944.

Williams, P. G.: A Reverse Flow Computation in the Theory of Self-Induced Separation, in R. D. Richtmyer (ed.), "Lecture Notes in Physics, vol. 35, Proceedings of the Fourth International Conference on Numerical Methods in Fluid Dynamics," pp. 445–451, Springer-Verlag, Berlin, 1975.

Winant, C. D., and F. K. Browand: Vortex Pairing: The Mechanism of Turbulent Mixing-Layer Growth at Moderate Reynolds Number, *J. Fluid Mech.*, vol. 63, no. 2, p. 237, 1974.

Zamir, M., and A. D. Young: Experimental Investigation of the Boundary Layer in a Streamwise Corner, *Aero. Quart.*, vol. 21, p. 313, 1970.

Index